AF598122

Nutrition and Health

Adrianne Bendich, Ph.D., FASN, FACN, Series Editor

More information about this series at http://www.springer.com/series/7659

Rajkumar Rajendram
Victor R. Preedy • Vinood B. Patel
Editors

Branched Chain Amino Acids in Clinical Nutrition

Volume 1

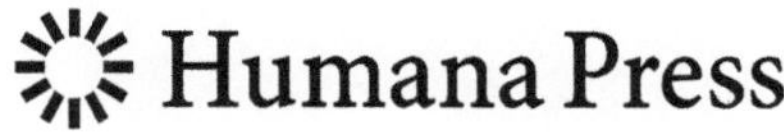

Editors
Rajkumar Rajendram
Intensive Care
Barnet and Chase Farm Hospitals, Royal Free London NHS Foundation Trust
London, United Kingdom

Victor R. Preedy
Diabetes and Nutritional Sciences
Kings College London
London, United Kingdom

Vinood B. Patel
Department of Biomedical Sciences
University of Westminster
London, United Kingdom

ISBN 978-1-4939-1922-2 ISBN 978-1-4939-1923-9 (eBook)
DOI 10.1007/978-1-4939-1923-9
Springer New York Heidelberg Dordrecht London

Library of Congress Control Number: 2014952897

Printed on acid-free paper

Humana Press is a brand of Springer
Springer is part of Springer Science+Business Media (www.springer.com)

Preface

In man, the branched chain amino acids (leucine, isoleucine, and valine) are essential amino acids and thus must be obtained from dietary components. The branched chain amino acids are not only necessary for the synthesis of proteins but also have other metabolic functions and roles. For example, over several decades' evidence has supported the notion that branched chain amino acids, particularly leucine, are important in ameliorating or restoring metabolic imbalance. Studies in the 1970s showed that leucine promoted protein synthesis in muscle in vitro. Later, in the 1980s, it was shown that the branched chain amino acids stimulated protein synthesis in vivo. Subsequently, studies showed that branched chain amino acids could potentially be used clinically in ameliorating muscle catabolism. More recently the branched chain amino acids have been added to performance-enhancing supplements. Although this is a simplistic synopsis of historical events, it is now evident that the branched chain amino acids have a variety of functions. In simple terms the knowledge base associated with the branched chain amino acids have now been successfully harvested to enhance human health. Branched chain amino acids, like some other amino acids, have an almost ubiquitous function and are important in maintaining the cellular milieu of virtually every organ in the human body. For example, branched chain amino acids have roles in carbohydrate and lipid metabolism, insulin release and resistance, proteolysis, formation of keto acids, obesity prevention, and cancer. This does not mean to say that branched chain amino acids are the universal panacea. Indeed the administration of high amounts of branched chain amino acids may be toxic. The science of branched chain amino acids is complex and finding all the relevant information in a single source has hitherto been problematic. This is, therefore, addressed in *Branched Chain Amino Acids in Clinical Nutrition.*

The book has seven major Parts in two volumes.

Volume I

Part I: Basic Processes at the Cellular Level
Part II: Inherited Defects in Branched Chain Amino Acid Metabolism
Part III: Experimental Models of Growth and Disease States: Role of Branched Chain Amino Acids

Volume II

Part I: Role of Branched Chain Amino Acids in Healthy Individuals
Part II: Branched Chain Amino Acids: Status in Disease States
Part III: Branched Chain Amino Acids and Liver Diseases
Part IV: Branched Chain Amino Acid Supplementation Studies in Certain Patient Populations

Coverage includes the individual branched chain amino acids, amino acid ratios, essential amino acids, metabolism, amino acids cocktails, aminotransferases, tRNA, PPAR, uncoupling proteins, insulin and insulin resistance, glucose and glycemic control, the hypothalamus, sirtuin, ammonia, cirrhosis, encephalopathy, apoproteins, maple syrup urine disease and oxidation disorders, mental retardation, fetal growth, skeletal and cardiac muscles, muscular dystrophy, amyotrophic lateral sclerosis, anorexia, obesity and weight loss, bladder carcinogenesis, tolerability, recovery, exercise, functional adaptations, psychomotor performance, whey protein, brain injury, obstructive pulmonary disease, ethanol oxidation, albumin, late evening snacks, organ transplantation, quality of life, and skin and radiotherapy. Finally there is a chapter on web-based material and additional reading.

Contributors are authors of international and national standing, leaders in the field, and trendsetters. Emerging fields of science and important discoveries are also incorporated in *Branched Chain Amino Acids in Clinical Nutrition.*

This book is designed for nutritionists and dietitians, public health scientists, doctors, epidemiologists, health care professionals of various disciplines, policy makers, and marketing and economic strategists. It is designed for teachers and lecturers, undergraduates and graduates, and researchers and professors.

London, UK

Rajkumar Rajendram
Victor R. Preedy
Vinood B. Patel

Series Editor Page

The great success of the Nutrition and Health Series is the result of the consistent overriding mission of providing health professionals with texts that are essential because each includes (1) a synthesis of the state of the science; (2) timely, in-depth reviews by the leading researchers and clinicians in their respective fields; (3) extensive, up-to-date fully annotated reference lists; (4) a detailed index; (5) relevant tables and figures; (6) identification of paradigm shifts and the consequences; (7) virtually no overlap of information between chapters, but targeted, interchapter referrals; (8) suggestions of areas for future research; and (9) balanced, data-driven answers to patient as well as health professional questions which are based upon the totality of evidence rather than the findings of any single study.

The series volumes are not the outcome of a symposium. Rather, each editor has the potential to examine a chosen area with a broad perspective, both in subject matter as well as in the choice of chapter authors. The international perspective, especially with regard to public health initiatives, is emphasized where appropriate. The editors, whose trainings are both research and practice oriented, have the opportunity to develop a primary objective for their book, define the scope and focus, and then invite the leading authorities from around the world to be part of their initiative. The authors are encouraged to provide an overview of the field, discuss their own research and relate the research findings to potential human health consequences. Because each book is developed de novo, the chapters are coordinated so that the resulting volume imparts greater knowledge than the sum of the information contained in the individual chapters.

Branched Chain Amino Acids in Clinical Nutrition, a two-volume book, edited by Rajkumar Rajendram, Victor R. Preedy, and Vinood B. Patel, is a very welcome addition to the Nutrition and Health Series and fully exemplifies the Series' goals. The first volume of *Branched Chain Amino Acids in Clinical Nutrition*, "Cellular Processes, Genetic Factors and Experimental Models of Branched Chain Amino Acid Functions and Metabolism," is organized into three relevant parts. The ten introductory chapters in the first part, entitled "Basic Processes at the Cellular Level," provide readers with the basics so that the more clinically related chapters can be easily understood. The first chapter provides a broad-based perspective on the protein requirements for humans and describes the 20 amino acids that are classified as essential for humans to consume as these cannot be synthesized de novo in the human body. Tables and figures included describe major food sources of essential amino acids and specific food sources of branched chain amino acids. The second chapter describes the structures and functions of the three branched chain amino acids, leucine, isoleucine, and valine, that are classified as essential amino acids. We learn that these three branched chain amino acids make up approximately one third of all the amino acids in the body. The majority of the three amino acids are found in skeletal muscle where these function as both structural elements and stores for systemic nitrogen. The dietary requirements are approximately 40, 20, and 19 mg/kg of body weight/day of leucine, valine, and isoleucine, respectively. Excellent sources of these amino acids are red meat, dairy, and soy protein-containing products. The typical Western diet provides sufficient protein to normally assure consumption of adequate levels of branched chain amino acids. The enzymes involved in the catabolism of these amino acids are described in detail and their locations within the body are reviewed with emphasis on their roles in muscle and the brain. The third chapter provides more detailed descriptions of the 15 enzymes responsible for the transamination and metabolism of the individual branched chain amino acids and describes the enzymatic reactions that share the same enzymes as well as those that differ in the metabolism of these amino acids. Peripheral as well as brain enzyme systems and shuttles are described in detail and illustrated in the included figures.

Chapter 4 describes in detail the role of isoleucine specifically in the functioning of certain adipocytes as well as its function in glucose metabolism. Laboratory animal studies are reviewed and the importance of isoleucine in the activation of liver and skeletal muscle free fatty acid uptake and oxidation is linked to a potential role in the development of obesity. The fifth chapter describes the unique metabolic activities of leucine. Leucine has been found to be a potent activator of the mammalian target of rapamycin (mTOR) pathway which is a critical nutrient-sensing pathway that governs cell metabolism, cell growth, and proliferation. Leucine also has a role in regulating insulin secretion and glucose utilization. The chapter includes an in-depth investigation of leucine regulation of several critical cellular processes in pancreatic β-cells, including metabolism, growth, proliferation, and insulin secretion, which ultimately influence overall glucose homeostasis. The multiple routes that are involved in acute regulation of insulin secretion are illustrated in the included figures.

Chapter 6 looks at the data from laboratory animal experiments that examine the effects of leucine and isoleucine on glucose metabolism and insulin regulation. These three chapters provide a comprehensive review of the roles of these amino acids in energy metabolism and suggest that this is an area where well-controlled clinical studies are needed. The seventh chapter provides additional detailed information concerning the importance of the hypothalamic metabolism of leucine in the brain and its effects on liver glucose production. The eighth chapter reviews new studies in cell culture and laboratory animals that are exposed to leucine and resveratrol. The rationale for this combination is that both leucine and resveratrol stimulate sirtuin-dependent pathways that are linked to enhancing longevity. The authors describe the synergistic actions of both substances on sirtuin-dependent downstream metabolic effects.

The next chapter in this part describes the role of branched chain amino acids in the metabolism of ammonia. This is a complex area of basic as well as clinical investigation and both are well described in this chapter. The author also describes the open questions concerning the interactions between muscle and liver metabolism of ammonia and the potential for unexpected effects when branched chain amino acids are given therapeutically. The last chapter in this part, Chapter 10, describes the use of a stable isotope form of leucine to calculate the in vivo rates of synthesis and catabolism of lipoproteins that are important in the determination of cardiovascular risk, survival of HIV, and other clinical manifestations of abnormal lipoprotein levels. This detailed chapter describes the calculations required to determine in vivo circulating HDL, VLDL, and LDL levels without exposing the patient to radiation.

Part II contains four chapters that review the inherited defects in branched chain amino acid metabolism and describes the resultant inborn errors of metabolism. The first chapter in this part focuses on genetic defects in the oxidative pathways involved in the breakdown of valine, leucine, and isoleucine. We learn that there are a total of 15 different enzyme reactions that are required for the breakdown of valine, leucine, and isoleucine. The chapter focuses on the 11 genetic defects and the enzymatic consequences of 11 of the pathways. Descriptions of the pathophysiological changes that are seen with each of the amino acid's associated genetic mutations in critical enzymes are described. Maple sugar urine disease is the most common manifestation of errors in the coding of the catabolic enzymes involved in the decomposition of the three branched chain amino acids and results in higher than normal levels of the amino acids in the blood and urine. Chapter 12 describes in detailed text and tables the 202 mutations described in 2013 that are associated with maple sugar urine disease. Because of the rapid onset of severe brain damage in neonates with this disease, newborn screening is done on all infants in the USA and in many other parts of the world. Clinically, treatments involve the provision of a diet low in branched chain amino acids. There is ongoing research concerning the biochemical mechanisms associated with the clinical presentations of the many of the genetic defects associated with this disease. Chapter 13 describes other adverse consequence of genetic mutations specifically to the isoleucine degradation enzymes. The authors have documented cases of mental retardation, motor dysfunctions, behavioral disorders, and other abnormal manifestations of physical and mental functions in children and young adults that carry these mutations. The X-linked inborn errors of isoleucine degradation are described in detail and the author clearly indicates the need for further research. The next chapter examines the results of laboratory animal studies that look at the effects of a genetic defect in the metabolism of valine and the consequences to eating behaviors. The authors posit that the resultant valine deficiency specifically can affect satiety centers in the brain resulting in anorexia. Translating this research into clinical studies has not as yet occurred.

Part III: Experimental Models

Part III includes seven chapters that describe laboratory studies with experimental models of growth and certain disease states where branched chain amino acids have been examined to determine the metabolic role of these amino acids; in several chapters the administration of one or all of the branched chain amino acids has been shown to provide some improvement. Chapters 15 and 16 review the importance of protein and specific amino acids in the growth of the fetus (Chapter 15) and the neonate (Chapter 16). Intrauterine growth retardation is associated with a reduction in organ growth and permanent changes in organ metabolism and/or structure. As discussed by the author, intrauterine growth retardation (IUGR) may result from maternal undernutrition that can increase the risk of lifelong adverse health effects. Experimental models have provided data indicating that IUGR causes changes in islet cells, in the hypothalamic-pituitary-adrenal axis, and in the secretion of prolactin, progesterone, estradiol, and insulin, as well as in the glucose uptake by muscles, body fat content, and mitochondrial function. IUGR increases the risk of cardiovascular diseases, diabetes, and obesity in adult life. IUGR causes a reduction in organ growth and permanent changes in organ metabolism and/or structure. The experimental models have shown that leucine supplementation in models of low protein intake may not be as valuable as increase in total protein intake. The chapter includes excellent tables and figures. Chapter 16 provides unique data from a neonatal pig model on the role of leucine in muscle metabolism. The authors have used parenteral leucine infusions and show that a physiological rise in leucine enhances protein synthesis in skeletal muscle and cardiac muscle. They have also shown that leucine supplementation of a meal acutely stimulated protein synthesis in the neonatal pig model. Two chapters examined the importance of branched chain amino acids in models of obesity. Chapter 17 reviews the data from rodent models concerning the effects of branched chain amino acid status on insulin resistance and alterations in adipose tissue. Chapter 18 examines the data from models of high fat intake that have shown that leucine functions as a nutrient signal to coordinately regulate three major signaling pathways in the liver, skeletal muscle, and adipose tissue. Dietary supplementation of leucine significantly ameliorated the deleterious effects of consumption of a high-fat diet including obesity, hepatic lipid accumulation, mitochondrial dysfunction, and insulin resistance. The metabolic benefits of leucine supplementation included upregulation of genes related to mitochondrial synthesis of critical enzymes, increases in metabolic rates, and suppression of inflammation in adipose tissue.

The last three chapters in this part discuss different models that reflect the functions of branched chain amino acids. Chapter 19 describes research underway to understand the etiology of amyotrophic lateral sclerosis (ALS), an adult-onset neurodegenerative disease, also known as Lou Gehrig disease, that is characterized by degeneration of neurons in the brain and spinal cord leading to progressive paralysis of respiratory and limb muscles. Both in vitro and laboratory animal studies are reviewed. There are indications of adverse effects of branched chain amino acids in certain of the models as a result of neurotoxicity that may be linked to oxidative stress. At present it is not known if these effects would be seen in humans, but caution is urged. The next chapter provides a unique perspective on the potential for leucine to enhance ethanol metabolism in the liver. The author describes experiments showing that leucine specifically (and not valine) accelerates ethanol clearance after acute ethanol administration by enhancing alcohol-metabolizing enzyme activities. Leucine treatment before alcohol intake also enhanced alcoholic enzyme activities and accelerates ethanol metabolism. It is well recognized that chronic alcohol intake leads to liver failure, such as hepatic inflammation and fatty liver, and induces liver cirrhosis. Accelerating ethanol oxidation may prevent liver failure. The last chapter in this part describes the development of a two-stage model of bladder cancer that involves the use of known cancer initiators and promoters. In this model, branched chain amino acid supplementation at very high doses compared to normal human dietary intakes enhanced tumor formation when certain basal diets were fed, but not with others. As indicated by the authors, "To date, however, there is no epidemiological data relevant to a relationship of dietary BCAA with the risk of bladder cancer."

Volume II

The second volume of *Branched Chain Amino Acids in Clinical Nutrition* concentrates on the role of these amino acids in healthy individuals, the effects of certain diseases on branched chain amino acid status, and finally, data from clinical studies that included therapeutic use of the amino acids in patients.

Part I: Role of Branched Chain Amino Acids in Healthy Individuals

The first part contains five chapters that begin with an in-depth examination of the safety of leucine supplementation in healthy adults. Often healthy adults use leucine supplements when they are exercising or attempting to build muscle. The next three chapters review the effects of leucine on muscle. The last chapter in the Part describes the potential for optimal surgery outcomes when amino acid ratios are used as an index of protein status. Chapter 1 describes the process used by the authors to determine the tolerable, and presumably safe, upper limit of intake of leucine in healthy adults. They define this value as the point at which the metabolic capacity to catabolize or oxidize the excess amino acid is exceeded because it represents the intake where the normal regulatory mechanisms are no longer sufficient to dispose of the excess. The amino acid intake corresponding to this inflection point does not represent a toxic intake level, but rather suggests that with increasing dietary intakes above this level the potential or risk for adverse events will increase. Also, amino acid intakes above this point are usually characterized by an increasing rate of accumulation in blood and excretion of the amino acid, and its secondary catabolites in urine. The figures included in this chapter help to illustrate these relationships. The experimental design used for leucine is reviewed and the authors report that with increasing intakes of leucine, a dose–response in leucine oxidative capacity was observed, with a breakpoint estimated at 550 mg/kg bw/day or 39 g/day for a 70 kg healthy adult. Simultaneous and significant increases in blood ammonia concentrations, plasma leucine concentrations, and urinary leucine excretion were observed with leucine intakes higher than 500 mg/kg bw/day. Thus, under acute dietary conditions, intakes greater than 500 mg leucine/kg bw/day may potentially increase the risk of adverse events, and is proposed as the tolerable upper safe intake (UL) for leucine in healthy adults.

The next chapter examines the requirements for protein and leucine intake in elite athletes to support skeletal muscle regeneration processes following endurance exercise in trained skeletal muscle. The authors present their rationale for adding free leucine to dietary protein to enhance the combined exercise-nutrient muscle response to the post-exercise regenerative processes and protein synthesis. Both the high dose (70 g whey protein/15 g leucine) and low dose (23 g whey protein, 5 g leucine) co-ingested with carbohydrate and fat over the first 90 min following intense cycling resulted in a proinflammatory transcriptome associated with increased leukocytes that reverted by 240 min to an anti-inflammatory signal in skeletal muscle. The other measurements also pointed to a positive effect of the post-exercise supplementation; however, in addition to the acute studies, long-term studies in trained athletes are needed. Chapter 3 reviews studies that used whey protein (from milk) plus leucine. The doses of L-leucine used in the studies reviewed varied from 2.24 to 7.5 g while the whey protein doses varied from 6.7 to 25 g. Outcomes also varied and included muscle responses post-exercise, in different aged populations and immune functions following exercise. The authors suggest that future studies assess dietary intakes and use consistent outcomes. The next chapter in this part reviews the data linking muscle atrophy in patients caused by several diseases to metabolic rationales for administering branched chain amino acids. The authors describe both animal models and patients with cancer-induced muscle loss (cachexia), glucocorticoid-induced muscle atrophy from Cushing's syndrome, as well as long-term therapeutic use of glucocorticoids, sarcopenia, and sepsis-induced muscle protein degradation. The final chapter, Chapter 5, describes the use of a ratio of branched chain amino acids to tyrosine as an index of presurgery health in patients with liver cancer; the index can also help to identify patients who would benefit with amino acid supplementation post surgery. Liver cancer patients undergoing surgical resection who had a low ratio before surgery had significantly more complications as well as more severe complications than patients with higher ratios. The authors suggest that liver cancer patients who have higher albumin levels (indicative of better liver function) and also have a higher ratio are at reduced risk for postsurgery complications. Patients with high albumin levels and low ratio may be the best candidates for supplementation with branched chain amino acids postsurgery.

Part II: Branched Chain Amino Acids: Status in Disease States

Part II contains five chapters that examine diseases of the heart, brain, and lungs and how these diseases affect the branched chain amino acid status of the patient. Branched chain amino acids, as discussed above, are critical for skeletal muscle integrity and are a major site of these amino acids' metabolism. Their role in cardiac muscle is the topic of Chapters 6 and 7. Chapter 6 examines clinical manifestations of genetic defects in branched chain amino acid metabolism as seen in propionic acidemia and methylmalonic acidemia that have been associated with dilated and hypertrophic cardiomyopathies. In addition to genetically related changes in branched chain amino acid metabolism, alterations in their metabolism are also seen in heart failure patients independent of a genetic cause. This is a new area of clinical research and a number of metabolic paths involving the catabolism of branched chain amino acids are hypothesized to potentially adversely affect cardiac tissue. Another new area that links branched chain amino acids with heart function, reviewed in Chapter 7, is the complex outcomes of mitochondrial cardiomyopathies (MCM) produced by mutational defects in energetic metabolism. Specifically, a mutation involving the branched chain amino acid valine has been identified and linked to impaired heart functions in the neonates that inherit this defect.

The next two chapters examine the effects of branched chain amino acids on brain functions, first as related to psychomotor skills and, in Chapter 9, their role in traumatic brain injury. With regard to psychomotor skills, we learn that these coordinated actions are used in everyday life, occupational work, and sport activities. Psychomotor performance depends mainly on cognitive function, attention, concentration, and decision-making. Branched chain amino acids, especially leucine, can cross the blood–brain barrier and influence these functions through their involvement in the synthesis of neurotransmitters. Ingestion of small doses of branched chain amino acids has been shown to improve psychomotor performance. However, higher doses can exert negative effects on some brain functions and may impair psychomotor performance. New research is underway to better understand the biochemical changes in the brain following traumatic brain injury. Clinical research has reported that immediately following the brain injury, there are phase-dependent changes in plasma amino acids with a different profile in the acute, subacute, and rehabilitation phase. During the acute phase, decreased plasma branched chain amino acid levels are associated with increased plasma aromatic amino acid concentrations. There are numerous other changes in the brain and the authors of Chapter 9 indicate that currently, it is not known if repletion of the branched chain amino acids is of benefit immediately or will be of more value during a different phase of recovery. The last chapter in this section reviews the effects of chronic obstructive pulmonary disease (COPD) on branched chain amino acid status and the potential to enhance the strength of these patients with nutritional addition of these amino acids. The authors point out that COPD patients with low body mass index (BMI) and/or severe airflow limitations exhibit reduced branched chain amino acid profiles, but those with a normal BMI and/or moderate airflow limitations do not. Nevertheless, both muscle and plasma levels of the amino acid levels are directly related to the levels of muscle wasting seen in patients with COPD.

Part III contains five chapters that examine the influence of branched chain amino acids in patients with liver diseases. Chapter 11 describes the effects of liver disease on branched chain amino acid status and then reviews the use of the amino acids as therapeutic supplements in patients with liver cirrhosis, liver cancer, and additional adverse consequences of liver disease. In patients with liver cirrhosis, levels of branched chain amino acids are decreased in the blood and the levels of aromatic amino acids and methionine are increased. The authors examine in depth the adverse effects of these changes on brain and muscle biochemistry and by using global gene expression analysis, the authors provide critical data on the molecular mechanisms by which hepatic damage can be reversed following branched chain amino acid supplementation. The next chapter discusses the importance of the liver in the synthesis of albumin. Albumin is the major protein produced in the liver and albumin represents over 50% of the total plasma proteins. Plasma albumin level is a standard index of nutritional status, liver function, and/or pathophysiological conditions. One of the most important factors regulating plasma albumin levels is ingestion of a protein-rich meal. The authors indicate that in Japan, pharmacological supplementation of branched chain amino acids is used to improve hypoalbuminemia in patients with liver cirrhosis. The chapter focuses on the effects of amino acid supplementation, including branched chain amino acids, on the regulation of albumin synthesis and provides excellent figures to help the reader. Chapter 13 describes the effects of liver cirrhosis caused by hepatitis C and nonalcoholic fatty liver disease and the consequent protein energy malnutrition and muscle wasting that can accompany these serious liver diseases. The author indicates that these major metabolic defects result in a state of starvation during the overnight sleep time and provides clinical evidence of the benefits of a nighttime snack that includes branched chain amino acids. Based upon the preliminary findings, it is suggested that well-controlled studies be undertaken.

The next chapter reviews the importance of stabilizing the nutritional status in patients with end-stage liver disease requiring liver transplantation. Accurate nutritional assessment and adequate perioperative nutritional treatment are essential for improving outcomes after liver transplant. The overall survival rate in patients with low skeletal muscle mass was found to be significantly lower than in patients with normal/high skeletal muscle mass. The authors report that perioperative nutritional therapy including branched chain amino acids is useful for patients with sarcopenia, whose prognosis is poor without nutritional therapy. The final two chapters in this part review the importance of branched chain amino acid supplementation in the treatment of the post-liver surgery patient. In Chapter 15, we learn that post-transplant bacteremia is one of the most serious complications following liver transplantation. One potential avenue for prevention of bacteremia that is being tested is nutritional support. It is known that in patients who have not been given supplementation prior to transplantation, serum levels of branched chain amino acids generally decrease and have been reported to be a risk factor for post-transplant bacteremia. This chapter reviews previous studies on the beneficial effects of branched chain amino acid supplementation in liver transplant patients: prevention of bacteremia, potential mechanisms of action, and avenues for future research. Chapter 16 examines the effects of liver surgery as a result of liver cancer and the data suggesting that branched chain amino acid supplementation may enhance the prognosis of these patients. The authors indicate that most liver cancers occur in patients with chronic liver disease. Advances in surgical technology and perioperative management have led to the standard use of hepatic surgical procedures for liver cancer and also metastatic liver tumors. Branched chain amino acid supplementation improves postoperative quality of life over the long term after hepatic resection by restoring and maintaining nutritional status and whole-body kinetics.

The last part in volume II examines the use of branched chain amino acid supplementation in several different patient populations that further attests to the growing interest in the clinical value of these amino acids. The final chapter provides an extensive and up-to-date listing of relevant websites and other resources. The first four chapters discuss the importance of these amino acids in patients who are losing muscle mass and may at the same time show symptoms of insulin resistance. Chapter 17 provides an overview of the most promising therapeutic uses of branched chain amino acids. Topics discussed include the effects of physical immobility in the elderly, cancer cachexia, and weight reduction in the obese patient. The common thread is the loss of muscle mass and the potential for leucine and/or branched chain amino acid-rich protein supplementation to reduce the muscle loss seen in these conditions. Chapter 18 reviews the data that link liver cirrhosis with insulin resistance. Insulin resistance increases in chronic liver disease and is a risk factor for the progression of liver disease, the potential for the development of liver cancer, and a decrease in long-term survival. The authors suggest that insulin resistance should be considered as an important therapeutic target in patients at any stage of chronic liver disease. They hypothesize that branched chain amino acids play a dual role in glucose metabolism in skeletal muscle, enhancing glucose uptake under normal insulin conditions while causing insulin resistance under high insulin conditions. The next chapter concentrates on the actions of leucine on glucose metabolism. Since its discovery, leucine has been shown to affect glucose homeostasis in liver, muscle, adipose tissue, and pancreatic β cells. In the β cells, leucine acutely stimulates insulin secretion by several mechanisms. These mechanisms can be modulated by factors that include, but are not limited to, the body weight of the patient, whether they are type 2 diabetics and whether or not they routinely exercise. Chapter 20 focuses on protein metabolism during insulin resistance and the effects of surgically implemented weight loss on branched chain amino acid status and requirements. Elevated plasma amino acid levels have been associated with obesity and insulin resistance and circulating branched chain amino acids have been identified as early biomarker predictors of diabetes risk in obese insulin-resistant subjects. The chapter includes detailed discussions of the results of bariatric surgery on branched chain amino acid levels and examines the data from twin studies to better understand the interactions between branched chain amino acid status and improvements in glucose control following acute, significant weight loss.

Chapter 21 examines the metabolism of branched chain amino acids within the skin and provides a unique perspective on the importance of these amino acids, especially leucine, in stimulating dermal collagen synthesis in wound healing. The chapter reviews the requirements for mixtures of amino acids to optimize skin collagen formation in the face of protein malnutrition, wound healing, and UV radiation and discusses the mechanism of action of these stressor to the skin. Chapter 22 reviews the potential for branched chain amino acid supplementation to reduce the muscle wasting seen in boys that have inherited the genetic disorder, Duchenne muscular dystrophy (DMD). The disease is characterized by progressive loss of muscle mass and accumulation of body fat. Steroids (e.g., prednisone) are the only treatment available at this time, but the drugs increase the accumulation of body fat and have other adverse side effects. The chapter provides a strong biochemically based rationale for supplementing DMD patients with branched chain amino acids; however as yet, there are no large, well-controlled and patient-monitored intervention studies.

Part III, above, contained chapters that consistently reported a decreased branched chain amino acid status in patients with liver diseases including liver cancer. The next two chapters investigate branched chain amino acid supplementation in patients with serious liver diseases. Chapter 23 examines the potential for branched chain amino acid supplementation to reduce the adverse effects of radiotherapy used in patients with liver cancer. The chapter reviews the recent development of radiotherapeutic technologies that permit their application in this patient population. The authors inform us that most of the patients undergoing radiation therapy have chronic liver disease and frequently also have protein-calorie malnutrition. In addition, radiation can result in general fatigue, nausea, and vomiting that further aggravate the patients' nutritional status. Preliminary clinical investigations suggest that supplementation with branched chain amino acids during radiation therapy helped to reduce the loss of these amino acids and may be of benefit in maintaining albumin levels. Chapter 24 describes the condition called hepatic encephalopathy, which is a metabolic neuropsychiatric syndrome of cerebral dysfunctions due to severe chronic or acute liver disease. The manifestations of hepatic encephalopathy range from minor symptoms with personality changes and altered sleep patterns to deep coma. The chapter includes a detailed review of all clinical trials and meta-analyses that included studies where patients have been given branched chain amino acids for the treatment of hepatic encephalopathy regardless of route of administration. The combined evidence supported the use of oral branched chain amino acids as treatment for patients with hepatic encephalopathy based upon the results from several large, high-quality randomized controlled trials. However, further research is required to determine the optimal dose of these amino acids.

The above descriptions of the two volumes' 46 chapters attest to the depth of information provided by the 156 well-recognized and respected editors and chapter authors. Each chapter includes complete definitions of terms with the abbreviations fully defined for the reader and consistent use of terms between chapters. Key features of the two comprehensive volumes include over 250 detailed tables and informative figures, an extensive, detailed index, and more than 1,900 up-to-date references that provide the reader with excellent sources of worthwhile information. Moreover, the final chapter contains a comprehensive list of web-based resources that will be of great value to the health provider as well as graduate and medical students.

In conclusion, *Branched Chain Amino Acids in Clinical Nutrition*, a two-volume book, edited by Rajkumar Rajendram, Victor R. Preedy, and Vinood B. Patel, provides health professionals in many areas of research and practice with the most up-to-date, well-referenced volume on the importance of branched chain amino acids in maintaining the nutritional status and overall health of the individual especially in certain disease conditions. The volumes will serve the reader as the benchmarks in this complex area of interrelationships between dietary protein intakes and individual amino acid supplementation, the unique role of the branched chain amino acids in the synthesis of brain neurotransmitters, collagen formation, insulin and glucose modulation, and the functioning of all organ systems that are involved in the maintenance of the body's metabolic integrity. Moreover, the physiological, genetic, and pathological interactions between plasma levels of branched chain amino acids and aromatic amino acids are clearly delineated so that students as well as practitioners can better understand the complexities of these interactions. Unique chapters examine the effects of branched chain amino acid status and the effects of genetic mutations from pre-pregnancy, during fetal development and birth, and infancy through the aging process. The editors are applauded for their efforts to develop the most authoritative and unique resource in the area of branched chain amino acids in health and disease to date and this excellent text is a very welcome addition to the Nutrition and Health Series.

Adrianne Bendich, Ph.D., F.A.C.N., F.A.S.N.
Series Editor

About the Series Editor

Adrianne Bendich, Ph.D., F.A.S.N., F.A.C.N. has served as the "Nutrition and Health" Series Editor for over 15 years and has provided leadership and guidance to more than 120 volume editors that have developed the 60+ well-respected and highly recommended volumes in the series.

In addition to ***Branched Chain Amino Acids in Clinical Nutrition*** **volume I and volume II, edited by Rajkumar Rajendram M.D., Victor R. Preedy Ph.D., and Vinood B. Patel Ph.D.**, **major new editions** in 2012–2014 include:

1. ***Glutamine in Clinical Nutrition,*** edited by Rajkumar Rajendram M.D., Victor R. Preedy Ph.D., and Vinood B. Patel Ph.D., 2014
2. ***Handbook of Clinical Nutrition and Aging, Third Edition,*** edited by Connie W. Bales Ph.D., RD, Julie L. Locher Ph.D., MSPH, and Edward Saltzman, M.D., 2014
3. ***Nutrition and Oral Medicine, Second Edition,*** edited by Dr. Riva Touger-Decker, Dr. Connie C. Mobley, and Dr. Joel B. Epstein, 2014
4. ***Fructose, High Fructose Corn Syrup, Sucrose and Health,*** edited by Dr. James M. Rippe, 2014
5. ***Nutrition in Kidney Disease, Second Edition,*** edited by Dr. Laura D. Byham-Gray, Dr. Jerrilynn D. Burrowes, and Dr. Glenn M. Chertow, 2014
6. ***Handbook of Food Fortification and Health, volume I*** edited by Dr. Victor R. Preedy, Dr. Rajaventhan Srirajaskanthan, and Dr. Vinood B. Patel, 2013
7. ***Handbook of Food Fortification and Health, volume II*** edited by Dr. Victor R. Preedy, Dr. Rajaventhan Srirajaskanthan, and Dr. Vinood B. Patel, 2013
8. ***Diet Quality: An Evidence-Based Approach, volume I*** edited by Dr. Victor R. Preedy, Dr. Lan-Ahn Hunter, and Dr. Vinood B. Patel, 2013
9. ***Diet Quality: An Evidence-Based Approach, volume II*** edited by Dr. Victor R. Preedy, Dr. Lan-Ahn Hunter, and Dr. Vinood B. Patel, 2013

10. ***The Handbook of Clinical Nutrition and Stroke,*** edited by Mandy L. Corrigan, MPH, RD, Arlene A. Escuro, MS, RD, and Donald F. Kirby, MD, FACP, FACN, FACG, 2013
11. ***Nutrition in Infancy, volume I*** edited by Dr. Ronald Ross Watson, Dr. George Grimble, Dr. Victor R. Preedy, and Dr. Sherma Zibadi, 2013
12. ***Nutrition in Infancy, volume II*** edited by Dr. Ronald Ross Watson, Dr. George Grimble, Dr. Victor R. Preedy, and Dr. Sherma Zibadi, 2013
13. ***Carotenoids and Human Health,*** edited by Dr. Sherry A. Tanumihardjo, 2013
14. ***Bioactive Dietary Factors and Plant Extracts in Dermatology,*** edited by Dr. Ronald Ross Watson and Dr. Sherma Zibadi, 2013
15. ***Omega 6/3 Fatty Acids,*** edited by Dr. Fabien De Meester, Dr. Ronald Ross Watson, and Dr. Sherma Zibadi, 2013
16. ***Nutrition in Pediatric Pulmonary Disease,*** edited by Dr. Robert Dumont and Dr. Youngran Chung, 2013
17. ***Magnesium and Health,*** edited by Dr. Ronald Ross Watson and Dr. Victor R. Preedy, 2012.
18. ***Alcohol, Nutrition and Health Consequences,*** edited by Dr. Ronald Ross Watson, Dr. Victor R. Preedy, and Dr. Sherma Zibadi, 2012
19. ***Nutritional Health, Strategies for Disease Prevention, Third Edition,*** edited by Norman J. Temple, Ted Wilson, and David R. Jacobs, Jr., 2012
20. ***Chocolate in Health and Nutrition,*** edited by Dr. Ronald Ross Watson, Dr. Victor R. Preedy, and Dr. Sherma Zibadi, 2012
21. ***Iron Physiology and Pathophysiology in Humans,*** edited by Dr. Gregory J. Anderson and Dr. Gordon D. McLaren, 2012

Earlier books included ***Vitamin D, Second Edition*** edited by Dr. Michael Holick; ***Dietary Components and Immune Function*** edited by Dr. Ronald Ross Watson, Dr. Sherma Zibadi, and Dr. Victor R. Preedy; ***Bioactive Compounds and Cancer*** edited by Dr. John A. Milner and Dr. Donato F. Romagnolo; ***Modern Dietary Fat Intakes in Disease Promotion*** edited by Dr. Fabien De Meester, Dr. Sherma Zibadi, and Dr. Ronald Ross Watson; ***Iron Deficiency and Overload*** edited by Dr. Shlomo Yehuda and Dr. David Mostofsky; ***Nutrition Guide for Physicians*** edited by Dr. Edward Wilson, Dr. George A. Bray, Dr. Norman Temple, and Dr. Mary Struble; ***Nutrition and Metabolism*** edited by Dr. Christos Mantzoros; and ***Fluid and Electrolytes in Pediatrics*** edited by Leonard Feld and Dr. Frederick Kaskel. Recent volumes include ***Handbook of Drug-Nutrient Interactions*** edited by Dr. Joseph Boullata and Dr. Vincent Armenti; ***Probiotics in Pediatric Medicine*** edited by Dr. Sonia Michail and Dr. Philip Sherman; ***Handbook of Nutrition and Pregnancy*** edited by Dr. Carol Lammi-Keefe, Dr. Sarah Couch, and Dr. Elliot Philipson; ***Nutrition and Rheumatic Disease*** edited by Dr. Laura Coleman; ***Nutrition and Kidney Disease*** edited by Dr. Laura Byham-Grey, Dr. Jerrilynn Burrowes, and Dr. Glenn Chertow; ***Nutrition and Health in Developing Countries*** edited by Dr. Richard Semba and Dr. Martin Bloem; ***Calcium in Human Health*** edited by Dr. Robert Heaney and Dr. Connie Weaver; and ***Nutrition and Bone Health*** edited by Dr. Michael Holick and Dr. Bess Dawson-Hughes.

Dr. Bendich is President of Consultants in Consumer Healthcare LLC, and is the editor of ten books including ***Preventive Nutrition: The Comprehensive Guide for Health Professionals, Fourth Edition*** co-edited with Dr. Richard Deckelbaum (www.springer.com/series/7659). Dr. Bendich serves on the Editorial Boards of the *Journal of Nutrition in Gerontology and Geriatrics*, and *Antioxidants*, and has served as Associate Editor for *Nutrition* the International Journal; served on the Editorial Board of the *Journal of Women's Health and Gender-Based Medicine*, and served on the Board of Directors of the American College of Nutrition.

Dr. Bendich was Director of Medical Affairs at GlaxoSmithKline (GSK) Consumer Healthcare and provided medical leadership for many well-known brands including TUMS and Os-Cal. Dr. Bendich had primary responsibility for GSK's support for the Women's Health Initiative (WHI) intervention study. Prior to joining GSK, Dr. Bendich was at Roche Vitamins Inc. and was involved with the groundbreaking clinical studies showing that folic acid-containing multivitamins significantly reduced major classes of birth defects. Dr. Bendich has coauthored over 100 major clinical research studies in the area of preventive nutrition. She is recognized as a leading authority on antioxidants, nutrition and immunity and pregnancy outcomes, vitamin safety, and the cost-effectiveness of vitamin/mineral supplementation.

Dr. Bendich received the Roche Research Award, is a *Tribute to Women and Industry* Awardee, and was a recipient of the Burroughs Wellcome Visiting Professorship in Basic Medical Sciences. Dr. Bendich was given the Council for Responsible Nutrition (CRN) Apple Award in recognition of her many contributions to the scientific understanding of dietary supplements. In 2012, she was recognized for her contributions to the field of clinical nutrition by the American Society for Nutrition and was elected a Fellow of ASN. Dr. Bendich is an Adjunct Professor at Rutgers University. She is listed in Who's Who in American Women.

About Volume Editors

Dr. Rajkumar Rajendram is an intensivist, anesthetist, and perioperative physician. He graduated in 2001 with a distinction from Guy's, King's, and St. Thomas Medical School, in London. As an undergraduate he was awarded several prizes, merits, and distinctions in preclinical and clinical subjects. This was followed by training in general medicine and intensive care in Oxford, during which period he attained membership of the Royal College of Physicians (MRCP) in 2004. Dr. Rajendram went on to train in anesthesia and intensive care in the Central School of Anaesthesia, London Deanery and became a fellow of the Royal College of Anaesthetists (FRCA) in 2009. He has completed advanced training in intensive care in Oxford and was one of the first intensivists to become a fellow of the faculty of intensive care medicine (FFICM) by examination. He coauthored the Oxford Case Histories in Cardiology which was published by the Oxford University Press in 2011. He is currently preparing the text for the Oxford Case Histories in Intensive Care. His unique training and experience has been tailored for a career in intensive care with a subspecialty interest in perioperative medicine.

Dr. Rajendram recognizes that nutritional support is a fundamental aspect of critical care. He has therefore devoted significant time and effort into nutritional science research. As a visiting research fellow in the Nutritional Sciences Research Division of King's College London he has published over 50 textbook chapters, review articles, peer-reviewed papers, and abstracts from his work.

Victor R. Preedy B.Sc., Ph.D., D.Sc., FSB, FRSH, FRIPHH, FRSPH, FRCPath, FRSC, is a senior member of King's College London (Professor of Nutritional Biochemistry) and King's College Hospital (Professor of Clinical Biochemistry; Hon). He is attached to both the Diabetes and Nutritional Sciences Division and the Department of Nutrition and Dietetics. He is also Director of the Genomics Centre and a member of the School of Medicine. Professor Preedy graduated in 1974 with an Honors Degree in Biology and Physiology with Pharmacology. He gained his University of London Ph.D. in 1981. In 1992, he received his Membership of the Royal College of Pathologists and in 1993 he gained his second doctoral degree, for his contribution to the science of protein metabolism in health and disease. Professor Preedy was elected as a Fellow of the Institute of Biology in 1995 and to the Royal College of Pathologists in 2000. Since then he has been elected as a Fellow to the Royal Society for the Promotion of Health (2004) and the Royal Institute of Public Health and Hygiene (2004). In 2009, Professor Preedy became a Fellow of the Royal Society for Public Health and in 2012 a Fellow of the Royal Society of Chemistry. In his career Professor Preedy worked at the National Heart Hospital (part of Imperial College London) and the MRC Centre at Northwick Park Hospital. He has collaborated with research groups in Finland, Japan, Australia, the USA, and Germany. He is a leading expert on biomedical sciences and has a longstanding interest in how nutrition and diet affects well-being and health. He has lectured nationally and internationally. To his credit, Professor Preedy has published over 500 articles, which include peer-reviewed manuscripts based on original research, reviews, and numerous books and volumes.

Dr. Vinood B. Patel is currently a Senior Lecturer in Clinical Biochemistry at the University of Westminster and honorary fellow at King's College London. He presently directs studies on metabolic pathways involved in liver disease, particularly related to mitochondrial energy regulation and cell death. In addition, research is being undertaken to study the role of nutrients, phytochemicals, and fatty acids in the development of fatty liver disease and iron homeostatic regulation. Another area includes identifying new biomarkers that can be used for diagnosis and prognosis of liver disease. Dr. Patel graduated from the University of Portsmouth with a degree in Pharmacology and completed his Ph.D. in protein metabolism from King's College London in 1997. His postdoctoral work was carried out at Wake Forest University Baptist Medical School studying structural–functional alterations to mitochondrial ribosomes, where he developed novel techniques for characterizing their biophysical properties. Dr. Patel is a nationally and internationally recognized alcohol researcher and was involved in several NIH-funded biomedical grants related to alcoholic liver disease. Dr. Patel has edited biomedical books in the area of nutrition and health prevention, autism, and biomarkers and has published over 160 articles.

Acknowledgments

The development of this book is built upon the foundation of the excellent work provided by the staff of Springer Science+Business Media, LLC and we are very thankful to them. In particular, we wish to acknowledge the outstanding support, advice, and patience of the Series Editor, Dr. Adrianne Bendich; the Developmental Editor, Michael Griffin; and Associate Editors, Amanda Quinn and Joanna Perey.

The Editors

Contents

List of Contributors

Niels Kristian Aagaard, M.D., Ph.D. Department of Hepatology and Gastroenterology, Aarhus University Hospital, Aarhus, Denmark

Asha V. Badaloo, B.Sc., M.Sc., Ph.D. Tropical Metabolism Research Unit, University of the West Indies, Kingston, Jamaica

Makoto Bannai, Ph.D. Institute for Innovation, Ajinomoto, Co., Inc., Kawasaki, Kanagawa, Japan

W. Ted Brown, M.D., Ph.D. New York State Institute for Basic Research in Developmental Disabilities, Staten Island, NY, USA

Antje Bruckbauer, M.D., Ph.D. NuSirt Sciences Inc., Knoxville, TN, USA

Gemma Calamandrei, Master Degree in Biology Unit of Neurotoxicology and Neuroendocrinology, Department of Cell Biology and Neurosciences, Istituto Superiore di Sanità, Rome, Italy

Kevin Carpenter, Ph.D. NSW Biochemical Genetics Service, The Children's Hospital at Westmead, University of Sydney, Westmead, NSW, Australia

Jeffrey T. Cole, Ph.D., M.S. Department of Neurology, Uniformed Services University of the Health Sciences, Bethesda, MD, USA

Annamaria Confaloni, Ph.D. Section of Clinic, Diagnostic and Therapy of Degenerative Diseases of the Central Nervous System, Department of Cell Biology and Neurosciences, Istituto Superiore di Sanità, Rome, Italy

David Conti Department of Transplant Surgery, Albany Medical College, Albany, NY, USA

Myra E. Conway, Ph.D. University of the West of England, Bristol, UK

Alessio Crestini, Ph.D., M.S., M.Sc. Res. Section of Clinic, Diagnostic and Therapy of Degenerative Diseases of the Central Nervous System, Department of Cell Biology and Neurosciences, Istituto Superiore di Sanità, Rome, Italy

Gitte Dam, M.D., Ph.D. Department of Hepatology and Gastroenterology, Aarhus University Hospital, Aarhus, Denmark

Teresa A. Davis, Ph.D. USDA/ARS Children's Nutrition Research Center, Baylor College of Medicine, Houston, TX, USA

Alessia De Felice, Master Degree in Biology Unit of Neurotoxicology and Neuroendocrinology, Department of Cell Biology and Neurosciences, Istituto Superiore di Sanità, Rome, Italy

Roberta De Simone, M.Sc. Section of Experimental Neurology, Department of Cell Biology and Neurosciences, Istituto Superiore di Sanità, Rome, Italy

Carl Dobkin, Ph.D. New York State Institute for Basic Research in Developmental Disabilities, Staten Island, NY, USA

Michael Dolinger Division of Gastroenterology, Department of Internal Medicine, Albany Medical College, Albany, NY, USA

Marinus Duran, Ph.D. Laboratory of Genetic Metabolic Diseases, Academic Medical Centre, University of Amsterdam, Amsterdam, AZ, The Netherlands

Chie Furuta, D.V.M., Ph.D. Institute for Innovation, Ajinomoto Co., Inc., Kawasaki, Kanagawa, Japan

Lise Lotte Gluud, M.D., D.M.Sc. Department of Medicine, Copenhagen University Hospital Gentofte, Copenhagen, Denmark

Xue-Ying He, Ph.D. New York State Institute for Basic Research in Developmental Disabilities, Staten Island, NY, USA

Yunfei Huang, M.D., Ph.D. Center for Neuropharmacology and Neuroscience, Albany Medical College, Albany, NY, USA

Susan M. Hutson, Ph.D. Department of Human Nutrition, Foods and Exercise, Virginia Tech, Blacksburg, VA, USA

Charles Isaacs, Ph.D. New York State Institute for Basic Research in Developmental Disabilities, Staten Island, NY, USA

Farook Jahoor, B.Sc., M.Sc., Ph.D. Baylor College of Medicine, Houston, TX, USA

Roger Gutiérrez-Juárez, M.D., Ph.D. Department of Medicine and Diabetes Research Center, Albert Einstein College of Medicine, Yeshiva University, Bronx, NY, USA

Hiroko Jinzu, R.D. Institute for Innovation, Ajinomoto, Co., Inc., Kawasaki, Kanagawa, Japan

Anna Kakehashi, Ph.D. Department of Pathology, Osaka City University Graduate School of Medicine, Osaka, Japan

Michio Komai, Ph.D. Tohoku University, Sendai, Miyagi, Japan

Ference Loupatty, Ph.D. Laboratory of Genetic Metabolic Diseases, Academic Medical Centre, University of Amsterdam, Amsterdam, AZ, The Netherlands

Department of Clinical Chemistry, Reinier de Graaf Groep, GA, The Netherlands

Fiorella Malchiodi-Albedi, M.D. Section of Molecular Neurobiology, Department of Cell Biology and Neurosciences, Istituto Superiore di Sanità, Rome, Italy

Takayuki Masaki, M.D., Ph.D. Department of Endocrinology and Metabolism, Faculty of Medicine, Oita University, Oita, Japan

Alberto Martire, Ph.D. Section of Central Nervous System Pharmacology, Department of Therapeutic Research and Medicines Evaluation, Istituto Superiore di Sanità, Rome, Italy

Andrea Matteucci, Ph.D., Biology. Section of Molecular Neurobiology, Department of Cell Biology and Neurosciences, Istituto Superiore di Sanità, Rome, Italy

Luisa Minghetti, Ph.D. Section of Experimental Neurology, Department of Cell Biology and Neurosciences, Istituto Superiore di Sanità, Rome, Italy

Hitoshi Murakami, Ph.D. Institute for Innovation, Ajinomoto Co., Inc., Kawasaki, Kanagawa, Japan

Kenji Nagao, Ph.D. Institute for Innovation, Ajinomoto Co., Inc., Kanagawa, Japan

Yasushi Noguchi, Ph.D. Institute for Innovation, Ajinomoto, Co., Inc., Kanagawa, Japan

Peter Ott, M.D., D.M.Sc. Department of Hepatology and Gastroenterology, Aarhus University Hospital, Noerrebrogade, Aarhus, Denmark

Lucas Carminatti Pantaleão, M.Sc. University of São Paulo, Avenida Professor Lineu Prestes, São Paulo, SP, Brazil

Patrizia Popoli, M.D. Section of Central Nervous System Pharmacology, Department of Therapeutic Research and Medicines Evaluation, Istituto Superiore di Sanità, Rome, Italy

Marvin Reid, M.B.B.S., Ph.D. Tropical Metabolism Research Unit, University of the West Indies, Kingston, Jamaica

Gabrielle Ritaccio Division of Gastroenterology, Department of Internal Medicine, Albany Medical College, Albany, NY, USA

Agus Suryawan, Ph.D. USDA/ARS Children's Nutrition Research Center, Baylor College of Medicine, Houston, TX, USA

Tetsuya Takimoto, Ph.D. Institute for Innovation, Ajinomoto, Co. Inc., Kawasaki, Kanagawa, Japan

Gabriela Fullin Resende Teodoro, M.Sc. University of São Paulo, Avenida Professor Lineu Prestes, São Paulo, SP, Brazil

Julio Tirapegui University of São Paulo, São Paulo, SP, Brazil

Aldina Venerosi, Ph.D. Section of Neurotoxicology and Neuroendocrinology, Department of Cell Biology and Neurosciences, Istituto Superiore di Sanità, Rome, Italy

Daiana Vianna, M.Sc. University of São Paulo, São Paulo, SP, Brazil

Hendrik Vilstrup, M.D., D.M.Sc. Department of Hepatology and Gastroenterology, Aarhus University Hospital, Aarhus, Denmark

Ronald J.A. Wanders, Ph.D. Laboratory of Genetic Metabolic Diseases, Academic Medical Centre, University of Amsterdam, Amsterdam, AZ, The Netherlands

Hideki Wanibuchi, M.D., Ph.D. Department of Pathology, Osaka City University Graduate School of Medicine, Osaka, Japan

Min Wei, M.D., Ph.D. Department of Pathology, Osaka City University Graduate School of Medicine, Osaka, Japan

Xiao-Li Xie, M.D., Ph.D. Department of Pathology, Osaka City University Graduate School of Medicine, Osaka, Japan

Yuran Xie, B.S. Section of Molecular Medicine, Department of Medicine, University of Oklahoma Health Sciences Center, Oklahoma City, OK, USA

Zhonglin Xie, M.D., Ph.D. Section of Molecular Medicine, Department of Medicine, University of Oklahoma Health Sciences Center, Oklahoma City, OK, USA

Shotaro Yamano, M.S. Department of Pathology, Osaka City University Graduate School of Medicine, Osaka, Japan

Jun Yang Center for Neuropharmacology and Neuroscience; Center for Cardiovascular Science; Division of Gastroenterology, Department of Internal Medicine, Albany Medical College, Albany, NY, USA

Song-Yu Yang, M.D., Ph.D. New York State Institute for Basic Research in Developmental Disabilities, Staten Island, NY, USA

Fumiaki Yoshizawa, Ph.D. Department of Agrobiology and Bioresources, Faculty of Agriculture, Utsunomiya University, Utsunomiya, Tochigi, Japan

Michael B. Zemel, Ph.D. NuSirt Sciences Inc., The University of Tennessee, Knoxville, TN, USA

Xinjun Zhu Center for Cardiovascular Science; Division of Gastroenterology, Department of Internal Medicine, Albany Medical College, Albany, NY, USA

Part I
Basic Processes at the Cellular Level

Chapter 1
Impact of Dietary Essential Amino Acids in Man

Kenji Nagao, Hiroko Jinzu, Yasushi Noguchi, and Makoto Bannai

Key Points

- During the course of evolution, animals lost their enzymatic capacity to synthesize certain amino acids that are required for growth and tissue function. These substances are called essential amino acids and include at least nine types of amino acids depending on the animal's pathophysiological conditions.
- Dietary essential amino acids have two distinct nutritional functions: (1) the substrates act as protein building blocks and (2) they act as nutritional signals to evoke physiological reactions.
- In the past, nitrogen balance studies suggested the essential amino acid requirements for humans, but these values have been upwardly revised by a number of studies using tracer techniques and approaches.
- Essential amino acids have unique physiological signaling roles in obesity-related metabolism.
- The supplementation or fortification of dietary essential amino acids or branched chain amino acids exerts beneficial effects on body weight, body fat, lean body mass, and insulin sensitivity in animals and humans.
- To enjoy the health benefits of essential amino acids, both as building blocks of proteins and as physiological signals, choosing food ingredients based on their amino acid composition would be a new approach, together with amino acid supplementation.

Keywords Essential amino acids • Branched chain amino acids • Requirements • Antiobesogenic effect • Anabolic effect • Food ingredients

Abbreviations

His	Histidine
Ile	Isoleucine
Leu	Leucine
Lys	Lysine
Met	Methionine

K. Nagao, Ph.D. (✉) • H. Jinzu, R.D. • Y. Noguchi, Ph.D. • M. Bannai, Ph.D.
Institute for Innovation, Ajinomoto Co., Inc. 1-1 Suzuki-cho, Kawasaki-ku, Kawasaki-shi, Kanagawa, 210-8681, Japan
e-mail: kenji_nagao@ajinomoto.com; hiroko_jinzu@ajinomoto.com; yasushi_noguchi@ajinomoto.com; makoto_bannai@ajinomoto.com

R. Rajendram et al. (eds.), *Branched Chain Amino Acids in Clinical Nutrition: Volume 1*, Nutrition and Health, DOI 10.1007/978-1-4939-1923-9_1,

Phe	Phenylalanine
Thr	Threonine
Trp	Tryptophan
Val	Valine
Glu	Glutamate
Gln	Glutamine
Ala	Alanine
Asp	Asparagine
BCAA	Branched chain amino acid
IAAB	Indicator amino acid balance
DAAB	Direct amino acid balance
BCAT	Branched chain aminotransferase
INTERMAP	International Study of Macro-/Micronutrients and Blood Pressure

Introduction

During the course of evolution, animals lost their enzymatic capacity to synthesize certain molecules that are required for growth and tissue function. These molecules are classified as essential dietary nutrients, including vitamins, specific fatty acids, and certain amino acids. Although all plants can synthesize the 20 types of amino acids commonly found in proteins, animals, from protozoa to mammals, are dependent on external sources for at least 9 amino acids: histidine (His), isoleucine (Ile), leucine (Leu), lysine (Lys), methionine (Met), phenylalanine (Phe), threonine (Thr), tryptophan (Trp), and valine (Val). Depending on the physiological and pathological conditions of the individual, in addition to these 9 amino acids, some of the other 11 common protein-bound amino acids may be required to maintain body protein homeostasis. These other amino acids are called "conditionally indispensable" amino acids.

For humans, the amino acid requirements of adults are stated in the 1985 FAO/WHO/UNU report [1], which was based on data in the 1973 FAO/WHO/UNU report [2] (Table 1.1). The amino acid values were primarily based on N balance studies by Rose [3]. However, the 1985 FAO/WHO/UNU report suggested that further research was needed on amino acid requirements because several new tracer techniques and approaches emerged in the early 1980s for estimating the efficiency of amino acid utilization and the requirements for specific essential amino acids. Essential amino acids not only play

Table 1.1 1985 FAO/WHO/UNU recommendations for essential amino acid requirements in adults [1]

	Requirement	
Amino acids	mg/kg/day	mg/g of protein
Histidine	8–12	16
Isoleucine	10	13
Leucine	14	19
Lysine	12	16
Methionine + Cystine	13	17
Phenylalanine + Tyrosine	14	19
Threonine	7	9
Tryptophan	3.5	5
Valine	10	13
Total (minus histidine)	83.5	111

Source: FAO/WHO/UNU, *Energy and Protein Requirements*, Report of a Joint FAO/WHO/UNU Expert Consultation, Technical Report Series 724, World Health Organization, Geneva, 1985, page 206 [1]

a role as building blocks of proteins but also act as physiological signals. There is clinical evidence suggesting that the supplementation of essential amino acids or branched chain amino acids (BCAAs) has beneficial effects on body weight, body fat, lean body mass, and insulin sensitivity. In the last part of this chapter, we will introduce variations in the amino acid composition of commonly used food ingredients.

Dietary Essential Amino Acids as Required Nutrition

Vernon R. Young defines the minimum physiological requirement for an essential amino acid as "the continuing dietary intake of an indispensable amino acid that is just sufficient to balance whole-body irreversible losses in an initially healthy individual at energy balance under conditions of moderate physical activity and as determined after a brief period of adjustment to a new quasi-steady state with a change in test amino acid intake. For pregnant and lactating women, the amino acid requirement is taken to also include the extra dietary need associated with the deposition of protein in tissues or secretion of milk at rates consistent with good health" [4].

Amino acid requirements for adult humans have been determined using one of the following methods: (1) the nitrogen balance method, (2) the plasma amino acid response method, (3) the indicator of amino acid balance (IAAB) approach, and (4) the direct amino acid balance (DAAB) approach [4]. The nitrogen balance method, a simple way to estimate amino acid requirements, examines the difference between nitrogen intake and total nitrogen excretion. The earlier estimates of the required values for adults proposed and used by the FAO/WHO/UNU were determined using the nitrogen balance method, but they are now considered significant underestimates (Table 1.1). It is technically difficult to accurately evaluate N intake and all of the N loss routes, including integumental and other minor routes of N loss. These additional unmeasured N losses can have a profound influence on the estimated amino acid requirements. Moreover, N balance estimates are affected by overestimates of intake and underestimates of losses via urine and feces, which are also affected by various dietary/metabolic factors.

Plasma amino acid concentrations change in response to various dietary factors, including the levels and sources of energy-yielding substrates and the amount and profile of amino acid intake [4, 5]. Excessive or inadequate intake of single essential amino acids is reflected in increases or decreases, respectively, in the plasma free amino acid concentration [6]. Studies in experimental animals suggest that there may be an even greater change in the plasma concentration of the free amino acid pool in body tissues, especially in skeletal muscle [7]. The qualitative relationship between the adequacy of dietary amino acid intake and plasma amino acid concentrations has been reasonably well established [4, 5]. Consequently, the quantitative relationship between the intake of a specific amino acid and the level of that amino acid in plasma was assessed in a series of studies [8, 9].

With advances in the mass spectrometric measurements of stable isotope enrichment in biological matrices and the expanded use of tracers enriched with these isotopes in human metabolic research, a series of experiments using tracer techniques was conducted in the early 1980s to revise the estimates of the amino acid requirements for adults. There are several methods for determining human amino acid requirements using tracer-based studies [4]. These methods can be divided into categories based on tracer choice and the protocol design:

1. Studies that use a tracer to analyze a dietary amino acid by measuring its rate of oxidation at various intake levels (the DAAO approach).
2. Studies that use an indicator tracer to assess the status of whole-body amino acid oxidation (IAAO) at varying levels of a dietary amino acid.
3. Kinetic studies designed to assess the retention of protein during the postprandial phase of amino acid metabolism using ^{13}C-leucine as a tracer (the postprandial protein utilization (PPU) approach).

Table 1.2 The essential amino acid requirements (mg/kg/day) in recent reports

	IOM/FNB (2002)	Young and Borgonha (2000)	Young and Tharakan's review (2004)
Histidine	11	–	8–12
Isoleucine	15	23	19
Leucine	34	40	40
Lysine	31	30	30
Methionine + Cystine	15	13	15
Phenylalanine + Tyrosine	27	39	39
Threonine	16	15	15
Tryptophan	4	6	5
Valine	19	20	25

Source: "Nutritional essentiality of amino acids and amino acid requirements in healthy adult" in "Metabolic and Therapeutic Aspects of Amino Acids in Clinical Nutrition" CRC Press, London and New York, 2004, page 463 [4] with permission

Each method has pros and cons for estimating the requirements. Despite some variability in the results, there is general agreement that the requirements for nitrogen balance considerably exceed the theoretical estimate for optimum amino acid intake, suggesting that human adults can efficiently use no more than approximately 60 % of proteins, even high-class types. The FAO/WHO/UNU report in 1985 [1] concluded that 52.5 g of first-class protein would meet the needs of 97 % of the adult population; however, Vermon R. Young insisted that some populations are unable to obtain nitrogen balance with such small intake amounts [4]. Young repeatedly argued that it was unsafe to accept the 1985 international standards for adult amino acid requirements (Table 1.2) based solely on the results of the nitrogen approach, which used an indirect measure of the balance of individual amino acids [4]. The general conclusion from such studies was that the adult requirements for individual essential amino acids appeared considerably greater than the standards derived from earlier balance studies. Because of the errors in balance studies and the complexity of isotope studies, final conclusions regarding amino acid requirements have remained controversial; however, there has been growing acceptance that the former estimates were too low.

Dietary Essential Amino Acids as Physiological Signals

Dietary essential amino acids have two distinct nutritional functions: (1) the substrates act as protein building blocks, as described above, and (2) they act as nutritional signals to evoke physiological reactions. Among the essential amino acids, BCAAs (Val, Leu, and Ile) are abundant in food, accounting for approximately 20 % of total protein intake [10, 11]. Because of the low hepatic expression of mitochondrial branched chain aminotransferase (BCAT2 or BCATm; EC 2.6.1.42), BCAAs from dietary protein bypass the initial metabolism in the liver; in contrast, the other 17 amino acids are predominantly metabolized in this organ [12, 13]. This bypass may contribute to the sharp elevation of plasma BCAA levels in response to a meal, thereby playing a signaling role in the peripheral tissues and the brain that is unique among amino acids. Thus, circulating BCAAs act as nutrient signals and regulate protein synthesis and degradation, as well as insulin secretion.

Studies have investigated the effects of dietary BCAA ingestion on a number of diseases and conditions, such as obesity and metabolic disorders, liver disease, muscle atrophy, cancer, impaired immunity, and a variety of injuries (postoperative injuries, trauma, burns, and sepsis) [14]. Of these, BCAAs play unique roles in obesity-related metabolism. For example, Leu in food improves muscle glucose uptake and whole-body glucose metabolism [15], mediates the postprandial leptin increase [16], and decreases food intake and body weight via central mTOR signaling. Macotela et al. showed that doubling dietary Leu reversed many of the metabolite abnormalities in diet-induced obesity models and caused a marked improvement in glucose tolerance and insulin signaling without altering food intake or weight status. Increased dietary Leu was also associated with a decrease in hepatic steatosis and a decrease in inflammation in adipose tissue, indicating that Leu acts on multiple tissues and at multiple levels of metabolism to impact insulin resistance [17]. Doi et al. reported a hypoglycemic effect of Ile ingestion that involves increases in muscle glucose uptake and whole-body glucose oxidation and a decrease in hepatic gluconeogenesis [18].

Recently, there have been some reports with the intriguing conclusion that specific mixtures of amino acids, rather than a single amino acid supplement, may be more efficacious in lowering the blood glucose response to a glucose challenge. For example, Noguchi et al. [19] designed a novel diet with an elevated ratio of essential to nonessential amino acids (high-E/N diet). Dietary proteins in the high-E/N diet were partially replaced with a mixture of five free essential amino acids (Leu, Ile, Val, Lys, and Thr) without altering the carbohydrate and fat content of the diet. This dietary amino acid manipulation improved glucose tolerance, decreased lipogenesis, and prevented hepatic steatosis in mice with diet-induced obesity; the approach was suggested as a novel preventive and therapeutic approach to nonalcoholic fatty liver disease in humans [20]. Similarly, another report found that rats orally administered an amino acid mixture (containing cysteine, Met, Val, Ile, and Leu) combined with a high-glucose solution showed improved glucose tolerance via an increase in skeletal muscle glucose uptake and intracellular disposal through enhanced intracellular signaling compared with nonsupplemented rats [21]. The other interesting effect of amino acid mixtures is that combinations of BCAAs and glutamine or proline play a role in restoring dermal collagen protein synthesis impaired by UV irradiation [22].

There have also been several clinical studies suggesting that supplementing essential amino acids or BCAAs has beneficial effects on body weight, body fat, and insulin sensitivity. Solerte et al. studied the effects of a balanced amino acid formula containing BCAAs in a long-term randomized study of elderly subjects with type 2 diabetes and found improved glucose control and insulin sensitivity [23]. From the perspective of nutritional epidemiology, the population-based International Study of Macro-/Micronutrients and Blood Pressure (INTERMAP) provided unique evidence to evaluate the effects of dietary BCAAs across different cultures. Higher dietary BCAA intake was associated with a lower prevalence of overweight and obesity in middle-aged individuals from East Asian and Western countries [24].

The anabolic effect after consuming a protein-containing meal, thus supplying essential amino acids to the blood, has been thoroughly studied. Protein synthesis increases and protein breakdown decreases after essential amino acid ingestion. In skeletal muscle, which is the major contributor to lean body mass in humans, eating appears to double the rate of muscle protein deposition [25], and much of this change can be attributed to the action of amino acids alone [26], without substantial insulin involvement [26]. Bohe et al. concluded that the rates of synthesis of all classes of muscle proteins are acutely regulated by the levels of essential amino acids in the blood. The stimulation of protein synthesis depends on the concentration of extracellular, rather than intramuscular, essential amino acids [27].

Furthermore, Dillon et al. [28] reported that 3 months of essential amino acid supplementation increased muscle IGF-1 levels and lean body mass in older women without affecting kidney function. The acute anabolic response to this supplementation (i.e., the increased muscle protein fractional synthesis rate) was maintained over time [28]. Moreover, many papers have reported that BCAA dietary supplements reduce sarcopenia in elderly subjects, and a variety of amino acid mixtures have been used to restore the protein content of defective tissues, especially skeletal muscles, in older subjects [28]. In a randomized trial involving 41 subjects with sarcopenia aged 66–84 years, consuming a special mixture of amino acids containing BCAAs increased muscle mass, reduced tumor necrosis factor-α, and improved insulin sensitivity [29]. As a result, The Society for Sarcopenia, Cachexia, and

Wasting Disease now recommends leucine-enriched balanced amino acid supplements in the nutritional recommendations for the management of sarcopenia [30]. Furthermore, in sarcopenic women, the combined effect of exercise and essential amino acids resulted in improvements in muscle strength, the combined variables of muscle mass and walking speed, and muscle mass [31].

In addition, the ingestion of essential amino acids, including BCAAs, is reported to have an influence on the incidence of infections. In geriatric long-term rehabilitation centers, the ingestion of BCAAs reduced the incidence of infections by 30 % and increased the serum albumin and total protein levels in hemodialysis patients while reducing inflammation markers and correcting anemia [32]. Three months of essential amino acid supplementation had comprehensive effects on nutritional status, muscle energy metabolism, blood oxygen tension, physical autonomy, cognitive function, and perceived health status in patients with severe COPD and secondary sarcopenia [33]. The parenteral supplementation of BCAAs also enhanced the cognitive recovery of patients with traumatic brain injury, even those in a vegetative or minimally conscious state [34].

In addition to BCAAs, the other essential amino acids are also associated with unique physiological signals. For example, Lys is one of the limiting amino acids (the essential amino acid found in the smallest quantity in a particular foodstuff) in most cereal grains. A 3-month randomized double-blind study was conducted to determine whether the Lys fortification of wheat reduced anxiety and stress responses in family members in poor Syrian communities that consumed wheat as a staple food [35]. In the Lys-fortified group, the plasma cortisol response to a blood draw as a cause of stress was reduced in females, as was sympathetic arousal in males, as measured by skin conductance. Lys fortification also significantly reduced chronic anxiety, as measured by the trait anxiety inventory in males. These results suggest that some stress responses can be improved by Lys fortification in economically weak populations that consume cereal-based diets [35]. Moreover, a series of studies has also shown that Lys fortification of wheat flour had a positive effect on diarrheal morbidity in women in Syria [35]. Lys is considered a partial 5-HT4 receptor antagonist; it suppresses 5-HT4 receptor-mediated intestinal pathologies and anxiety [36].

Variations in the Amino Acid Content of Food Ingredients

To date, although the amino acid content is known to vary considerably among various food ingredients, few analyses have been performed. Using data on the amino acid composition of food ingredients [11] published by the Ministry of Education, Culture, Sports, Science and Technology in Japan, we determined the unique amino acid composition of commonly used Japanese food ingredients. Hierarchical cluster analyses were conducted to obtain a visual representation of the amino acid content (Fig. 1.1). Ward's linkage method was performed on Euclidean distances from the standardized amino acid compositions. A cluster analysis of the amino acid composition of food ingredients (Fig. 1.1) showed three clusters. The first cluster mainly consisted of grain-based ingredients, the second cluster mainly consisted of meat and fish, and the third cluster included the remaining food ingredients. There were smaller amounts of Lys, Thr, Ala, and Asp in the first cluster, which contained grain-based ingredients, whereas this cluster had greater amounts of Glx (sum of Glu and Gln) and Pro. The second cluster, which contained meat and fish, had higher amounts of Lys, Thr, and Met (essential amino acids) and smaller amounts of Pro, Glx, and Ser.

Using this database, it is possible to determine not only the total amount of protein but also the amount of each essential amino acid. Figure 1.2 shows the food ingredients with the highest BCAA compositions for each 100 g edible portion according to the "Standard Tables of Food Composition in Japan: Amino Acid Composition of Foods, 2010" [11]. The highest values were observed in special proteins that are rarely used in home cooking, such as casein or isolated soy protein. In addition to these special proteins, there were certain types of fish (e.g., skipjack (bonito) and tuna), soybeans, cheese, chicken, and turkey but no beef or pork. Figure 1.3 shows the differences between the amino

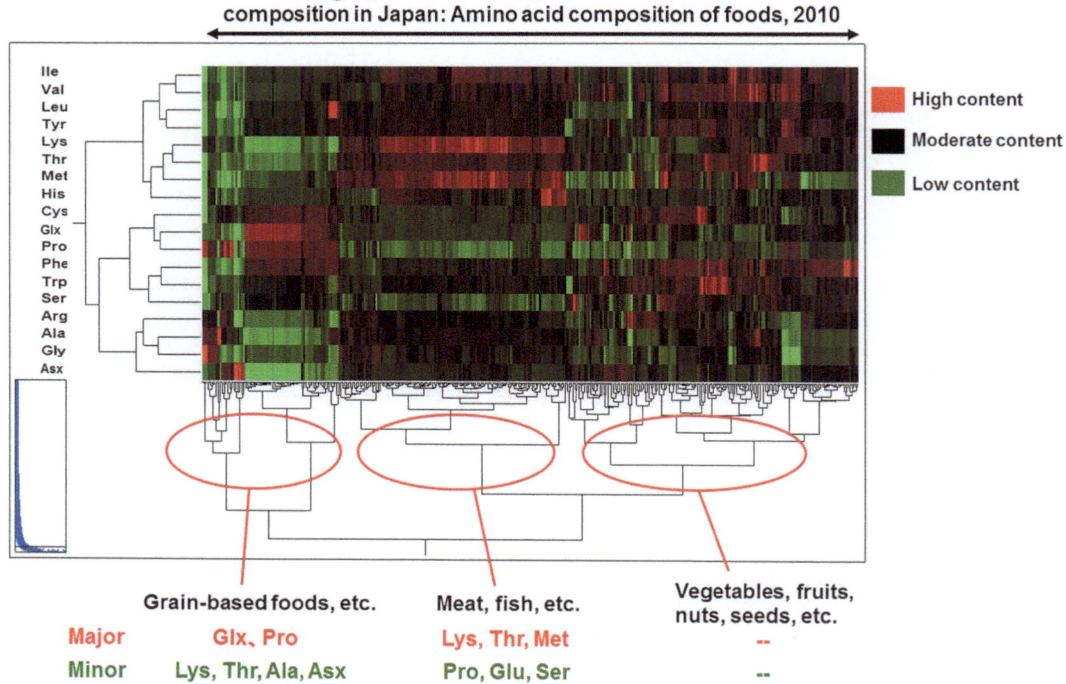

Fig. 1.1 Cluster analysis of the amino acid composition of food ingredients. A hierarchical cluster analysis using Ward's linkage method was performed to obtain a visual representation of the amino acid content of food ingredients. The analysis included 337 food ingredients that were listed in the "Standard Tables of Food Composition in Japan: Amino Acid Composition of Foods, 2010" [11]

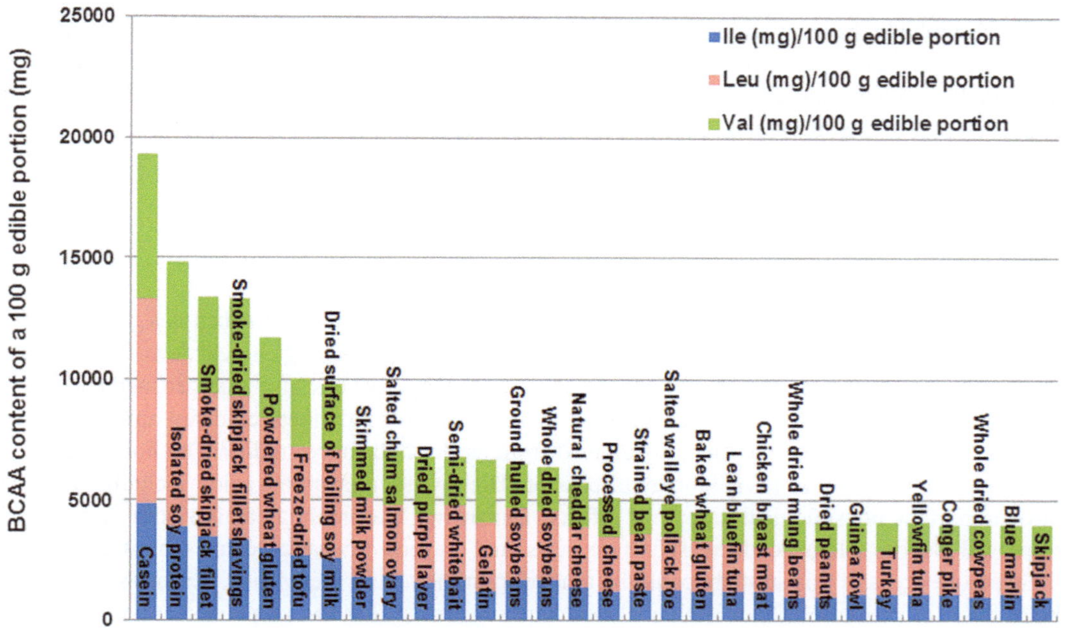

Fig. 1.2 The 30 food ingredients with the highest BCAA content among commonly used Japanese food ingredients. The BCAA content is shown for a 100 g edible portion. The food ingredients with high BCAA content were taken from the "Standard Tables of Food Composition in Japan: Amino Acid Composition of Foods, 2010" [11]

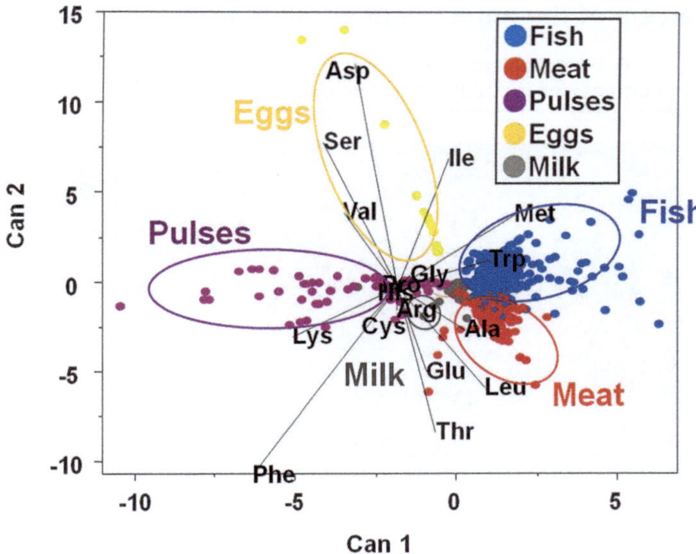

Fig. 1.3 Discriminant analysis using the amino acid composition of major dietary protein sources (meat, fish and shellfishes, pulses, eggs, and milk). A visual representation of the differences in the amino acid composition of major dietary protein sources (meat, fish and shellfish, pulses, eggs, and milk) that was obtained by conducting a discriminant analysis. The food ingredients were chosen from the meat, fish and shellfish, pulses (beans), eggs, and milk (dairy products) categories in the "Standard Tables of Food Composition in Japan: Amino Acid Composition of Foods, 2010" [11]

acid compositions of major dietary protein sources (meat, fish and shellfish, pulses, eggs, and milk), as assessed with discriminant analysis. Intriguingly, pulses (beans), eggs, and milk (dairy products) were different from fish and meat. Although the amino acid score is frequently used to evaluate the amino acid content of a diet, focusing on the amino acid content of foods (e.g., the content of Leu or BCAAs in a 100 g edible portion) would be a new way to consider dietary amino acid intake that could be applied whenever amino acid supplementation is necessary. To reap the health benefits of a specific dietary amino acid composition, selecting appropriate food ingredients can be an effective tool in addition to amino acid supplementation. Further studies should be conducted to advance the understanding of the importance of food selection with regard to amino acid composition.

Conclusions

Dietary essential amino acids have two distinct nutritional functions: (1) substrates act as protein building blocks, and (2) they act as nutritional signals to evoke physiological reactions. To determine amino acid requirements, a number of tracer techniques and approaches have emerged, and the general conclusion from these studies is that the adult requirements for individual essential amino acids are considerably greater than the standards derived from earlier N balance studies. The final conclusion regarding each requirement is still under discussion. Dietary essential amino acids, including BCAAs, have important physiological signaling effects. Enhancing the essential amino acid composition improved glucose tolerance, decreased lipogenesis, and prevented hepatic steatosis in mice with diet-induced obesity, and this strategy could be a novel preventive and therapeutic approach to nonalcoholic fatty liver disease. There is clinical evidence suggesting that the supplementation of essential amino acids or BCAAs has beneficial effects on body weight, body fat, lean body mass, and insulin sensitivity. BCAAs are known to be abundant in food, accounting for approximately 20 % of total protein intake, and they bypass initial metabolism in the liver, in contrast to the other 17 amino acids. These characteristics may contribute to the sharp elevation in plasma BCAA levels in response to a meal; BCAAs may therefore play a signaling role in the peripheral tissues and brain that is unique among amino acids. It would be reasonable to hypothesize that the signaling role of BCAAs developed as a result of evolutionary adaptations to organisms' metabolism. To reap the health benefits of essential amino acids (both as building blocks of proteins and as physiological signals), choosing food ingredients based on their amino acid composition is a potential new approach that could be combined with amino acid supplementation.

References

1. FAO/WHO/UNU. Energy and protein requirements. Geneva 1985.
2. FAO/WHO. Energy and protein requirements. Geneva 1973.
3. Rose WC. The amino acid requirements of adult man. Nutr Abstr Rev Ser Hum Exp. 1957;27(3):631–47.
4. Young VR, Tharakan JF. Metabolic and therapeutic aspects of amino acids in clinical nutrition. Boca Raton, FL: CRC Press; 2004.
5. Young VR, Scrimshaw NS. The nutritional significance of plasma and urinary amino acids. New York: Pergamon Press; 1972.
6. Harper AE, Benevenga NJ, Wohlhueter RM. Effects of ingestion of disproportionate amounts of amino acids. Physiol Rev. 1970;50(3):428–558.
7. Pion R. The relationship between the levels of free amino acids in blood and muscle and the nutritive value of proteins. New York: Academic; 1973.
8. Young VR, Tontisirin K, Ozalp I, Lakshmanan F, Scrimshaw NS. Plasma amino acid response curve and amino acid requirements in young men: valine and lysine. J Nutr. 1972;102(9):1159–69.
9. Tontisirin K, Young VR, RAND WM, Scrimshaw NS. Plasma threonine response curve and threonine requirements of young men and elderly women. J Nutr. 1974;104:495–505.
10. Layman DK. The role of leucine in weight loss diets and glucose homeostasis. J Nutr. 2003;133(1):261S–7.
11. MEXT: Ministry of Education C, Sports, Science and Technology. STANDARD TABLES OF FOOD COMPOSITION IN JAPAN AMINO ACID COMPOSITION OF FOODS 2010.
12. Ichihara A, Koyama E. Transaminase of branched chain amino acids. I Branched chain amino acids-alpha-ketoglutarate transaminase. J Biochem. 1966;59(2):160–9.
13. Wahren J, Felig P, Hagenfeldt L. Effect of protein ingestion on splanchnic and leg metabolism in normal man and in patients with diabetes mellitus. J Clin Invest. 1976;57(4):987–99.
14. Cynober L, Harris RA. Symposium on branched-chain amino acids: conference summary. J Nutr. 2006;136(1 Suppl):333S–6.
15. Zhang Y, Guo K, LeBlanc RE, Loh D, Schwartz GJ, Yu YH. Increasing dietary leucine intake reduces diet-induced obesity and improves glucose and cholesterol metabolism in mice via multimechanisms. Diabetes. 2007;56(6): 1647–54.
16. Lynch CJ, Gern B, Lloyd C, Hutson SM, Eicher R, Vary TC. Leucine in food mediates some of the postprandial rise in plasma leptin concentrations. Am J Physiol Endocrinol Metab. 2006;291(3):E621–30.
17. Macotela Y, Emanuelli B, Bang AM, et al. Dietary leucine–an environmental modifier of insulin resistance acting on multiple levels of metabolism. PLoS One. 2011;6(6):e21187.
18. Doi M, Yamaoka I, Nakayama M, Sugahara K, Yoshizawa F. Hypoglycemic effect of isoleucine involves increased muscle glucose uptake and whole body glucose oxidation and decreased hepatic gluconeogenesis. Am J Physiol Endocrinol Metab. 2007;292(6):E1683–93.
19. Noguchi Y, Nishikata N, Shikata N, et al. Ketogenic essential amino acids modulate lipid synthetic pathways and prevent hepatic steatosis in mice. PLoS One. 2010;5(8):e12057.
20. Theytaz F, Noguchi Y, Egli L, et al. Effects of supplementation with essential amino acids on intrahepatic lipid concentrations during fructose overfeeding in humans. Am J Clin Nutr. 2012;96(5):1008–16.
21. Bernard JR, Liao YH, Hara D, et al. An amino acid mixture improves glucose tolerance and insulin signaling in Sprague-Dawley rats. Am J Physiol Endocrinol Metab. 2011;300(4):E752–60.
22. Murakami H, Shimbo K, Inoue Y, Takino Y, Kobayashi H. Importance of amino acid composition to improve skin collagen protein synthesis rates in UV-irradiated mice. Amino Acids. 2012;42(6):2481–9.
23. Solerte SB, Fioravanti M, Locatelli E, et al. Improvement of blood glucose control and insulin sensitivity during a long-term (60 weeks) randomized study with amino acid dietary supplements in elderly subjects with type 2 diabetes mellitus. Am J Cardiol. 2008;101(11A):82E–8.

24. Qin LQ, Xun P, Bujnowski D, et al. Higher branched-chain amino acid intake is associated with a lower prevalence of being overweight or obese in middle-aged East Asian and Western adults. J Nutr. 2011;141(2):249–54.
25. Rennie MJ, Edwards RH, Halliday D, Matthews DE, Wolman SL, Millward DJ. Muscle protein synthesis measured by stable isotope techniques in man: the effects of feeding and fasting. Clin Sci (Lond). 1982;63(6):519–23.
26. Bennet WM, Connacher AA, Scrimgeour CM, Rennie MJ. The effect of amino acid infusion on leg protein turnover assessed by L-[15 N]phenylalanine and L-[1-13C]leucine exchange. Eur J Clin Invest. 1990;20(1):41–50.
27. Bohe J, Low A, Wolfe RR, Rennie MJ. Human muscle protein synthesis is modulated by extracellular, not intramuscular amino acid availability: a dose-response study. J Physiol. 2003;552(Pt 1):315–24.
28. Dillon EL, Sheffield-Moore M, Paddon-Jones D, et al. Amino acid supplementation increases lean body mass, basal muscle protein synthesis, and insulin-like growth factor-I expression in older women. J Clin Endocrinol Metab. 2009;94(5):1630–7.
29. Solerte SB, Gazzaruso C, Bonacasa R, et al. Nutritional supplements with oral amino acid mixtures increases whole-body lean mass and insulin sensitivity in elderly subjects with sarcopenia. Am J Cardiol. 2008; 101(11A):69E–77.
30. Morley JE, Argiles JM, Evans WJ, et al. Nutritional recommendations for the management of sarcopenia. J Am Med Dir Assoc. 2010;11(6):391–6.
31. Kim HK, Suzuki T, Saito K, et al. Effects of exercise and amino acid supplementation on body composition and physical function in community-dwelling elderly Japanese sarcopenic women: a randomized controlled trial. J Am Geriatr Soc. 2012;60(1):16–23.
32. Aquilani R, Zuccarelli GC, Dioguardi FS, et al. Effects of oral amino acid supplementation on long-term-care-acquired infections in elderly patients. Arch Gerontol Geriatr. 2011;52(3):e123–8.
33. Dal Negro RW, Aquilani R, Bertacco S, Boschi F, Micheletto C, Tognella S. Comprehensive effects of supplemented essential amino acids in patients with severe COPD and sarcopenia. Monaldi Arch Chest Dis. 2010;73(1):25–33.
34. Aquilani R, Boselli M, Boschi F, et al. Branched-chain amino acids may improve recovery from a vegetative or minimally conscious state in patients with traumatic brain injury: a pilot study. Arch Phys Med Rehabil. 2008;89(9):1642–7.
35. Smriga M, Ghosh S, Mouneimne Y, Pellett PL, Scrimshaw NS. Lysine fortification reduces anxiety and lessens stress in family members in economically weak communities in Northwest Syria. Proc Natl Acad Sci U S A. 2004;101(22):8285–8.
36. Smriga M, Torii K. L-Lysine acts like a partial serotonin receptor 4 antagonist and inhibits serotonin-mediated intestinal pathologies and anxiety in rats. Proc Natl Acad Sci U S A. 2003;100(26):15370–5.

Chapter 2
Metabolism of BCAAs

Jeffrey T. Cole

Key Points

- Branched chain amino acids are essential amino acids that cannot be synthesized in the body.
- Leucine, isoleucine and valine share enzymes for catabolizing the first two steps in their metabolism.
- The three branched chain amino acids are unique among amino acids in that their first catabolic step cannot occur in the liver.
- Disturbances in branched chain amino acids metabolism are primarily genetic in origin.
- Traumatic brain injury appears to uniquely influence the concentration of branched chain amino acids in the brain.

Keywords Traumatic brain injury • BCATm • BCKD • Astroglial-neuronal nitrogen cycle • Transamination • Michaelis constant • Nitrogen

Introduction

Leucine, isoleucine and valine are collectively known as the branched chain amino acids (BCAAs), a descriptive term derived from the structure of their side chains.

In addition to the structural similarity, and the metabolic synchronization discussed below, the plasma levels of the three BCAAs are remarkably consistent with one another in a variety of physiological conditions, including both Type 1 and Type 2 diabetes [1]. However, despite these remarkable similarities, there are subtle but important differences in their metabolism and function. Despite their similar structure, the side chains differ in the structural conformation and hydrophobicity. These differences account for the variability of roles the three BCAAs play in structural elements. For example, leucine is more common in α-helices, while valine and isoleucine tend to be found primarily in β-sheets. These slight differences in structure also suggest an underlying reason for why leucine is so often the BCAA involved in helical zippers. Additionally, these differences emphasize why the

J.T. Cole, Ph.D., M.S. (✉)
Department of Neurology, Uniformed Services University of the Health Sciences, Bethesda, MD, USA
e-mail: Jeffrey.cole@usuhs.edu

R. Rajendram et al. (eds.), *Branched Chain Amino Acids in Clinical Nutrition: Volume 1*, Nutrition and Health, DOI 10.1007/978-1-4939-1923-9_2,

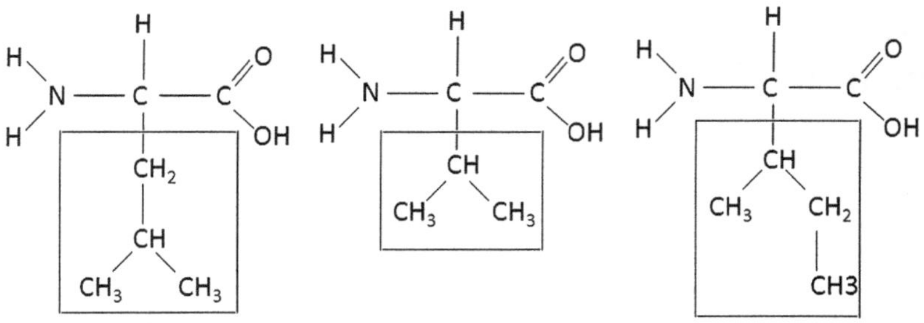

Fig. 2.1 The chemical structure of the three branched chain amino acids—leucine, valine and isoleucine. For each amino acid, the square outlines the R-group which gives all amino acids, not just BCAAs, their structural uniqueness and specificity. The slightly different structure of the BCAAs also accounts for their different formula weights (131 for leucine and isoleucine, 117 for valine). Unpublished figure from Cole

BCAAs should be considered, and researched, as separate entities. Often leucine, isoleucine and valine are treated as interchangeable amino acids and frequently, results gained from research into one of the three amino acids is thought to apply automatically to the other two BCAAs.

Structure and Physical Characteristics of BCAAs

The side chains of amino acids are what give each of the more than 20 amino acids their unique characteristics. In the case of leucine, isoleucine and vlaine, these side chains referred to as aliphatic, a general term which designates organic compounds composed of hydrogen and carbon atoms, that do not contain a benzene-like ring. The 3 BCAAs (Fig. 2.1) contain a 3 or 4 carbon side chain and are of virtually the same molecular weight (leucine 131.17 g/mol; isoleucine 131.17 g/mol; valine 117.15 g/mol). Collectively, the three branched chain amino acids make up approximately 33 % of all the amino acids in the body. A great proportion of these three amino acids are found in skeletal muscle, where they act as both a structural element and as a store for systemic nitrogen.

These amino acids are "essential" amino acids, meaning that mammals cannot de novo synthesize them, necessitating their intake from external sources. As essential amino acids, the human body must consume approximately 40, 20 and 19 mg/kg/day of leucine, valine and isoleucine, respectively. Collectively, this amounts in total to approximately 5.5 g/day for a 70 kg adult. The best sources of these amino acids are red meat or dairy products. However, it is possible for vegans to consume sufficient amounts to meet their needs by judicious consumption of soy protein and other vegetarian sources. In the typical Western diet, approximately 20 % of all dietary protein consists of the three BCAAs which makes deficiency an exceptionally rare occurrence. In fact, metabolic and physiological disorders associated with BCAAs typically involve the genetic disruptions in their metabolism or as secondary to other primary health problems and not due to deficiency (see Chap. 12 on MSUD and below for further discussion)[2–5].

Catabolism of BCAAs

To completely catabolize BCAAs it requires a number of enzymatic steps (Fig. 2.2), most of which occur in the mitochondria [6–8]. As with most amino acids, the final catabolic step results in the production of acetyl-CoA, propionyl-CoA and succinyl-CoA. However, it should be remembered that at

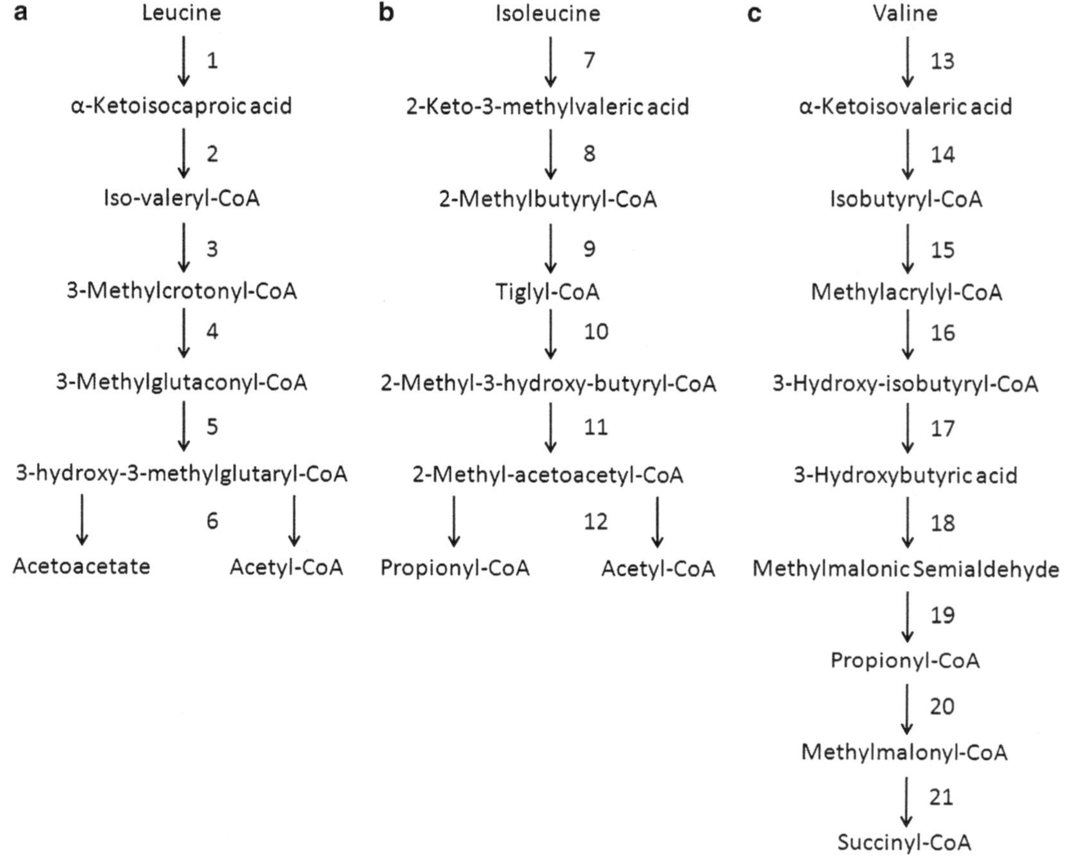

Fig. 2.2 Schematic depicting the complete metabolism of leucine, isoleucine and valine. The numbers correspond to the appropriate enzymes for each reaction. 1. Branched Chain AminoTransferase; 2. Branched Chain Keto-Acid Dehydrogenase; 3. Isovaleryl-CoA Dehydrogenase; 4. 3-Methylcrotonyl-CoA Carboxylase; 5. 3-Methylglutaconyl-CoA Hydratase; 6. 3-Hydroxy-3-Methylglutaryl-CoA lyase; 7. Branched Chain AminoTransferase; 8. Branched Chain Keto-Acid Dehydrogenase; 9. 2-Methylbutyryl-CoA Dehydrogenase; 10. Enol-CoA Dehydrogenase; 11. 2-Methylhydroxybutyryl CoA Thiolase;12. 3-Methylacetoacetyl-CoA Thiolase; 13. Branched Chain AminoTransferase; 14. Branched Chain Keto-Acid Dehydrogenase; 15. 2-Methylbutyryl-CoA Dehydrogenase; 16. Enol-CoA Dehydrogenase; 17. 3-Hydroxyisobutyryl-CoA Deacylase; 18. 3-Hydroxyisobutyryl-CoA Dehydrogenase; 19. Methylmalonic semialdehyde Dehydrogenase; 20. Propionyl-CoA Carboxylase; 21. Methylmalonyl-CoA Mutase. Unpublished figure from Cole

virtually every catabolic step, the metabolite can be diverted for other uses, such as fatty acid or cholesterol synthesis. Additionally, while some reactions are reversible, leucine, isoleucine and valine cannot be synthesized de novo by the body.

Two shared enzymes catabolize the first step: BCAA catabolism begins with a transamination reaction catalyzed by the branched chain aminotransferase (BCATs). The distribution of these enzymes will be discussed in greater depth in Chap. 3; briefly however, there are two isoforms—mitochondrial BCATm and cytosolic BCATc. In fact, the cytosolic isoform catalyzes the only catabolic step in BCAA metabolism that occurs outside of the mitochondria [9–12]. These isozymes transfer the α-amino group from leucine, isoleucine or valine to α-ketoglutarate [13, 14]. This metabolic step is unique to BCAA metabolism as the BCAT isozymes (and the second step, catabolized by BCKD) are common to three different amino acids, while no other amino acids share a primary first step enzyme. The products resulting from this reaction are glutamate, which now contains a nitrogen-containing group provided

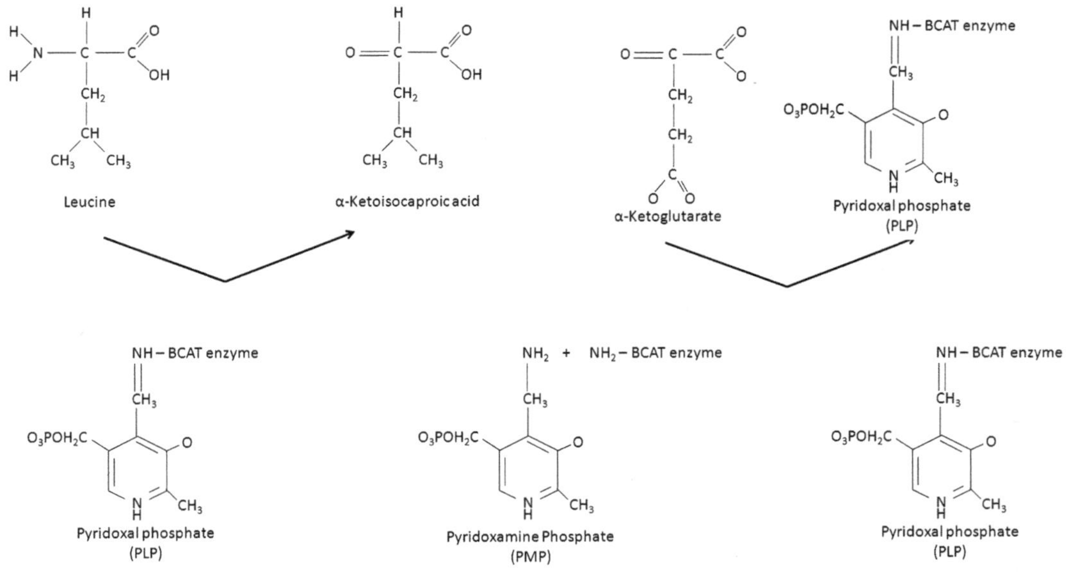

Fig. 2.3 An example of the first step of BCAA catabolism, using leucine as an example. In the first half of this "ping-pong" reaction, leucine is deaminated, forming α-ketoisocaproic acid, with the amino group being temporarily transferred to the BCAT isozyme. Both the mitochondrial and cytosolic isozymes function identically in this reaction. Accepting the amino group converts BCAT-PLP to BCAT-PMP. In the second half of the reaction, BCAT-PMP donates the amino group to an acceptor molecule, typically α-ketoglutarate, to form the amino acid glutamate. The BCAT-PMP is thus returned to its initial BCAT-PLP state. PLP—pyridoxal 5′ phosphate. PMP—pyridoxamine 5′ phosphate. Unpublished figure from Cole

by the BCAAs and a branched chain keto-acid (BCKA) [15–17]. Once deaminated, leucine, isoleucine and valine form α-ketoisocaproic acid, α-keto-β-methylvaleric acid and α-ketoisovaleric acid, respectively.

Both BCATm and BCATc use Vitamin B-6 cofactors to catalyze the reaction. During the first half of reaction, pyridoxal-5-phosphate (PLP) acts as a temporary acceptor of the α amino group being donated from the BCAA and becomes pyridoxamine-5-phosphate, which converts BCAT-PLP to BCAT-PMP. In the second half of the reaction, BCAT-PMP donates the amino group to α-ketoglutarate. This converts the BCAT-PMP back to BCAT-PLP, while simultaneously transaminating α-ketoglutarate to glutamate (Fig. 2.3). This is often referred to as "ping-pong metabolism", as the enzyme and cofactor revert back to their initial state following release of the temporarily attached amino group.

As mentioned, studies of the BCAAs frequently "lump" the three amino acids together and use one BCAA, usually leucine, as a representative for the other two BCAAs. However, there is a marked substrate preference for the BCAT enzymes for isoleucine, followed by leucine and then valine. Typically, α-ketoglutarate is presumed to be the default amino group acceptor although other ketoacids can serve this function. Indeed, kinetic data indicates that BCKAs themselves can act as substrates for the second half of the reaction (i.e. the donation of amino group from BCAT-PMP), with K_m values indicating that KIC is preferred to KIV and KMV, and even α-ketoglutarate. The Michaelis constant (K_m) characterizes the affinity of an enzyme for a substrate. In an enzymatic reaction, the K_m value is the substrate concentration at which the reaction proceeds at half its maximum speed. A low K_m value means that the enzyme reaction has a high affinity for its substrates, with just a small amount of substrate needed to make the reaction proceed at its maximum rate. However, despite this marked preference, Islam and Hutson clearly demonstrate that specificity constant (k_{cat}/K_m) for BCATm determine that α-ketoglutarate is the major substrate in vivo for the second half reaction. A specificity constant is the efficiency of the enzyme reaction and is determined by the K_{cat} (maximum rate at which an enzyme can function) divided by the K_m (the rate of substrate-enzyme interaction) [18].

Enzymatic characteristics of BCATc and BCATm: There are many unique features of BCAT enzymes, one of which is a redox-sensitive CXXC center that plays a major role in catalytic reactions. In this case, the C's represent the amino acid cysteine, while the X's can be any amino acid. Both isozymes of BCAT are reversible, and it is this redox center that permits this reversibility. In most cells, BCAT enzymes operate near equilibrium, with both substrate and product concentrations being at or below their K_m values. This allows BCAT isozymes to quickly respond to changes in tissue concentrations of substrate or product, which allows BCAAs to be an ideal reserve for both carbon skeletons and nitrogen for glutamate synthesis.

However, this near equilibrium status also means that for the reaction to proceed, rather than cycle between BCAAs and BCKAs, BCKAs must be eliminated. This can occur via simple removal from the cell which is not typical as demonstrated by the exceptionally low circulating levels of BCKAs in the serum [19]. However, the movement of BCKAs between tissues (or cell types, as seen in the brain, discussed above) would provide a mechanism for the transfer of BCAA-derived nitrogen between organs and tissues. The other, more prevalent, mechanism by which BCKAs can be eliminated is through simple oxidation within the cells. Removal of the BCKA carbon skeleton following transamination results in the net transfer of the amino group from BCAAs to glutamate.

BCKD catabolizes the second step: The oxidation of BCKAs following BCAA deamination is regulated by the second enzyme in BCAA catabolism. As with the BCAT isozymes, this enzyme is also shared among the three BCAAs. The mitochondrial branched chain α-keto acid dehydrogenase (BCKD) irreversibly catabolizes BCKAs and regulates the rate of BCAA metabolism. Interestingly, the rate at which BCAA carbon skeletons are oxidized mirrors the rate of net transfer of nitrogen to glutamate from BCAAs. BCKD, also sometimes referred to as BCKDC (the "C" designating "complex"), is one of a family of macromolecular proteins which include pyruvate dehydrogenase (PDH) and α-ketoglutarate dehydrogenase (α-KDH).

Structural arrangement of BCKD: BCKD contains within its complex 3 enzymes, each of which also has multiple copies. The three subunits have been identified as branched chain α-keto acid decarboxylase (E1), a dihydrolipoyl transacylase (E2) and a dihydrolipoyl dehydrogenase (E3). Ultimately, BCKD oxidatively decarboxylates any of the three BCKAs, forming isovaleryl-CoA from KIC, isobutyryl-CoA from KIV, and alpha-methylbutyryl-CoA from KMV, with NADH being also produced. This reaction is both rate limiting and highly regulated, with BCKD kinase (BCKD-K) mediating the reaction by associating and dissociating from the complex. BCKD-K association with BCKD results in phosphorylation and inactivation of E1 enzyme. The regulation of BCKD-K is in turn regulated by changes in the local concentrations of KIC. Elevated levels of KIC inactivate BCKD-K, which facilitates oxidation of all three BCKAs. When levels of KIC drop, the BCKD-K is reactivated, signaling it to inhibit BCKD activity. Some research groups have speculated about the existence of a BCKD phosphatase. However, it has yet to be identified and the exact function and role of this enzyme has not been determined, nor have the possible protein regulators of this phosphatase been identified [20–22].

All three enzymes catalyze multi-step reactions and irreversibly oxidatively decarboxylate α-keto acids. Due to its complexity, the structure and function of BCKD enzyme has only been partially elucidated [18]. However, the enzyme complex is assembled around a scaffolding consisting of a dihydrolipoamide transacylase (E2 subunit). This 24-meric core is the noncovalent attachment site for multiple copies each of the BCKA dehydrogenase (E1 subunit) and the dihydrolipoamide dehydrogenase (E3 subunit). It is the E1 subunit that catalyzes the decarboxylation of BCKAs, using as a cofactor thiamine diphosphate (ThDP). The reaction produces carbanion ThDP, which is then reduced by E1, transferring the carbanion to the lipoyl moiety bound to the E2 subunit. The acyl group is transferred to the active site of E2 by the sacyldihydrdolipoamide domain. The FAD moiety of E2 reoxidizes the dihydrolipoyl residue, while NAD+serves as the electron acceptor. In turn, the mitochondria will oxidize the NADH to provide energy (Fig. 2.4).

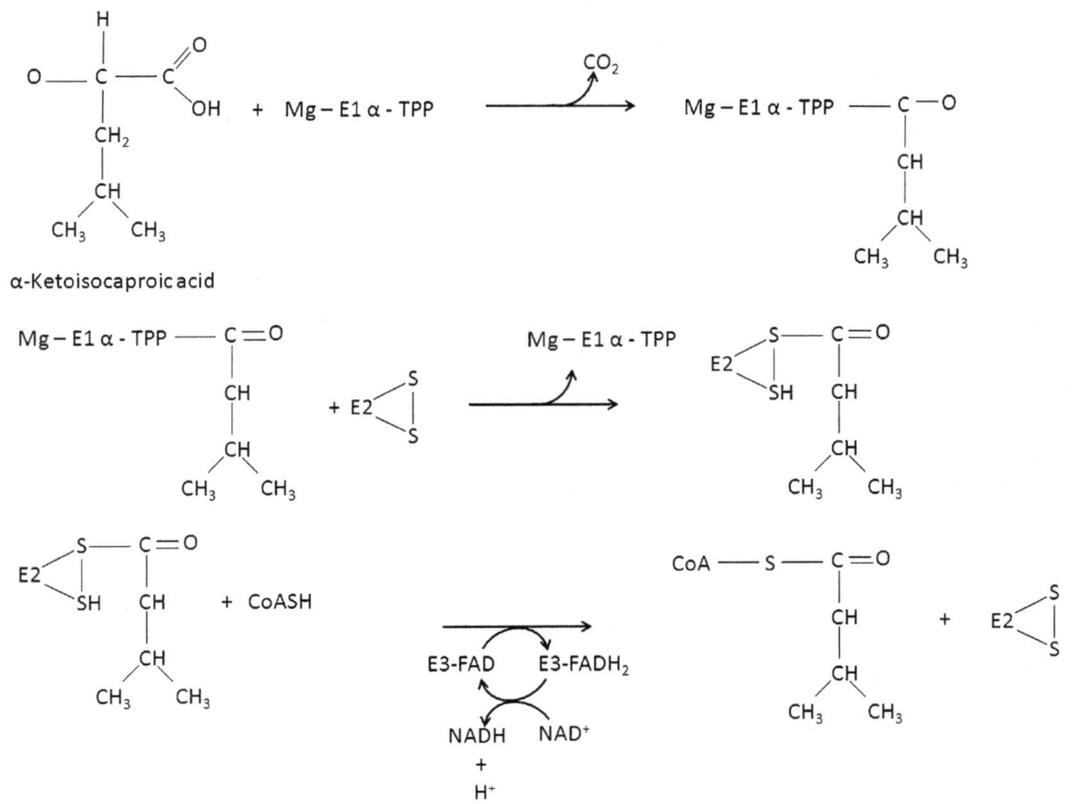

Fig. 2.4 Branched chain ketoacid dehydrogenase catabolizes the second, irreversible, step in BCAA catabolism. While any of the three ketoacids generated during BCAA catabolism can act as a substrate, this figure uses α-ketoisocaproic acid as an example. In step A, the E1 subunit of BCKD, in a thiamine pyrophosphate-mediated reaction temporarily binds to the BCKA, releasing a carbon dioxide (CO_2). This complex then interacts with the E2 subunit (step B), transferring the decarboxylated BCKA and releasing the E1 subunit. This E2 complex then transfers the BCKA derivative to CoASH, forming the third product in BCAA metabolism (after the BCAA and BCKA), which are isovaleryl-CoA, 2-methylbutyryl-CoA and isobutyryl-CoA. During this release, the FAD-linked E3 subunit acts as a temporary proton acceptor, mediating the conversion of NAD^+ to NADH. FAD—flavin adenine dinucleotide. NAD—nicotinamide adenine dinucleotide. Unpublished figure from Cole

BCATm and BCKD– a metabolon?: The efficiency of BCATm and BCKD is enhanced not just by the close physical proximity of the enzymes, but through an actual structural relationship referred to as a metabolon. Islam and Hutson demonstrated channeling of the BCKAs formed via BCATm activity directly to the E1 component of the BCKD complex [18, 23]. This direct channeling enhanced the capacity of BCKD to decarboxylate BCKA and the overall complex activity. Using isothermal calorimetry, Islam et al. determined that BCATm and the E1 subunit of BCKD directly, structurally, interact. However, this association, with a binding constant of 6 mM, is relatively weak. Despite this weak association, when a BCAA donates its amino group to BCATm-PLP, the resultant BCKA is channeled directly to the active site of E1, while the BCATm-PMP, having transaminated alpha-ketoglutarate, is released. These results suggest that not only is the activity of BCATm and BCKD dependent on substrate availability, but also on the ratio of BCATm to BCKD E1. This close relationship drives transamination and glutamate synthesis while providing additional carbon skeletons for energy generation.

Downstream BCAA Catabolism: The remaining catabolic steps are, like BCKD, mitochondrial in location. Two different dehydrogenases (isovaleryl-CoA dehydrogenase for leucine products and 2-methyl-butyryl-CoA dehydrogenase for isoleucine and valine products) oxidize the branched chain acyl-CoA products to form 3-methylcrotonyl CoA, tigloyl-CoA and methylacryl-CoA [24]. Following this step, the catabolic pathways of leucine, isoleucine and valine, which had shared enzymes and pathways, diverge (Fig. 2.2) The ketogenic leucine product Isovaleryl-CoA is ultimately catabolized to form acetyl-CoA and acetoacetate. Its metabolism involves β-methylcrotonyl-CoA carboxylase, which is both ATP and biotin dependent. In contrast, both valine and isoleucine are considered to be glucogenic, with both BCAAs being metabolized to propionyl-CoA (and Acetyl-CoA in the case of isoleucine) then succinate via methylmalonyl-CoA. Additionally, valine catabolism results in the production of 3-hydroxybutyrate (3-HB) which can be released from tissue. As a ketone body, 3-HB can be used as an alternative fuel source when glucose is depleted.

Distribution of BCAA Metabolism

Unique characteristics of BCAA metabolism: The metabolism of BCAAs is relatively well-understood and straightforward. However, it is the cellular and tissue distribution of their primary metabolizing enzymes that gives leucine, isoleucine and valine such an important role in nitrogen metabolism [25]. Significantly, neither isoform of BCAT is expressed in the liver. BCATc is expressed primarily in the brain, testes and ovaries, while BCATm is found throughout the body. The distribution of BCAT and BCKD will be discussed further in Chap. 3. This makes BCAAs somewhat unique in comparison to the other amino acids in that not only is their first catabolic step not occurring in the liver, but that its primary catabolism occurs in the skeletal muscle [26, 27]. Despite the fact that dietary amino acids typically contain about 20 % BCAAs, amino acids in the blood exiting the splanchnic bed typically consist of about 50 % BCAAs. In fact, approximately 50 % of all skeletal muscle amino acid uptake consists of BCAAs, since leucine, isoleucine and valine all avoid "first pass metabolism". The phrase "first pass metabolism" refers to hepatic metabolism of substrates immediately following their absorption from the intestine [25]. Indeed, following consumption of a protein rich meal (beef) in which BCAAs make up about 20 % of the amino acids, the splanchnic output consisted of about 50 % BCAAs [28–30]. This caused speculation that about half the capacity to catabolize BCAAs is found in the skeletal muscle, with a substantial portion also being seen in adipose tissue. This prompted experiments in which forearm metabolism was monitored after the ingestion of a steak. Following consumption of this high protein meal, BCAA uptake into the muscle was significantly increased. However the release of BCKAs was only minimally affected. This suggests that BCAAs are being completely catabolized in the muscle, or contributing to protein deposition, which may account for the popularity of these amino acids as muscle building supplements.

A key function of the BCAAs in the skeletal muscle is to provide the nitrogen needed to maintain muscle pools of glutamate, alanine and glutamine. In the skeletal muscle, BCAT activity is substantially higher than BCKD activity. The high ratio of BCAT:BCKD activity indicates a tissue capacity for efficiently transferring nitrogen from BCAAs to alpha-ketoglutarate and also possibility of a high release of BCKAs from the tissue instead of their being oxidized in situ. Despite the extensive release of BCKAs from the skeletal muscle, the concentrations of circulating BCKAs are exceptionally low. This is due to the complete absence of BCAT isozymes in the liver, while the hepatocytes demonstrate an extremely high level of BCKD activity. This allows virtually complete oxidation of BCKAs and maintains low levels.

Interorgan BCAA metabolism: Since muscle is not a gluconeogenic tissue, valine and isoleucine cannot be completely oxidized in this tissue if they are to be converted to glucose. Studies in cardiac muscle have demonstrated that KIC is almost completely oxidized, while KIV is not. In fact, in both cardiac and skeletal muscle, the primary end product of valine metabolism is not complete oxidation but instead beta-hydroxyisobutyrate. This metabolite is a key indicator for the fate of valine in the muscle. While in a fed state, plasma beta-hydroxyisobutyrate levels are approximately 20 mM, there is a fivefold increase in fasted humans. This is due to the role of β-hydroxyisobutyrate as a key inter-organ gluconeogenic substrate, and is in fact, an ideal gluconeogenic substrate in hepatocytes and renal cortical tubules.

BCAA metabolism in the brain: The metabolism of BCAAs in the brain represents a somewhat unique case, in comparison to the rest of the body. This is due to the unique distribution of the BCAA metabolizing enzymes, with the brain being the only tissue (along with the ovaries and testes) that express BCATc, BCATm and BCKD. In the central nervous system (CNS), BCATm and BCKD are expressed in different cell types, despite convincing evidence that elsewhere the two enzymes form a metabolon to facilitate BCAA catabolism. This suggests, perhaps, a slightly different role for BCAAs in the brain than in the rest of the body. In the CNS, BCATc is expresses primarily in neurons, as is BCKD, while in BCATm it appears to be limited to astrocytes [31, 32]. This will be discussed more in depth in Chap. 3.

Metabolic Disorders

While BCAA deficiency is exceptionally rare, there are several disorders in their metabolism that can cause significant health problems. Perhaps the most well-known is Maple Syrup Urine Disease, which is caused by mutations in the BCKD enzyme complex. This will be discussed completely in Chap. 12. There is another, equally rare, autosomal recessive disease known as Isovaleric Acidemia. This disorder is caused by a mutation in the enzyme isovaleryl-CoA dehydrogenase, which converts isovaleryl CoA to 3-methylcrotonyl-CoA. The accumulation of isovaleryl-CoA results in many of the symptoms of this disorder, which may prove fatal. As with MSUD, this disorder can best be treated by strict regulation of the dietary intake of BCAAs.

Extra-metabolic functions of BCAAs in the brain: In the brain, the BCAAs play a key role in maintaining the supply of the neurotransmitter glutamate. To understand this contribution, a brief discussion of the glutamate:glutamine cycle and the astroglial:neuronal nitrogen hypothesis are in order. In the brain, glutamate serves as an excitatory neurotransmitter and is released from the synaptic cleft. After interacting with receptors on the post-synaptic neuron, it is removed from the synapse by astrocytes. Collectively, the astrocyte and pre- and post-synaptic neuron are referred to as a "tripartite synapse". The glutamate is removed from the synaptic cleft to prevent unrestrained, uncontrolled hyperactivation of the post-synaptic neuron. The astrocyte aminates the glutamate in a glutamine synthetase-mediated reaction to form glutamine. This amino acid is then returned to the pre-synaptic neuron where it can be deaminated by glutaminase to re-enter the pool of neurotransmitter glutamate. This process has been named the "glutamate:glutamine cycle", and allows for the conservation of excitatory neurotransmitters within the brain. However, as with any physiological processes, the glutamate:glutamine cycle is not 100 % efficient [15–17]. At each step of the process, glutamate can be lost or diverted for other purposes. In the astrocyte, glutamate can be aminated to form glutamine. However, astrocytes also convert a significant fraction of the glutamate to alpha-ketoglutarate via glutamate dehydrogenase, which then enters the TCA cycle and is catabolized eventually to lactate. In fact, the flow of TCA cycle intermediates to glycolytic products means that significant glutamate synthesis must occur to replace the lost glutamate. Additionally, the glutamine produced in the astrocyte can be used both within the astroycte or neuron for purine and pyrimidine synthesis, or for nitrogen fixation. When glutamine has been returned to the pre-synaptic neuron, the glutamate produced by its deamination by glutaminase can similarly be scavenged for other uses, notably for energy

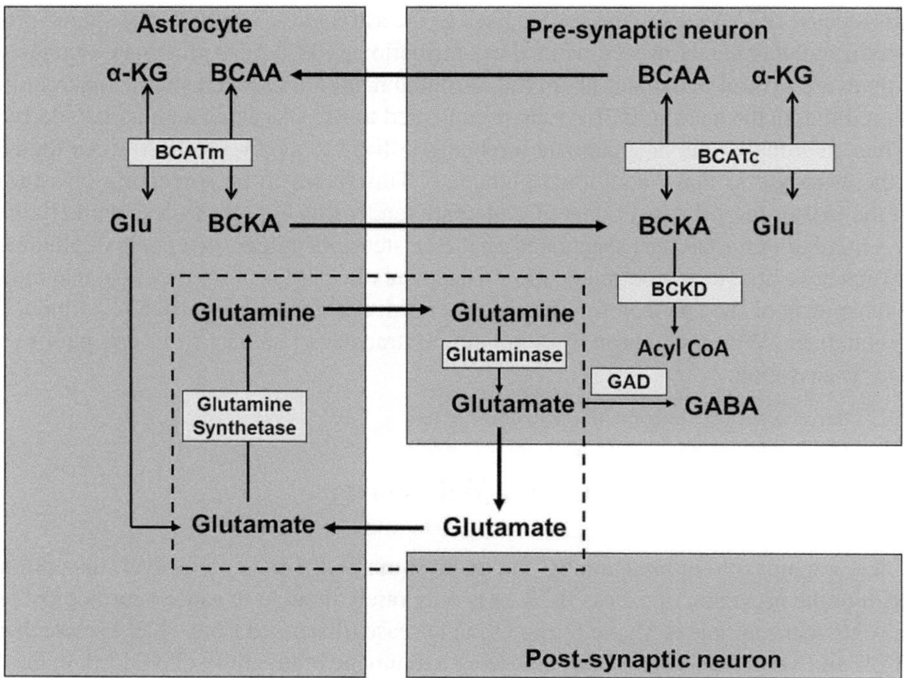

Fig. 2.5 Scheme for the Astroglial-Neuronal Shuttle depicts the current model for BCAA function in the brain. The glutamate:glutamine cycle (within the *dotted line box*) depicts the uptake of glutamate from the synaptic cleft following release from the pre-synaptic neuron. After re-amination, to form glutamine, the synaptically inert glutamine is returned from the astrocyte to the pre-synaptic neuron. It is deaminated to form the excitatory neurotransmitter glutamate. Since this reaction is not 100 % efficient, it requires a continual input of new glutamate (and nitrogen), which led to the development of the astroglial-neuronal nitrogen shuttle hypothesis. Astrocytic BCATm transaminates BCAAs to form glutamate. The resultant BCKA carbon skeleton is returned to the neuron for re-amination, prior to return to the astrocyte. Alternatively, the BCKAs are disposed of in the neuron by irreversible catabolism by branched chain keto acid dehydrogenase (BCKD). BCAT- branched chain aminotransferase. BCKD- branched chain ketoacid dehydrogenase. GAD- glutamic acid decarboxylase. BCAA, branched chain amino acid. Glu- glutamate. α-kg—ketoglutarate. Unpublished figure from Cole

production via the TCA cycle, as in the astrocyte. Put simply, there is no guarantee that the glutamate released from the pre-synaptic neuron will be returned to the pre-synaptic neuron as a neurotransmitter without being used for other physiological purposes.

Astroglial-neuronal nitrogen shuttle: While the glutamate:glutamine pool is well-documented, less well understood is the means by which the glutamate pool is maintained despite losses for other purposes. This led to the creation of the "astroglial-neuronal nitrogen shuttle hypothesis" based on the contribution of BCAAs to glutamate metabolism in the brain (Fig. 2.5). Approximately 40 % of glutamate in the brain contains an amino group that was derived from a BCAA. This hypothesis was originally developed from results obtained in cultured astrocytes in which leucine oxidation is very slow, while release of alpha-ketoisocaproic acid is high, as well as studies using radio-labeled leucine to track the nitrogen label.

In the astroglial-neuronal nitrogen shuttle, BCATm-mediated BCAA transamination occurs in the astrocyte, with alpha-ketoglutarate being the amino group receptor resulting in glutamate production. The glutamate can then be further aminated as it enters the glutamate:glutamine cycle. However, the BCKAs are not primarily catabolized in the astrocyte. In fact, the BCKAs themselves are also putatively transferred to the neuron. Once in the neuron, the BCKAs are re-aminated by BCATc.

The re-capitulated BCAAs are then shuttled back to the astrocyte to complete the shuttle process and further produce glutamate. However, while the contribution of BCAAs to glutamate synthesis is indisputable, there are several paradoxes about the astroglial neuronal nitrogen shuttle that require further elaboration. First, in the neuron, BCKAs are re-aminated to BCAAs using a glutamate-derived amino group. Thus, neuronal pools of glutamate are being utilized to synthesis BCAAs for the return trip back to the astrocyte to make additional glutamate. This seems to be something of a futile cycle, although the existence of distinct pools of glutamate render this hypothesis less unlikely than it first appears. A number of researchers speculated on the existence of at least two pools of glutamate in the neuron which have little to no communication. The metabolic pool is many orders of magnitude larger than the synaptic pool and can therefore provide the needed amino group to BCKAs without depleting synaptic glutamate. Why the neuron does not simply transfer glutamate from one pool to the other remains to be answered.

Traumatic Brain Injury and BCAA Metabolism

BCAAs play a major role in maintaining glutamate stores in the brain. However, disruption of their metabolism or the provision of excess BCAAs is only rarely thought to cause neurological problems. While the classical example is Maple Syrup Urine Disease (discussed Chap. 12), a research effort led by Drs. Cole and Verma determined that following a traumatic brain injury (TBI) [33], in the damaged region of the brain, there is a significant decrease in the concentration of BCAAs. Decreases in BCAAs correlated very strongly to impairments in electrophysiological activity and neurological/cognitive function. Interestingly, application of BCAAs restored brain function in the injured areas and performance on memory tasks used to assess cognitive function was improved. Aquilani et al. (2005, 2008) conducted a clinical trial in which severely brain injured patients are given IV BCAAs. Using a Disability Rating Score, these researchers demonstrated a significant improvement in the neurological health of these individuals. See chapter regarding "The branched chain amino acids in the context of other amino acids in traumatic brain injury" for more details.

Conclusions

Leucine, isoleucine and valine are key amino acids in a number of systemic functions, most particularly nitrogen homeostasis and neurological function. In comparison to other amino acids, the three BCAAs have a unique distribution of enzymes, and, perhaps more uniquely, the property of sharing key enzymes. Understanding these structural, biochemical, and metabolic properties of BCAAs will shed significant light onto the following chapters.

References

1. Marchesini G, Bianchi GP, Vilstrup H, Capelli M, Zoli M, Pisi E. Elimination of infused branched-chain amino-acids from plasma of patients with non-obese type 2 diabetes mellitus. Clin Nutr. 1991;10(2):105.
2. Lang CH, Lynch CJ, Vary TC. BCATm deficiency ameliorates endotoxin-induced decrease in muscle protein synthesis and improves survival in septic mice. Am J Physiol Regul Integr Comp Physiol. 2010;299(3):R935.
3. Perez-Villasenor G, Tovar AR, Moranchel AH, Hernandez-Pando R, Hutson SM, Torres N. Mitochondrial branched chain aminotransferase gene expression in AS-30D hepatoma rat cells and during liver regeneration after partial hepatectomy in rat. Life Sci. 2005;78(4):334.

4. Richardson MA, Small AM, Read LL, Chao HM, Clelland JD. Branched chain amino acid treatment of tardive dyskinesia in children and adolescents. J Clin Psychiatry. 2004;65(1):92.
5. Rossi Fanelli F, Cangiano C, Capocaccia L, Cascino A, Ceci F, Muscaritoli M, Giunchi G. Use of branched chain amino acids for treating hepatic encephalopathy: clinical experiences. Gut. 1986;27 Suppl 1:111.
6. Berkich DA, Ola MS, Cole J, Sweatt AJ, Hutson SM, LaNoue KF. Mitochondrial transport proteins of the brain. J Neurosci Res. 2007;85(15):3367.
7. Bixel M, Shimomura Y, Hutson S, Hamprecht B. Distribution of key enzymes of branched-chain amino acid metabolism in glial and neuronal cells in culture. J Histochem Cytochem. 2001;49(3):407.
8. Cole JT, Sweatt AJ, Hutson SM. Expression of mitochondrial branched-chain aminotransferase and alpha-keto-acid dehydrogenase in rat brain: implications for neurotransmitter metabolism. Front Neuroanat. 2012;6:18.
9. Bixel MG, Hutson SM, Hamprecht B. Cellular distribution of branched-chain amino acid aminotransferase isoenzymes among rat brain glial cells in culture. J Histochem Cytochem. 1997;45(5):685.
10. Castellano S, Casarosa S, Sweatt AJ, Hutson SM, Bozzi Y. Expression of cytosolic branched chain aminotransferase (BCATc) mRNA in the developing mouse brain. Gene Expr Patterns. 2007;7(4):485.
11. Garcia-Espinosa MA, Wallin R, Hutson SM, Sweatt AJ. Widespread neuronal expression of branched-chain aminotransferase in the CNS: implications for leucine/glutamate metabolism and for signaling by amino acids. J Neurochem. 2007;100(6):1458.
12. Sweatt AJ, Garcia-Espinosa MA, Wallin R, Hutson SM. Branched-chain amino acids and neurotransmitter metabolism: expression of cytosolic branched-chain aminotransferase (BCATc) in the cerebellum and hippocampus. J Comp Neurol. 2004;477(4):360.
13. Brosnan JT, Brosnan ME. Branched-chain amino acids: enzyme and substrate regulation. J Nutr. 2006;136(1 Suppl):207S.
14. Drown PM, Torres N, Tovar AR, Davoodi J, Hutson SM. Use of sulfhydryl reagents to investigate branched chain alpha-keto acid transport in mitochondria. Biochim Biophys Acta. 2000;1468(1–2):273.
15. Yudkoff M, Daikhin Y, Grunstein L, Nissim I, Stern J, Pleasure D. Astrocyte leucine metabolism: significance of branched-chain amino acid transamination. J Neurochem. 1996;66(1):378.
16. Yudkoff M, Daikhin Y, Lin ZP, Nissim I, Stern J, Pleasure D. Interrelationships of leucine and glutamate metabolism in cultured astrocytes. J Neurochem. 1994;62(3):1192.
17. Yudkoff M, Daikhin Y, Nissim I, Horyn O, Luhovyy B, Lazarow A. Brain amino acid requirements and toxicity: the example of leucine. J Nutr. 2005;135(6 Suppl):1531S.
18. Islam MM, Wallin R, Wynn RM, Conway M, Fujii H, Mobley JA, Chuang DT, Hutson SM. A novel branched-chain amino acid metabolon. Protein-protein interactions in a supramolecular complex. J Biol Chem. 2007;282(16):11893.
19. She P, Van Horn C, Reid T, Hutson SM, Cooney RN, Lynch CJ. Obesity-related elevations in plasma leucine are associated with alterations in enzymes involved in branched-chain amino acid metabolism. Am J Physiol Endocrinol Metab. 2007;293(6):E1552.
20. Lang CH, Frost RA, Deshpande N, Kumar V, Vary TC, Jefferson LS, Kimball SR. Alcohol impairs leucine-mediated phosphorylation of 4E-BP1, S6K1, eIF4G, and mTOR in skeletal muscle. Am J Physiol Endocrinol Metab. 2003;285(6):E1205.
21. Lynch CJ. Role of leucine in the regulation of mTOR by amino acids: revelations from structure-activity studies. J Nutr. 2001;131(3):861S.
22. Lynch CJ, Halle B, Fujii H, Vary TC, Wallin R, Damuni Z, Hutson SM. Potential role of leucine metabolism in the leucine-signaling pathway involving mTOR. Am J Physiol Endocrinol Metab. 2003;285(4):E854.
23. Islam MM, Nautiyal M, Wynn RM, Mobley JA, Chuang DT, Hutson SM. The branched chain amino acid (BCAA) metabolon: interation of glutamate dehydrogenase with the mitochondrial branched chain aminotransferase (BCATm). J Biol Chem. 2010;285(1):265–76.
24. Sweatt AJ, Wood M, Suryawan A, Wallin R, Willingham MC, Hutson SM. Branched-chain amino acid catabolism: unique segregation of pathway enzymes in organ systems and peripheral nerves. Am J Physiol Endocrinol Metab. 2004;286(1):E64.
25. Goichon A, Chan P, Lecleire S, Coquard A, Cailleux AF, Walrand S, Lerebours E, Vaudry D, Dechelotte P, Coeffier M. An enteral leucine supply modulates human duodenal mucosal proteome and decreases the expression of enzymes involved in fatty acid beta-oxidation. J Proteomics. 2013;78:535.
26. DeSantiago S, Torres N, Suryawan A, Tovar AR, Hutson SM. Regulation of branched-chain amino acid metabolism in the lactating rat. J Nutr. 1998;128(7):1165.
27. Dillon EL. Nutritionally essential amino acids and metabolic signaling in aging. Amino Acids. 2013;45(3):431–41.
28. Faure M, Glomot F, Papet I. Branched-chain amino acid aminotransferase activity decreases during development in skeletal muscles of sheep. J Nutr. 2001;131(5):1528.
29. Pelletier V, Marks L, Wagner DA, Hoerr RA, Young VR. Branched-chain amino acid interactions with reference to amino acid requirements in adult men: leucine metabolism at different valine and isoleucine intakes. Am J Clin Nutr. 1991;54(2):402.

30. Purpera MN, Shen L, Taghavi M, Munzberg H, Martin RJ, Hutson SM, Morrison CD. Impaired branched chain amino acid metabolism alters feeding behavior and increases orexigenic neuropeptide expression in the hypothalamus. J Endocrinol. 2012;212(1):85.
31. Castellano S, Macchi F, Scali M, Huang JZ, Bozzi Y. Cytosolic branched chain aminotransferase (BCATc) mRNA is up-regulated in restricted brain areas of BDNF transgenic mice. Brain Res. 2006;1108(1):12.
32. Mersey BD, Jin P, Danner DJ. Human microRNA (miR29b) expression controls the amount of branched chain alpha-ketoacid dehydrogenase complex in a cell. Hum Mol Genet. 2005;14(22):3371.
33. Cole JT, Mitala CM, Kundu S, Verma A, Elkind JA, Nissim I, Cohen AS. Dietary branched chain amino acids ameliorate injury-induced cognitive impairment. Proc Natl Acad Sci U S A. 2010;107(1):366.

Chapter 3
The Cytosolic and Mitochondrial Branched Chain Aminotransferase

Myra E. Conway and Susan M. Hutson

Key Points

- The branched chain aminotransferases (BCAT) are PLP-dependent proteins which catalyze the transfer of an amino group from the donor amino acid to α-ketoglutarate, forming glutamate and the respective keto acids.
- Structurally the BCAT proteins are homodimers, where the active site between each isoform is largely conserved.
- The cytosolic and mitochondrial isoforms show cell and tissue specific expression where the aminotransferase proteins play an integrated role in shuttling metabolites between cells and tissues.
- These anaplerotic shuttles interface with core metabolic pathways and protein complexes such as the branched chain α-keto acid dehydrogenase complex and glutamate dehydrogenase, respectively, indicating a role in the regeneration of key metabolites such as the primary neurotransmitter glutamate.
- Leucine is a nutrient signal and involved in mTOR signalling, which controls the synthesis of cellular protein levels.
- Moreover, the BCAT proteins have a unique redox-active CXXC motif regulated through changes in the redox environment, likely to play a key role in this signalling mechanism.
- Site-directed mutagenesis studies have identified that the N-terminal cysteine acts as the 'redox sensor' and the C-terminal cysteine as its resolving partner, which permits reversible regulation.
- Oxidation, S-nitrosation and S-glutathionylation are important redox regulators of BCAT activity and are reversibly controlled through the glutaredoxin/glutathione system.
- Biochemical and X-ray crystallography studies of the redox-active mutant proteins describe the importance of the N-terminal cysteine in the orientation of the substrate and its interaction with key residues of the interdomain loop.

Keywords BCATm • BCATc • Brain-glutamate metabolism • Metabolite shuttling • BCKDC • Redox-regulation • S-nitrosylation and X-ray crystallography

M.E. Conway, Ph.D. (✉)
University of the West of England, Bristol, UK
e-mail: Myra.conway@uwe.ac.uk

S.M. Hutson, Ph.D.
Department of Human Nutrition, Foods and Exercise, Virginia Tech, Blacksburg, VA, USA
e-mail: susanh5@vt.edu

R. Rajendram et al. (eds.), *Branched Chain Amino Acids in Clinical Nutrition: Volume 1*, Nutrition and Health, DOI 10.1007/978-1-4939-1923-9_3,

Abbreviations

PLP	Pyridoxal phosphate
PMP	Pyridoxamine
BCKA	Branched chain α–keto acids
KIC	α-Ketoisocaproate
KIV	α-Ketoisovalerate
KMV	α-Keto-β-methylvalerate
BCAT	Branched chain aminotransferase
ALT	Alanine aminotransferase
AST	Aspartate aminotransferase
BCAAs	Branched chain amino acids
BCKDC	Branched chain α-keto acid dehydrogenase complex
E2	Dihydrolipoamide acyltransferase
E1	Branched chain α-keto acid dehydrogenase
E3	Dihydrolipoamide dehydrogenase
ThDP	Thiamine diphosphate
BCATm	Mitochondrial BCAT
BCATc	Cytosolic BCAT
GDH1	Glutamate dehydrogenase
mTORC1	Mammalian target of rapamycin complex 1
GSNO	S-nitrosoglutathione
NO	Nitric oxide
eBCAT	*E. coli* branched chain aminotransferase
4MeVa	4-Methylvalerate
BCATm-ox	Oxidized BCATm

Introduction

Transamination reactions are catalysed by pyridoxal phosphate-dependent (PLP) enzymes [1]. The key aminotransferase proteins have two predominant isoforms, a mitochondrial and cytosolic isoform. These include the branched chain aminotransferases (BCAT) [E.C. 2.6.1.42], the alanine aminotransferases (ALT) [glutamate pyruvate tranaminase or alanine 2-oxo-glutarate E.C. 2.6.1.2] and the aspartate aminotransferase proteins (AST) [glutamic oxaloacetic transaminase or L-aspartate:2-oxo-glutarate aminotransferase, E.C. 2.6.1.1]. These enzymes all require the cofactor PLP for activity and have the same basic transamination mechanism. The PLP-dependent aminotransferases are largely placed with the fold-type I or L-aspartate aminotransferase family [2, 3]. However, the BCAT proteins fall into the fold-type IV class of proteins, in part due the fact that the proton is abstracted from the C4′ atom of the coenzyme-imine or external aldimine on the *Re* face instead of the *Si* face of the PLP cofactor [4].

The BCAT proteins are key metabolic proteins responsible for the metabolism of the branched chain amino acids (BCAAs). Products of their metabolism are essential nutrient signals and glutamate is one of the major neurotransmitters in the brain. These proteins are distributed throughout the body but show tissue and cell specific compartmentalization which facilitates the interplay of highly regulated pathways. Knowledge of these pathways is not only important to our understanding of normal physiological mechanisms but more so how they are altered and contribute to the pathogenesis of disease. This chapter details the whole body distribution of the BCAT aminotransferases with particular focus on their role in anaplerotic pathways to generate and maintain the pool of brain and whole body glutamate through their involvement in metabolic shuttles that channel nitrogen. Furthermore, the structure/function relationship of these proteins is discussed in detail with particular emphasis on how these proteins are regulated through changes in the redox environment.

Transamination

The first step in BCAA catabolism is a nitrogen transfer reaction transamination catalyzed by the BCAT isozymes. Deamindation of amino acids was first observed by H. A. Krebs in the 1930s in tissue slices, and subsequently it was observed that amino acid transformations could occur without changes in total nitrogen, i.e., nitrogen transfer [5]. The identity of amino group acceptors such as α-ketoglutarate, pyruvate and oxaloacetate soon followed. Thus, transamination is a nitrogen transfer reaction with the α-amino group of one amino acid transferred to a receptor α-keto acid to form the respective amino acid and α-keto acids [5].

Purification of the first BCAT enzyme was reported in 1966 [6, 7]. Subsequent research in the 1980s and 1990s by Hutson and co-workers established that there are only two mammalian BCAT (for review see [8])—mitochondrial BCATm (*BCAT2)* and cytosolic BCATc (*BCAT1*). Like other aminotransferases, BCAT use vitamin B_6-(PLP) as a cofactor and exhibit ping-pong kinetic mechanisms (Fig. 3.1). In the first half-reaction, the PLP form of the enzyme reacts with the α-amino group of a BCAA, leucine, isoleucine or valine. Then the reaction proceeds to the pyridoxamine (PMP) form of the enzyme releasing the respective branched chain α-keto acids (BCKA), α-ketoisocaproate (KIC), α-keto-β-methylvalerate (KMV) or α-ketoisovalerate (KIV) (Reaction 1). The PMP enzyme then reaminates a second α-keto acid, usually α-ketoglutarate, followed by reversal of the first half reaction to reform the PLP enzyme and release glutamate (Reaction 2).

$$\mathrm{BCAT \bullet PLP + R_1 - CH(NH_2) - COOH \rightleftarrows BCAT \bullet PMP + R_1 - CO - COOH} \tag{3.1}$$

$$\mathrm{BCAT \bullet PMP + R_2 - CO - COOH \rightleftarrows BCAT \bullet PLP + R_2 - CH(NH_2) - COOH} \tag{3.2}$$

The mammalian enzymes are very specific for BCAA and glutamate with substrate preferences for isoleucine ≥ leucine > valine >> glutamate [9–11]. Once BCAA nitrogen enters the large glutamate pool, it is available for the synthesis of amino acids that transaminate with glutamate such as alanine and aspartate and synthesis of glutamine via glutamine synthetase [12]. Thus, through transamination, BCAA contribute to the synthesis of nonessential amino acids (glutamate, alanine and glutamine), nevertheless, net nitrogen transfer into the nonessential amino acid pool only occurs when the BCKA products are oxidized.

Transamination is a reversible reaction and net transfer of BCAA nitrogen into the nonessential amino pool requires oxidation of the BCAA carbon skeleton. The committed step in BCAA oxidation is catalyzed by the second enzyme in the catabolic pathway, the branched chain α-keto acid dehydrogenase complex (BCKDC) (Reaction 3). The enzyme complex catalyzes multistep reactions that lead to the irreversible oxidative decarboxylation of the BCKA giving rise to branched chain acyl-CoAs.

$$\mathrm{R - CO - COOH + CoA - SH + NAD^+ \rightarrow R - CO - S - CoA + CO_2 + NADH + H^+} \tag{3.3}$$

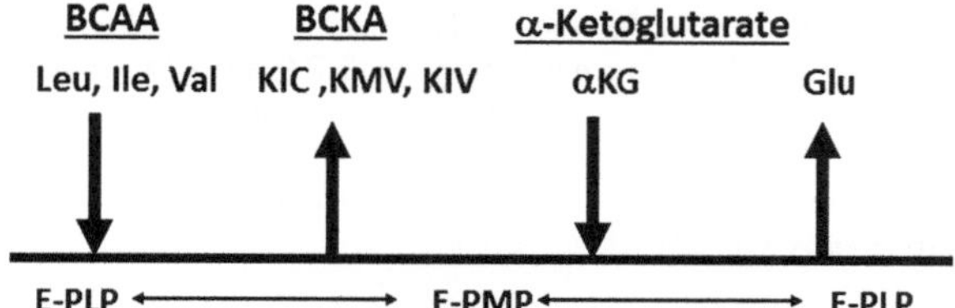

Fig. 3.1 Transamination of the BCAAs. Transamination of the BCAA with α-keto glutarate forming glutamate and the α-keto acids by the BCAT isozymes via a ping-pong kinetic mechanism

BCKDC is organized around a 24-meric core scaffold consisting of dihydrolipoamide acyltransferase (E2) subunits to which multiple copies of the branched chain α-keto acid dehydrogenase (E1) and dihydrolipoamide dehydrogenase (E3) are noncovalently attached [13]. The E1 enzyme binds thiamine diphosphate (ThDP) and catalyzes the ThDP-mediated decarboxylation of BCKAs [14]. The product of the reaction, carbanion-ThDP, undergoes the E1-catalyzed reductive transfer of carbanion to the lipoyl moiety that is covalently bound to dihydrolipoyl transacylase (E2) [15]. The dihydrolipoyl residue is re-oxidized by the FAD moiety of E3, and NAD^+ is the electron acceptor which is oxidized in the mitochondria to provide energy. These first two steps in BCAA catabolism are common to all three BCAA and regulation of BCKDC is achieved by phosphorylation/dephosphorylation; however, BCATm and BCKDC activities may also be regulated by complex formation.

Substrate Channelling Between Mitochondrial BCAT and the BCKDC Complex

The mitochondria BCAT (BCATm) isozyme has been found to form a metabolon protein complex with the BCKDC and other proteins [16–18]. Metabolon formation promoted channelling of the BCATm transamination product to the E1 component of BCKDC which enhanced E1-catalyzed decarboxylation of the BCKAs and overall BCKDC activity [16–18]. The reaction cycle is depicted schematically in Fig. 3.2. The interaction between BCATm and the E1 enzyme of BCKDC was confirmed in vitro with human recombinant enzymes to identify the protein partner for PLP-BCATm as the E1 dehydrogenase of BCKDC with a binding constant of 6 μM, indicating a weak association. The highest activity was obtained when α-ketoglutarate was also added, because it facilitated

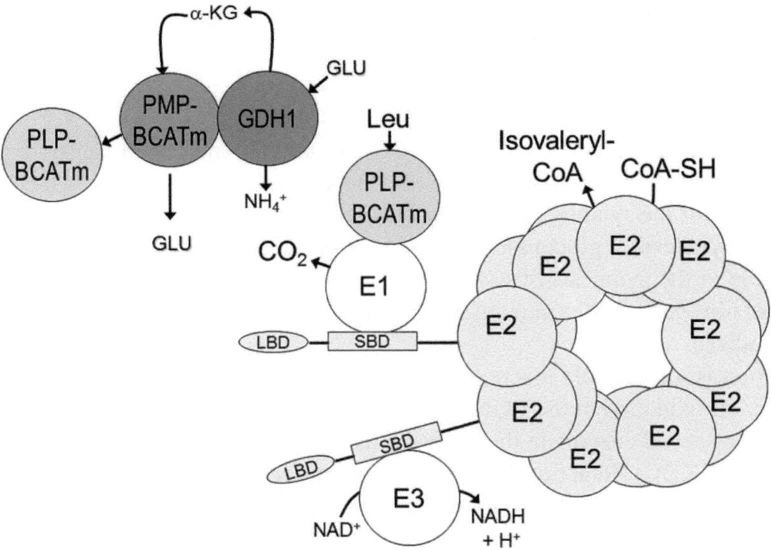

Fig. 3.2 Metabolon formation between BCATm, the BCKDC complex and GDH. Human BCATm binds directly to the E1 subunit of BCKDC, forming the BCAA metabolon. Binding of BCATm to the E1 component promotes channelling of substrate from BCATm to E1. The PMP form of BCATm is released from the E1 component. α-KG then binds to the PMP form of BCATm regenerating the PLP form of BCATm, initiating a new reaction cycle. GDH1, binds to the PMP-form of human BCATm facilitating reamination of the α-KG product of the GDH1 oxidative deamination (Adapted from Islam MM, et al., J Biol Chem. 2007;282(16):11893–903, and Hutson SM, Islam MM, Zaganas I. Interaction between glutamate dehydrogenase (GDH) and L-leucine catabolic enzymes: intersecting metabolic pathways. Neurochem Int. 2011;59(4):518–24)

converting PMP-BCATm back to PLP-BCATm which can then initiate another cycle, i.e., cycling versus a single turnover reaction [16]. Binding of BCATm to BCKDC is blocked by phosphorylation of E1.

More recently, GDH1 was identified as a potential metabolon component in rat tissues [17, 18]. GDH1, but not GDH2, binds to the PMP form of human BCATm (PMP-BCATm) but not to PLP-BCATm in vitro. This protein interaction facilitates reamination of the α-ketoglutarate product of the GDH1 oxidative deamination reaction. Another metabolic enzyme also found in the metabolon is pyruvate carboxylase. Kinetic results suggest that PC binds to the E1 decarboxylase of BCKDC but does not affect BCAA catabolism. The protein interaction of BCATm and GDH1 promotes regeneration of PLP-BCATm which then binds to BCKDC resulting in channelling of the BCKA products from BCATm first half reaction to E1 and promoting BCAA oxidation and net nitrogen transfer from BCAAs (Fig. 3.2). The cycling of nitrogen through glutamate via the actions of BCATm and GDH1 releases free ammonia. Formation of ammonia may be important for astrocyte glutamine synthesis in the central nervous system of rodents.

Shuttling of BCAA Metabolites and Their Role in Brain Metabolism

BCAA metabolite shuttling has been observed for astrocytes and neurons in rodent brain [19]. In the rodent brain BCATm and BCKDC do not co-localize, but in human brain the localization of BCKDC is not yet known. Immunolocalization of BCATm and BCKDC in rats revealed that BCATm is present in astrocytes in white matter and in neuropil, while BCKDC is essentially found in neuronal populations. In the rat, BCATm appears uniformly distributed in astrocyte cell bodies throughout the brain. The segregation of BCATm to astrocytes and BCKDC to neurons provides further support for the existence of a BCAA-dependent astroglial-neuronal nitrogen shuttle since the data show that BCKAs produced by glial BCATm must be exported to neurons and support the hypothesis that BCATm association with GDH1 in astrocytes could provide ammonia for glutamine synthesis. Thus, in rodent brain the hypothetical astrocyte neuronal shuttle involving BCKA produced from BCATm transamination in astrocytes shuttling to neurons where the cytosolic isoform, BCATc, reaminates BCKA transferring BCAA nitrogen between the two cell types and providing nitrogen for glutamate/glutamine synthesis is supported by enzyme localization [20, 21]. Because BCKDC is also localized in neurons, though BCKDC and BCATc are not necessarily found in the same neurons, net nitrogen transfer occurs when the BCKA product is oxidized. The localization of these key enzymes provides the potential for significant shuttling of BCAA metabolites between different cell types in the rodent brain.

On the other hand, the localization of BCAT isozymes in human brain has been reported recently. Like the rat/mouse, BCATc is primarily found in neurons that are GABAergic or glutamatergic with more immunopositive GABAergic neurons [22]. As in rodents, neurons that were neither glutamatergic nor GABAergic were also labeled (serotonergic, dopaminergic) with some neuroendocrine cells also showing BCATc immunolabeling. BCATc labeling was found in the paraventricular nucleus of the hypothalamus and substantia nigra among other structures in both human and rodent brain [22, 23]. The expression of BCATm in the vasculature and not in astrocytes in human brain presents a clear difference between humans and rodents. The localization of BCKDC in human brain cells is not known nor are there reports that GDH1 is localized in the vascular endothelium. It is clear that there is the potential for shuttling of BCAA and their metabolites in human brain, but how BCAA nitrogen and other metabolites contribute to human brain metabolism is not known at this time. Based on human tissue BCAT and BCKDC activities which represent about 20 % of human whole body BCAA metabolic capacity, the potential for brain oxidation of BCAAs suggests brain may contribute significantly to human BCAA metabolism.

Metabolite Shuttling in Peripheral BCAA Metabolism

Metabolite shuttling is also a hallmark of peripheral BCAA metabolism [24, 25]. The high ratio of transaminase to oxidative capacity in skeletal muscle of the rat, absence of BCAT and high oxidative capacity in liver led to the prediction of extensive interorgan shuttling of BCAA metabolites, first postulated by Harper and collaborators [24–26]. The majority of peripheral tissues contain both BCATm and BCKDC [27] whereas GDH1 is found at varying levels in tissues, and it is not clear if BCATm and GDH1 co-localize, certainly not in liver. This led to the hypothesis that BCAAs play an important role in body nitrogen metabolism, particularly as nitrogen donors for the major nitrogen carriers alanine and glutamine [28, 29]. BCATm is not expressed nor it is expressed at low levels in liver, and at higher levels in other organs, leaving the initial step in BCAA degradation to tissues other than liver [26, 30].

BCAAs represent about 50 % of skeletal muscle amino acid uptake, and skeletal muscle is not the primary site for catabolism of most of the other plasma amino acids. The BCAAs can be used for maintenance of muscle pools of glutamate, alanine and glutamine and to provide nitrogen for alanine and glutamine release from muscle with subsequent uptake by the liver, kidney and intestines. In human skeletal muscle, BCAT activity is high compared to BCKDC activity (ratio of measured BCAT/actual BCKDC activities and for total dephosphorylated BCKDC activity) [31]. The high ratios of BCAT/BCKDC activities favor (1) release of BCKAs rather than their oxidation [24, 25, 30] and (2) efficient transfer of BCAA nitrogen. In hepatocytes, the absence of the BCATm isozyme and high activity of BCKDC favors liver oxidation of circulating BCKAs. This would account for the low concentrations of BCKAs found in liver [32]. The flexibility of metabolic sites also explains the success of liver transplantation in patients with the inborn error of BCAA metabolism at the BCKDC step, Maple Syrup Urine Disease. On the other hand, not all tissues actively metabolize BCAAs. A recent study suggests that adipose tissue is not a site of BCAA disposable due to the limited uptake of BCAAs by this tissue [33].

Another unique feature of BCAA metabolism is the compartmentalization of BCAT isozymes and BCKDC within organs and tissues. The mitochondrial BCATm is expressed ubiquitously whereas the cytosolic isozyme BCATc is found almost exclusively in nervous tissue (Fig. 3.3). Sweatt et al. [23, 27] have shown that in the rat, BCATc is expressed in peripheral nerves. Neither BCATm nor BCKDC were found in peripheral nerves. In peripheral tissues BCATm is by far the predominant isoform expressed. Its expression, however, is cell specific. It is often localized in secretory epithelial cells [23]. Surprisingly the enzyme is highly expressed in the acinar cells of the exocrine pancreas and parietal and chief cells of the stomach whereas the enzyme was not found in the secretory epithelial cells of the intestine. The physiological significance of BCAA metabolism in the digestive system is not known, but it may be important for glutamine synthesis.

With a few exceptions (liver, kidney and brain), the BCKDC complex appears to co-localize with BCATm in rat and primate tissues. Major advantages of metabolite shuttling are: (1) BCAA nitrogen can be transferred within and between organs and tissues; (2) oxidation of BCAA carbon skeleton can occur throughout the body in a tissue and cell-specific manner depending on metabolic state; (3) leucine signalling can be regulated within a specific cell-type within a tissue. Because leucine activates mammalian target of rapamycin complex 1 (mTORC1) leading to increased cellular protein synthesis, reduced autophagy and regulation of lipid homeostasis [34, 35], the ability to adjust intracellular leucine concentrations and nutrient activation of mTORC1 is advantageous. Studies with transgenic mice with deletion of BCATm [36] show that blocking BCAA catabolism at the transamination step throughout the body has profound effects on mTOR signalling. The BCATm knockout mouse has elevated BCAAs without accumulation of BCKAs. The mouse is lean and resistant to high fat diet-induced obesity, and it appears to be hypermetabolic as a result of increased protein

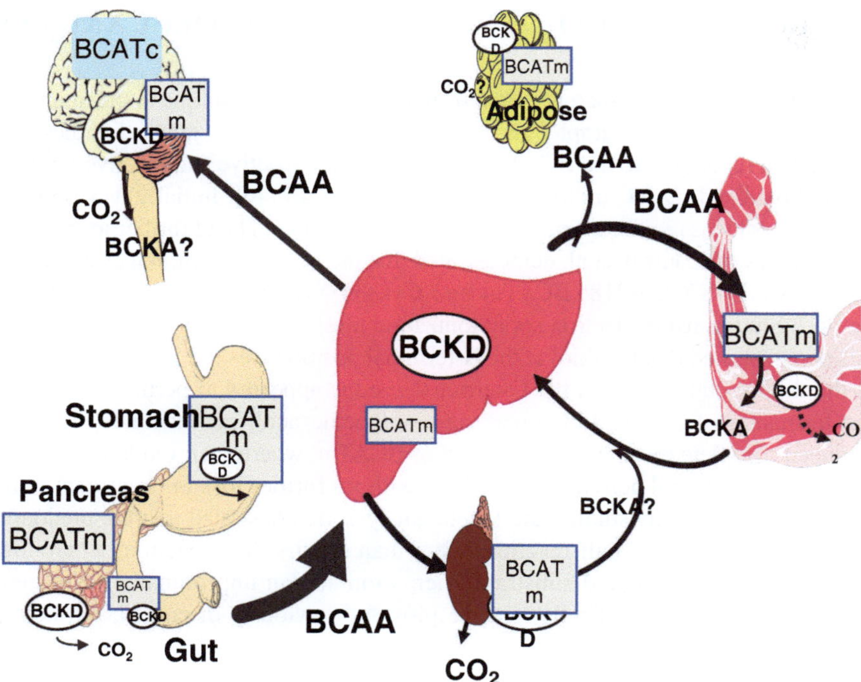

Fig. 3.3 Whole body distribution of the BCAT and BCKDC proteins. The mitochondrial isoform is ubiquitously expressed whereas the cytosolic isoform is largely expressed in neuronal cells of the brain and peripheral nervous system

turnover (increased synthesis and degradation) [36]. The BCATc (unpublished data) knockout mouse does not exhibit an obvious phenotype, however, BCATc may modulate mTOR signalling in cells of the immune system (unpublished data). Whether or not BCATc impacts mTOR signalling, through modulation of leucine/KIC concentrations or influencing α-ketoglutarate concentrations, remains to be determined. A recent publication suggests that leucine and not leucine metabolites are key regulators [37]. However, the lack of BCATc in T lymphocytes, which prevents induction of BCATc during T cell receptor engagement, strongly influences mTOR signalling leading to higher mTORC1 activation (unpublished data). The hypothesis that BCATc is an oncogene was first postulated in the 1970s by Ichihara [38]. Tonjes et al. have just reported that BCATc promotes cell proliferation in gliomas and expression is highly correlated with proliferation. The mechanism appears to be BCATc's contribution to glioma glutamate excretion which is essential for glioma invasion [39].

The response of the BCAT proteins to cellular events that regulate BCAA catabolism (oxidation) and impact on intracellular leucine is likely to be governed by its regulatory CXXC motif [40, 41], which has been shown to be important for binding of BCATm to the E1 subunit of the BCKDC, facilitating substrate channelling as previously discussed [16]. In a separate study [42], using extracts from the neuronal IMR32 cells, overexpressed mammalian BCATc was shown to have specific peroxide related redox associations with several proteins, including septin 4, kalirin rho GEF, β-tubulin, myosin-6 and the sodium channel type 10 α subunit. These proteins have either known reactive cysteine residues or CXXC motifs with phosphorylation sites and/or are directly involved or controlled by G protein cell signalling, which is known to be modulated by peroxide. These associations were abolished when the environment became more oxidising. Thus, this indicates a potential novel role for human BCATc in neuronal cell signalling.

The Redox Regulation of the BCAT Proteins: Impact of the CXXC Motif

The redox-active CXXC motif, which is ~10 Å from the active site, is unique to the mammalian BCAT proteins [43, 44]. X-ray crystallography and biochemical studies have shown that these thiol groups confer redox-linked regulation to the BCAT proteins and that the active and inactive forms result from the reduced and oxidized CXXC motif, respectively [41, 42, 45, 46]. Initially it was observed that a reducing environment was required for maximum BCAT activity [11] and further mapping of the reactive cysteine residues by Conway et al. demonstrated that the N- and C-terminal cysteine groups of the CXXC motif (Cys315-XX-Cys318; BCATm and Cys335-XX-Cys338; BCATc) were the residues responsible [41]. Studies using electron spray ionization mass spectrometry analysis and site-directed mutagenesis demonstrated that the thiol at the N-terminal position was the 'redox sensor' and the most reactive of the thiol groups whereas, the C-terminal residue appeared to permit reversible regulation through the formation of a disulphide bond via a sulphenic acid intermediate (Fig. 3.4) [41, 46]. Peroxide induced complete reversible inactivation of BCATm, whereas air oxidation alone resulted in a loss of 40–45 % functional activity for BCATc, with no further loss on treatment with peroxide [40, 42]. X-ray crystallography studies and kinetic analysis demonstrated that the predominant effect of oxidation was on the second half reaction rather than the first half reaction, where disruption of the CXXC center results in altered substrate orientation and an unprotonated PMP amino group, thus rendering the enzyme catalytically inactive [46]. As previously discussed, physiologically, the

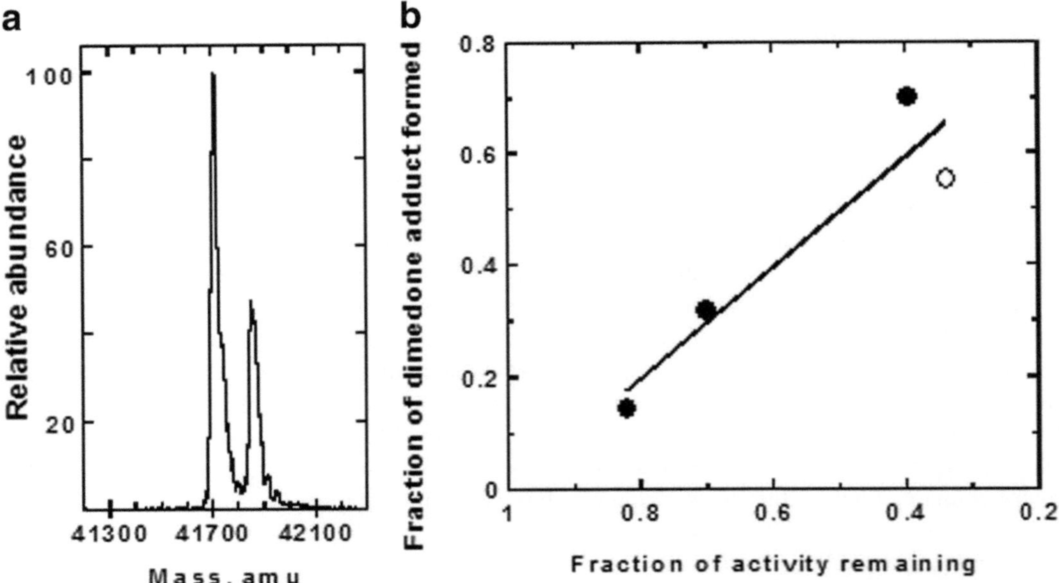

Fig. 3.4 Trapping of the sulphenic acid intermediate using dimedone and activity loss for the H_2O_2-treated C318A mutant of hBCATm. Dimedone adduct formation and activity loss for the H_2O_2-treated C318A mutant of hBCATm. Samples (4.5 nmol each) were incubated with 1.5 mM (*open circle*) or 2 mM H_2O_2 (*closed circles*) and 5 mM dimedone at 25 °C prior to removal of aliquots, addition of catalase and BCAT activity and electrospray ionization mass spectrometery analyses, as described in Materials and Methods. Panel (**a**) shows the transformed mass spectrometry data for the sample after 427 min treatment, with peaks of maximal abundance at 41,708 and 41,848 atomic mass units for the free and dimedone-adducted protein, respectively. Panel (**b**) illustrates the correlation between the fraction of dimedone adduct formed and the BCAT activity of the C318A samples from two independent experiments, after treatment for 224, 427, 1180 and 1375 min (in order of increasing adduct formation; $r=0.936$ and slope $=-0.987$). (Reprinted with permission from Conway ME, Poole LB, Hutson SM. Roles for cysteine residues in the regulatory CXXC motif of human mitochondrial branched chain aminotransferase enzyme. Biochemistry. 2004;43(23), 7356–7364. Copyright (2004) American Chemical Society.)

reduced state of the CXXC motif of the BCAT proteins has been shown to be essential for substrate channelling and redox associations with proteins involved in G-protein cell signalling.

The reactive cysteines of both isoforms are also targets for the S-nitrosylating agent, S-nitroso glutathione (GSNO) (a reaction transferring a NO group to the reactive cysteine of this protein). Nitric oxide (NO) plays numerous important roles in maintaining cell function including inflammation and cell signalling cascades. However, when intracellular levels of NO become elevated a series of pathologies associated with prolonged exposure have been identified, including diabetes, malignancy, atherosclerosis and neurodegeneration. Recent studies demonstrated that human BCATc and BCATm were differentially inactivated through S-nitrosation suggesting alternative mechanisms of regulation in response to levels of NO in the mitochondria and the cytosol [47]. In these studies, low concentrations of the physiologically relevant NO donor, GSNO, caused a reversible time-dependent loss of 50 % activity, characterized predominantly through S-nitrosation [47]. However, increased exposure resulted in a shift towards S-glutathionylation (addition of GSH to the reactive thiol), a marker of oxidative stress, which was considered to play a role in protecting BCAT, pointing to an adaptive response to cellular stress (Fig. 3.5) [47]. Moreover, the BCAT transaminase activity was recovered with the physiologically important reducing system, the glutaredoxin/glutathione system, supporting a role for BCAT in cellular redox control. Recent work by Conway's group has established that the BCAT proteins have endogenous dithiol disulphide isomerase activity that catalyse the refolding of proteins, indicating new redox substrates for the BCAT proteins (unpublished data). Further studies are currently ongoing to investigate the interactions of these proteins with redox systems in cells.

Substrate Orientation and the Structural Importance of the Interdomain Loop in Catalysis

The analysis of the 3D-structure of the BCAT proteins has provided invaluable insight into their structure/function relationship in particular with respect to the importance of the redox-active CXXC motif. The human BCAT share 58 % sequence homology and although they share only 29 % homology with eBCAT, similarities exist in the active sites [43–45, 48, 49]. Both BCAT isoforms are homodimers with each monomer existing as a large and small domain and have a molecular weight of 41,730 Da (BCATm) and 43,400 Da (BCATc), respectively [11, 43, 49]. Although they share the same substrate specificity they have subtle catalytic differences and as previously described are differentially regulated through changes in the redox environment. The steady-state kinetics are summarised in Table 3.1. In brief, for deamination of the BCAA and reamination for KIC and KIV the k_{cat}/K_m for BCATc was found to range between 2 and 5 times faster than its mitochondrial isoform [11]. Although the catalytic turnover is different between isozymes the key active-site residues involved in substrate and cofactor binding were found to be largely identical in amino acid sequence in both structures [43, 49]. Thus, this variation in steady-state kinetics was not readily explained by the active-site binding pocket. Rather the residues that surround the active site have been considered to play a more important role. Moreover, BCATc but not BCATm is inhibited by the neuroactive drug Gabapentin, a drug used in the treatment of epilepsy but more recently used to reduce neuropathic pain. The interaction of BCATc with gabapentin facilitated the previously elusive 3D-structure.

The Structure of BCATm Complexed with Isoleucine

Using molecular replacement methods with eBCAT [48] as a search model the crystal structure of BCATm has been resolved in several forms; the PLP (PDB codes 1EKP, 1EKF), ketimine (PDB code 1KT8) and PMP (PDB code 1KTA) form [49]. In addition a TRIS-inhibited structure has also

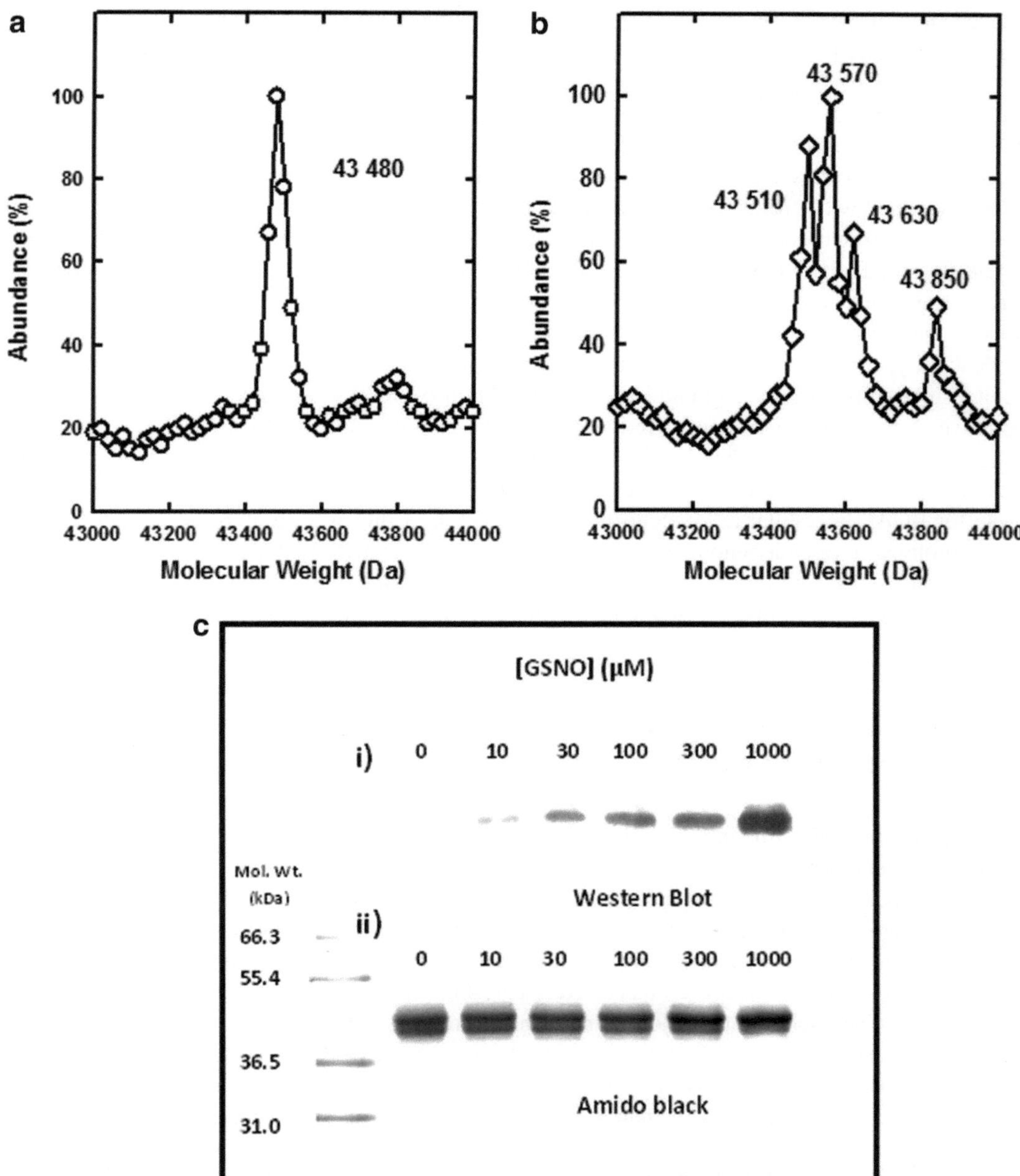

Fig. 3.5 GSNO-mediated S-nitrosylation and S-glutathionylation of hBCATc. Twenty nmol of hBCATc was exchanged into 10 mM ammonium hydrogen carbonate (pH 7.5), and incubated with 300 μM GSNO for 30 min at 37 °C prior to Q-TOF MS analysis as described in Experimental Procedures. Panel (**a**). The principal peak has a mass of 43,480 corresponding to control hBCATc. Panel (**b**). A total of 4 peaks were identified with GSNO-modified protein, with masses of 43,510, 43,570, 43,630 and 43,850, corresponding to hBCATc with 1 NO, 3 NO, 5 NO and 2 NO plus 1 GSH adducts, respectively. In addition, 4 nmol of hBCATc in 50 mM HEPES (pH 7.2) and 1 mM EDTA was incubated for 30 min at 37 °C with increasing concentrations of GSNO. Aliquots were removed for Western Blot analysis using the anti-GSH antibody (Virogen). Panel (**c**). Lanes 1–6, hBCATc incubated with 0 μM, 10 μM, 30 μM, 100 μM, 300 μM and 1000 μM of GSNO respectively (i) Western blot analysis of these NO-modified proteins using anti-GSH (ii) SDS-PAGE on a 12 % resolving gel. The extent of thiol modification was monitored using F5M as described in Experimental Procedures. (Reprinted with permission from Coles SJ, Easton P, Sharrod H, et al. S-Nitrosoglutathione inactivation of the mitochondrial and cytosolic BCAT proteins: S-nitrosation and S-thiolation. Biochemistry. 2009;48(3):645–656 Copyright (2009) American Chemical Society)

Table 3.1 Kinetic constants for the reactions of human branched chain aminotransferase isoenzymes

Variable substrate	Fixed substrate	BCATm k_{cat} s^{-1}	BCATm K_m mM	BCATm k_{cat}/K_m $\times 10^3$ M^{-1} s^{-1}	BCATc k_{cat} s^{-1}	BCATc K_m mM	BCATc k_{cat}/K_m $\times 10^3$ M^{-1} s^{-1}
BCAA[a]							
Leu	α-KG[b]	105±11	1.21±0.12	88	132±7	0.60±0.04	220
Ile	α-KG	80±1	0.60±0.02	165	172±9	0.77±0.02	223
Val	α-KG	68±2	6.10±0.11	11	122±8	2.40±0.09	51
BCKA[c]							
KIC	Glu	244±1	0.16±0.004	1525	309±11	0.063±0.002	4,905
	Ile	476±29	0.12±0.006	3967	381±5	0.057±0.002	6,684
	Ile[d]	426±34	0.50±0.080	852	482±31	0.120±0.005	4,017
KIV	Glu	147±4	0.50±0.050	294	221±3	0.150±0.003	1,473
	Ile	186±9	0.20±0.010	930	217±4	0.070±0.004	3,100
	Ile[d]	121±8	0.42±0.010	288	295±6	0.180±0.010	1,639

The fixed substrates were $10\times K_m$. Glutamate was fixed at 100 mM K_m and V_{max} values (used to calculate k_{cat}) were obtained from hyperbolic plots fit to the Michaelis-Menten equation. Means and S.E. values are from 3 to 5 separate determinations. (Adapted from Davoodi et al. 1998)

[a]Branched chain amino acids

[b]α-KG, α-ketoglutarate

[c]Branched chain α-keto acids

[d]Assays were performed in the presence of 150 mM KCl

been reported (PDB code 1EKV). The structure showed that each homodimer of BCATm has two monomers linked by an interdomain loop, 11 amino acids in length. Each monomer has 365 amino acids with one PLP cofactor attached, and is composed of a small domain (residues 1–175) and a large domain (residues 176 to 365), where the active site is close to the dimer interface. Each domain has four α-helical structures and the small subunit has a funnel shaped twisted β-pleated sheet whereas the large domain has a ten-stranded antiparallel β-pleated sheet in each of the monomers [43, 44].

In the PLP form, the 4′-aldehyde of the cofactor is covalently linked to the ε-amino group of the active site Lys202, which lies at the bottom of the *Re*-face, via a Schiff-base linkage. Most of the active-site residues at the PLP end are conserved or are similar to those found in eBCAT [44, 48]. However, structural differences in these fold-type IV proteins occur around the O3 of PLP together with altered hydrogen–bonding interactions due to the presence of water molecules. The substrate binding pocket was characterized by three binding surfaces, noted as A, B and C that surround the BCAAs [45]. The key residues in the substrate binding pocket are predominantly hydrophobic including Phe75, Tyr207 and Thr240 that form surface A and Phe30, Tyr141 plus Ala314 form surface B with Tyr70*, Leu153* and Val155* (*from the opposite subunit) forming surface C. These surfaces together with van der Waals interactions confer substrate specificity. Domain closure is considered to be governed by Tyr173, which is essential for catalysis. Binding of the substrate initiates movement of the interdomain loop, increasing access to the active site, controlling the entry or exit of substrates and products [43–45].

The thiols of the CXXC motif of BCATm (C315, Val336, Val337, C318) in its reduced state form a thiol-thiolate hydrogen bond and are 3.17–3.46 Å apart. Structural determinants were prepared for C315A (PDB code 2HGX), C315A-substrate mimic N-methyl leucine (PDB 2HG8), C318A (PDB code 2HGW), C315/8A (PDB 2HDK) and oxidized BCATm (BCATm-ox) (PDB code 2HHF) and all resolved to <2.4 Å [46]. These studies determined that the redox reactive thiol, C315, was identified as key for optimal substrate active-site orientation and for critical van der Waals forces influencing enzyme catalysis [46]. Moreover, the peptide dipole alignment between residues 312 and 315 was also important in stabilizing the carboxylate binding site. Here, the hydrogen bonding at the CXXC centre and at the β-turn preceding it results in the rotation of the peptide plane following C315 (Fig. 3.6i). Mutation of both cysteines also changed the hydrogen bonding at the β-turn and influenced substrate binding with KIC (Fig. 3.6ii) [46].

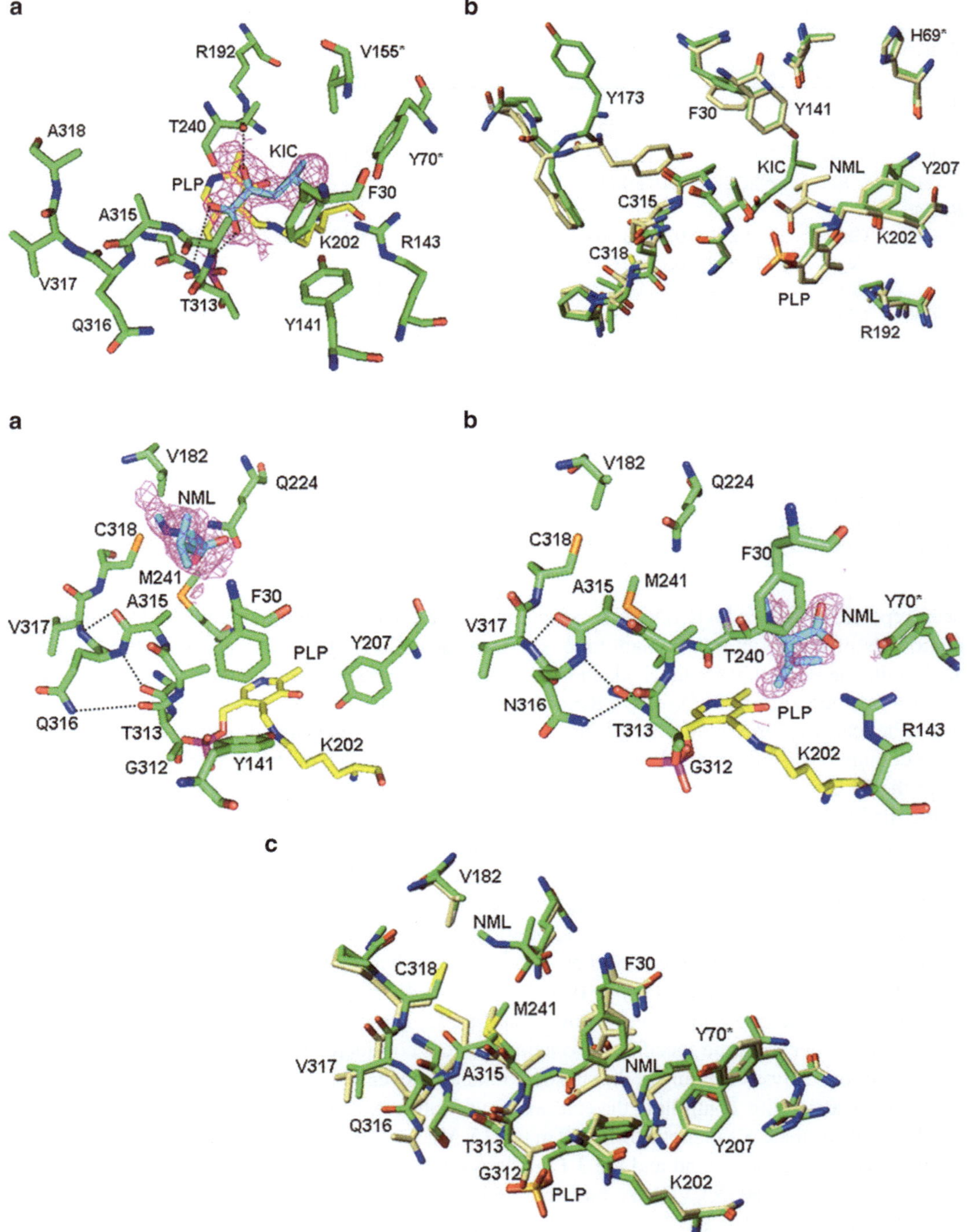

Fig. 3.6 Active-site structures of C315A mutant BCATm in complex with *N*-methylleucine (C315a-NML) (**i**) and active-site structures of C315A/C318A double mutant BCATm complexed with α-ketoisocaproate (**ii**). (I). *A* and *B* show the $2F_o$–F_c difference Fourier electron density maps of the active site in monomer A and monomer B, respectively. *C*, the superimposed model of the *N*-methylleucine-bound (NML) WT-BCATm structure and structure of NML-C315A mutant. (II). *A*, $2F_o$–F_c difference Fourier electron density map of the active site in monomer B of the double mutant in complex with KIC. *B*, the superimposed model of NML-WT and the structure of KIC-bound double mutant BCATm. The carbon skeletons are in khaki in WT-BCATm and in *green* in the single and double mutant structures, respectively. (Cited with permission: This research was originally published in J Biol Chem. Yennawar NH, Islam MM, Conway M, et al. Human mitochondrial branched chain aminotransferase isozyme: Structural role of the CXXC center in catalysis. 2006;281(51):39660–71)

Furthermore, Tyr173 of the interdomain loop, which is essential to catalysis, forms a weak S-h...π hydrogen bonding interaction with C315 in the native conformation [46]. In the oxidized protein this becomes abolished causing a flip in the orientation of Tyr173 away from the active site. Other key residues such as Phe30, Arg143, Tyr207 and Arg141 are also reoriented weakening the side chain binding of the substrate. Moreover, mutation of the thiol at C318 did not alter the structural integrity of the protein with the refined structure showing similarities to WT protein explaining why the C318 mutant has similar kinetic parameters to WT BCAT [42, 46].

The Structure of BCATc with the Neuroactive Drug, Gabapentin

The crystal structure of BCATc was elusive for quite some time until Goto et al. 2008 crystallized BCATc complexed with Gabapentin both in the oxidized (PDB code: 2coi) and reduced states (PDB code: 2coj) at 1.9 and 2.4 Å resolution, respectively and also with the leucine analog 4-methylvalerate (4MeVa) (PDB code: 2coq) at 2.1 Å [49]. Although larger in molecular weight (43,400 Da) with 395 amino acids, each monomeric form of BCATc, like BCATm, consists of a small domain (residues 1–188), an interdomain loop (pro-189-Pro-201) and a large domain (residues 202 to the C terminus). The small and large domain is folded into an open α/β and a pseudo-barrel structure, respectively with PLP showing similar structural alignment with its catalytic Lys-222 as observed for BCATm.

The residues of the substrate binding pocket of BCATc are identical to those found in BCATm with exception of one residue (Val-336 in BCATc is Gln in BCATm) [43, 44, 49]. The main groups of amino acids that surround and enclose gabapentin include the OH group of Tyr-90*, the guandino group of Arg-163, the OH group of Tyr-161, water molecules W1 and W2, the main chain C=O of Gly-97, the carboxylate of gabapentin and the phosphate O1 atom of PLP that form a hydrogen bond network. At the *Si*-face of PLP gabapentin fits in a hollow, where the α-carboxylate is coordinated to residues in the small and large sites. These small and large sites are surrounded by Phe-49, Phe-95, Tyr-161, Tyr-227, Thr-260, Thr-333, Ala-334, Tyr-90*, Leu-173* and Val-175* and with Phe-49, Tyr-161 and Thr-260 located at the boundary region between these areas (Fig. 3.7) [49]. Additionally, the side

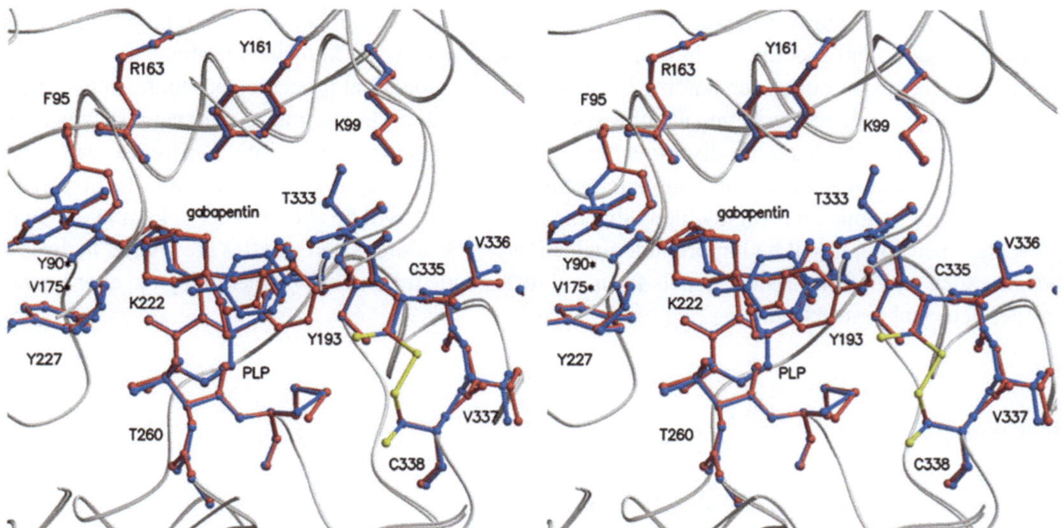

Fig. 3.7 Stereo diagram of the superimposed active sites of BCATc-ox-gabapentin and BCATc-gabapentin. The residues and gabapentin, which represent BCATc-ox-gabapentin and BCATc-gabapentin, are shown in *brown* and *deep blue*, respectively. (Cited with permission: This research was originally published in J Biol Chem. Goto M, Miyahara I, Hirotsu K, et al. Structural determinants for branched-chain aminotransferase isozyme-specific inhibition by the anticonvulsant drug gabapentin. 2005;280 (44), 37246–37256)

chain of Tyr193 of the interdomain loop approaches the aminomethyl group from the solvent side. As a result, in this oxidised form gabapentin is almost shielded in the protein inside by access of Tyr193, suggesting that BCATc-ox-gabapentin has a closed form of the active site indicating a weaker binding with gabapentin as opposed to that observed for the reduced form of the complex. When superimposed the active site for both forms are identical suggesting that changes to the redox environment should not alter gabapentin inhibition. However, in the oxidized and reduced BCATc-gabapentin complex, the interdomain loop shield gabapentin differentially where the interdomain loop of the reduced enzyme is closer to the central part of the active site than in the oxidized form. Here, Tyr-193 has closer access to gabapentin facilitating a stronger interaction with the cyclohexane ring of gabapentin, indicating that binding would be weaker in the oxidised form [49]. With exception to the aminomethyl group the active-site structure for 4MeVa was similar to BCATc-ox-gabapentin [49]. Thus, as observed with BCATm the Tyr residue of the interdomain loop is central to dictating access of substrates to the active site.

Conclusions

The BCAT proteins are key metabolic proteins that play a central role in shuttling metabolites between cells and tissues, regenerating nitrogen and key amino acids. Data from studies with human enzymes, evidence for protein-protein interactions with the BCKDC and GDH1 demonstrate that the BCAT proteins are part of a wider metabolon, supporting a role for substrate channelling. Clear evidence supports shuttling of metabolites between astrocytes and neurons in rat brain but in humans' expression of BCATm specifically to the vascular endothelium offers intriguing new frontiers of investigation to evaluate if similar shuttles exist and with equivalent metabolic partners in the human brain. Furthermore, the BCATc isozyme is a target of the neuroactive drug, gabapentin (Neurontin), which is used in the treatment of epilepsy and neuropathic pain.

The ability of the BCAT to interact with other proteins and structural changes that influence catalysis and protein-protein interactions may form the basis for a role for the BCAT redox regulatory switch that may influence BCAA metabolism and signalling. Evidently, despite significant advances much is still unknown about these metabolic pathways in particular in the human brain and the need for more sensitive models, and specific inhibitors, would further enhance our knowledge. This is important not just from the point of our understanding of the normal physiological processes but more so under diseased conditions, where biochemical pathways are altered. Of particular interest are neurodegenerative conditions such as Alzheimer's disease where glutamate toxicity features into the disease pathology. One would expect that as the aminotransferase proteins play a key role in facilitating the anaplerotic generation of glutamate it is highly likely that their metabolism will be altered where glutamate toxicity presents. Therefore, it will be important to understand if these suggested shuttles are altered in disease, because they may offer possible targets for novel therapeutic treatment to either delay onset or prevent further neuronal destruction.

References

1. Christen P, Metzler DE. Transaminases. New York: Wiley; 1975. p. 323–6.
2. Jansonius JN. Structure, evolution and action of vitamin B6-dependent enzymes. Curr Opin Struct Biol. 1998;8(6):759–69.
3. Schneider G, Kack H, Lindqvist Y. The manifold of vitamin B6 dependent enzymes. Structure. 2000;8(1):R1–6.
4. Yoshimura T, Jhee KH, Soda K. Stereospecificity for the hydrogen transfer and molecular evolution of pyridoxal enzymes. Biosci Biotechnol Biochem. 1996;60(2):181–7.

5. Braunstein AE. Transamination and transminases. New York: John Wiley & Sons; 1985.
6. Ichihara A, Koyama E. Transaminase of branched chain amino acids. I. Branched chain amino acids-alpha-ketoglutarate transaminase. J Biochem. 1966;59(2):160–9.
7. Taylor RT, Jenkins WT. Leucine aminotransferase. II. Purification and characterization. J Biol Chem. 1966; 241(19):4396–405.
8. Hutson S. Structure and function of branched chain aminotransferases. Prog Nucleic Acid Res Mol Biol. 2001;70:175–206. Review.
9. Wallin R, Hall TR, Hutson SM. Purification of branched chain aminotransferase from rat heart mitochondria. J Biol Chem. 1990;265(11):6019–24.
10. Hall TR, Wallin R, Reinhart GD, et al. Branched chain aminotransferase isoenzymes. Purification and characterization of the rat brain isoenzyme. J Biol Chem. 1993;268(5):3092–8.
11. Davoodi J, Drown PM, Bledsoe RK, et al. Overexpression and characterization of the human mitochondrial and cytosolic branched-chain aminotransferases. J Biol Chem. 1998;273(9):4982–9.
12. Hutson SM, Sweatt AJ, Lanoue KF. Branched-chain [corrected] amino acid metabolism: implications for establishing safe intakes. J Nutr. 2005;135(6 Suppl):1557S–64. Review.
13. Pettit FH, Hamilton L, Munk P, et al. Alpha-keto acid dehydrogenase complexes. XIX. Subunit structure of the Escherichia coli alpha-ketoglutarate dehydrogenase complex. J Biol Chem. 1973;248(15):5282–90.
14. Damuni Z, Lim Tung HY, Reed LJ. Specificity of the heat-stable proteininhibitor of the branched-chain alpha-keto acid dehydrogenase phosphatase. Biochem Biophys Res Commun. 1985;133(3):878–83.
15. Brautigam CA, Wynn RM, Chuang JL, et al. Structural insight into interactions between dihydrolipoamide dehydrogenase (E3) and E3 binding protein of human pyruvate dehydrogenase complex. Structure. 2006;3:611–21.
16. Islam MM, Wallin R, Wynn RM, Conway M, Fujii H, Mobley JA, Chuang DT, Hutson SM. A novel branched-chain amino acid metabolon. Protein-protein interact.ions in a supramolecular complex. J Biol Chem. 2007;282(16): 11893–903.
17. Hutson SM, Islam MM, Zaganas I. Interaction between glutamate dehydrogenase (GDH) and L-leucine catabolic enzymes: intersecting metabolic pathways. Neurochem Int. 2011;59(4):518–24.
18. Islam MM, Nautiyal M, Wynn RM, et al. Branched-chain amino acid metabolon: interaction of glutamate dehydrogenase with the mitochondrial branched-chain aminotransferase (BCATm). J Biol Chem. 2010;285(1):265–76.
19. Lieth E, LaNoue KF, Berkich DA, et al. Nitrogen shuttling between neurons and glial cells during glutamate synthesis. J Neurochem. 2001;76(6):1712–23.
20. Hutson SM, Berkich D, Drown P, et al. Role of branched-chain aminotransferase isoenzymes and gabapentin in neurotransmitter metabolism. J Neurochem. 1998;71(2):863–74.
21. Hutson SM, Lieth E, LaNoue KF. Function of leucine in excitatory neurotransmitter metabolism in the central nervous system. J Nutr. 2001;131(3):846S–50.
22. Hull J, Hindy ME, Kehoe PG, et al. Distribution of the branched chain aminotransferase proteins in the human brain and their role in glutamate regulation. J Neurochem. 2012;123(6):997–1009.
23. Sweatt AJ, Garcia-Espinosa MA, Wallin R, et al. Branched-chain amino acids and neurotransmitter metabolism: expression of cytosolic branched-chain aminotransferase (BCATc) in the cerebellum and hippocampus. J Comp Neurol. 2004;477(4):360–70.
24. Hutson SM, Cree TC, Harper AE. Regulation of leucine and alpha-ketoisocaproate metabolism in skeletal muscle. J Biol Chem. 1978;253(22):8126–33.
25. Shinnick FL, Harper AE. Branched-chain amino acid oxidation by isolated rat tissue preparations. Biochim Biophys Acta. 1976;437(2):477–86.
26. Harper AE, Benjamin E. Relationship between intake and rate of oxidation of leucine and alpha-ketoisocaproate in vivo in the rat. J Nutr. 1984;114(2):431–40.
27. Sweatt AJ, Wood M, Suryawan A, et al. Branched-chain amino acid catabolism: unique segregation of pathway enzymes in organ systems and peripheral nerves. Am J Physiol Endocrinol Metab. 2004;286(1):E64–76.
28. Odessey R, Khairallah EA, Goldberg AL. Origin and possible significance of alanine production by skeletal muscle. J Biol Chem. 1974;249(23):7623–9.
29. Garber AJ, Karl IE, Kipnis DM. Alanine and glutamine synthesis and release from skeletal muscle. II The precursor role of amino acids in alanine and glutamine synthesis. J Biol Chem. 1976;251(3):836–43.
30. Hutson SM, Zapalowski C, Cree TC, et al. Regulation of leucine and alpha-ketoisocaproic acid metabolism in skeletal muscle. Effects of starvation and insulin. J Biol Chem. 1980;255(6):2418–26.
31. Suryawan A, Hawes JW, Harris RA, et al. A molecular model of human branched-chain amino acid metabolism. Am J Clin Nutr. 1998;68(1):72–81.
32. Hutson SM, Harper AE. Blood and tissue branched-chain amino and alpha-keto acid concentrations: effect of diet, starvation, and disease. Am J Clin Nutr. 1981;34(2):173–83.
33. Lackey DE, Lynch CJ, Olson KC, et al. Regulation of adipose branched chain amino acid catabolism enzyme expression and cross-adipose amino acid flux in human obesity. Am J Physiol Endocrinol Metab. 2013; 304(11):E1175–87.

34. Atherton PJ, Smith K, Etheridge T, et al. Distinct anabolic signalling responses to amino acids in C2C12 skeletal muscle cells. Amino Acids. 2010;38(5):1533–9.
35. Ricoult SJ, Manning BD. The multifaceted role of mTORC1 in the control of lipid metabolism. EMBO Rep. 2013;14(3):242–51.
36. She P, Reid TM, Bronson SK, et al. Disruption of BCATm in mice leads to increased energy expenditure associated with theactivation of a futile protein turnover cycle. Cell Metab. 2007;6(3):181–94.
37. Schriever SC, Deutsch MJ, Adamski J, et al. Cellular signaling of amino acids towards mTORC1 activation in impaired human leucine catabolism. J Nutr Biochem. 2013;24(5):824–31.
38. Ichihara A. Isozyme patterns of branched-chain amino acid transaminase during cellular differentiation and carcinogenesis. Ann N Y Acad Sci. 1975;259:347–54.
39. Tönjes M, Barbus YJ, Wang W, et al. BCAT1 promotes cell proliferation via amino acid catabolism in IDH1 wildtype gliomas. Nat Med. 2013;19(7):901–8.
40. Conway ME, Yennawar N, Wallin R, et al. Identification of a peroxide-sensitive redox switch at the CXXC motif in the human mitochondrial branched chain aminotransferase. Biochemistry. 2002;41(29):9070–8.
41. Conway ME, Poole LB, Hutson SM. Roles for cysteine residues in the regulatory CXXC motif of human mitochondrial branched chain aminotransferase enzyme. Biochemistry. 2004;43(23):7356–64.
42. Conway ME, Coles SJ, Islam MM, et al. Regulatory control of human cytosolic branched-chain aminotransferase by oxidation and S-glutathionylation and its interactions with redox sensitive neuronal proteins. Biochemistry. 2008;47(19):5465–79.
43. Yennawar N, Dunbar J, Conway M, et al. The structure of human mitochondrial branched-chain aminotransferase. Acta Crystallogr D Biol Crystallogr. 2001;57(Pt 4):506–15.
44. Yennawar NH, Conway ME, Yennawar HP, et al. Crystal structures of human mitochondrial branched chain aminotransferase reaction intermediates: ketimine and pyridoxamine phosphate forms. Biochemistry. 2002;41(39):11592–601.
45. Conway ME, Yennawar N, Wallin R, et al. Human mitochondrial branched chain aminotransferase: structural basis for substrate specificity and role of redox active cysteines. Biochim Biophys Acta. 2003;1647(1–2):61–5.
46. Yennawar NH, Islam MM, Conway M, et al. Human mitochondrial branched chain aminotransferase isozyme: Structural role of the CXXC center in catalysis. J Biol Chem. 2006;281(51):39660–71.
47. Coles SJ, Easton P, Sharrod H, et al. S-Nitrosoglutathione inactivation of the mitochondrial and cytosolic BCAT proteins: S-nitrosation and S-thiolation. Biogeosciences. 2009;48(3):645–56.
48. Okada K, Hirotsu K, Sato M, et al. Three-dimensional structure of Escherichia coli branched-chain amino acid aminotransferase at 2.5 A resolution. J Biochem. 1997;121(4):637–41.
49. Goto M, Miyahara I, Hirotsu K, et al. Structural determinants for branched-chain aminotransferase isozyme-specific inhibition by the anticonvulsant drug gabapentin. J Biol Chem. 2005;280(44):37246–56.

Chapter 4
Isoleucine, PPAR and Uncoupling Proteins

Takayuki Masaki

Key Points

- Adipocytokines regulate whole body homeostasis and inflammatory state
- Ile regulates the levels of adipocytokine leptin and adiponectin
- Ile and glucose homeostasis
- Uncoupling protein and peroxisome proliferator-activated receptors are involved in fatty acid oxidation.
- Ile is involved in tissue fatty acid oxidation

Keywords Isoleucine • Obesity • Adiponectin • Leptin • Peroxisome proliferator-activated receptors

Abbreviations

Ile	Isoleucine
PPAR	Peroxisome proliferator-activated receptors
UCP	Uncoupling protein

Introduction

Isoleucine (Ile) is an essential branched chain amino acid (BCAA) that plays a variety of roles in the body. Ile has been shown to increase free fatty acid oxidation in skeletal muscle and liver. In addition, Ile prevents the accumulation of tissue triglycerides and up-regulates the expression of peroxisome proliferator-activated receptor (PPAR)-alpha and uncoupling protein (UCP) in diet-induced obese mice. These results provide Ile may find applications in the treatment of obesity and related metabolic disorder.

T. Masaki, M.D., Ph.D. (✉)
Department of Endocrinology and Metabolism, Faculty of Medicine,
Oita University, Idaigaoka 1-1, Yufu-hasama, Oita 879-5593, Japan
e-mail: masaki@oita-u.ac.jp

R. Rajendram et al. (eds.), *Branched Chain Amino Acids in Clinical Nutrition: Volume 1*, Nutrition and Health, DOI 10.1007/978-1-4939-1923-9_4,

Adipocyte as an Endocrine Organ

Dietary overload and an inactive lifestyle are modern phenomena leading to excess accumulation of body fat that is connected to insulin resistance.

A primary role for adipocyte is energy storage. Adipose tissue is an endogenous source of circulating lipids as well as the site of the production and secretion of several hormones and cytokines, including leptin, adiponectin and tumor necrosis factor-alpha [1, 2] (Fig. 4.1). Indeed, several adipocytokines regulate whole body homeostasis and inflammatory state [3].

The inflammation does not rely on the classic instigators of immune responses. Rather, it is an immunological response to adipose tissue malfunction. This type of inflammation is referred to as para-inflammation and is dependent on white adipose tissue infiltration by macrophages [3]. Mechanisms involved in the self-maintenance of this inflammatory state in response to chronic caloric overload are being investigated. Evidence suggests that adipose tissue resident macrophages secrete bioactive molecules including inflammatory and anti-inflammatory cytokines that could be related to the development of obesity and insulin resistance. The adipose-derived signaling molecules such as adiponectin and leptin are likely to play a key role in the complex network modulating obesity, inflammation and related disorders [4].

Ile Regulates the Levels of Adipocytokine Leptin and Adiponectin

Leptin is the product of the *ob* gene, and it is secreted by adipocytes with circulating levels proportional to fat mass. Levels of circulating leptin have a diurnal and pulsatile pattern. Leptin has several effects on the central nervous system [5]. Leptin is transported across the blood brain barrier by a saturable transporter system and it exerts its anorectic effect via the arcuate nucleus, where both neuropeptide Y (NPY)/agouti related protein (AgRP) and pro-opiomelanocortin (POMC)/cocaine- and amphetamine-regulated transcript (CART) neurons express leptin receptors. Leptin inhibits NPY/AgRP neurons and activates POMC/CART neurons, resulting in reduced food intake and increased energy expenditure. There are three types of leptin receptors identified: long, short and secreted form. Among those, Ob-Rb receptor, which is highly expressed in the hypothalamus is thought to act as the main receptor involved in appetite control [6]. The *db/db* mouse, with an inactivating mutation in the Ob-Rb receptor, has an obese phenotype and leptin-deficient *ob/ob* mice exhibit hyperphagia and obesity, which can be reversed by leptin administration.

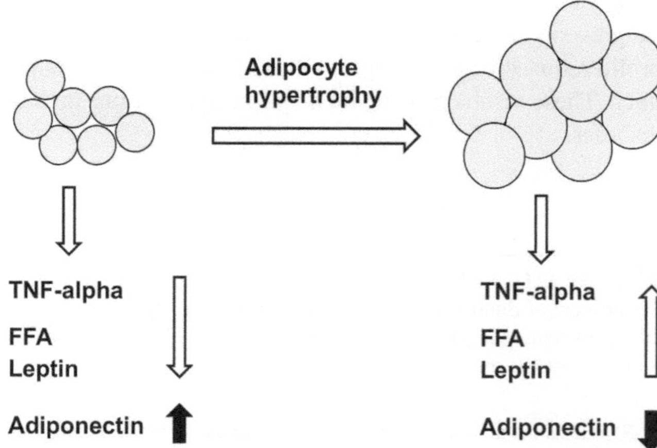

Fig. 4.1 Adipocyte as an endocrine organ. Several cytokine and hormone are secreted from adipocyte

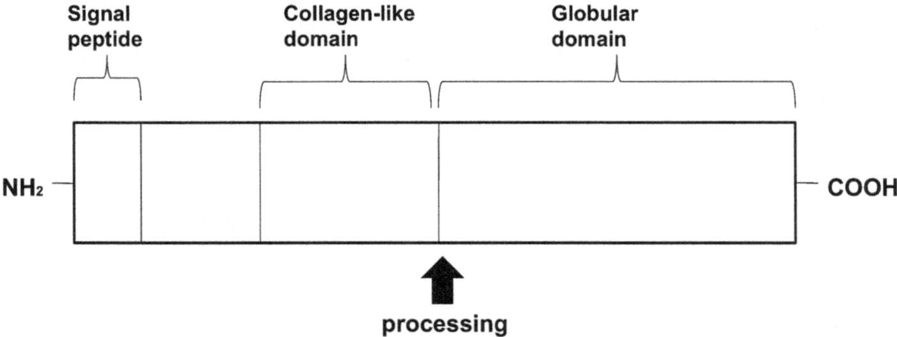

Fig. 4.2 Structure and domains of adiponectin. Adiponectin is composed of an N-terminal collagen-like sequence and a C-terminal globular region

Subcutaneous administration of recombinant leptin reduces fat mass, hyperinsulinaemia, and hyperlipidaemia in obese patients with congenital leptin deficiency. However, obese individuals often have high leptin levels which result in a failure to respond to exogenous leptin.

Previous study revealed the relationship between Ile and adipocytokine leptin [7]. High-fat diet-induced leptin elevation that related to leptin resistance in obesity. The levels of circulating leptin in the Ile-treated diet-induced obese mice were lower than in the controls. In addition, Ile decreased body adiposity and the size of adipose tissue enlargement in the obese mice. The data thus suggest that supplementation with Ile might be useful in the treatment of obesity and leptin resistance.

Adiponectin was identified in adipose tissue by screening for adipose-specific genes and have been shown to regulate metabolic syndrome [8]. Adiponectin mRNA is expressed in adipose tissue and the protein is abundant in the plasma of humans and rodents. In addition, patients with type 2 diabetes and/or obesity showed lower levels of plasma adiponectin, indicating the involvement of this adiponectin in glucose and fat metabolism. Serum adiponectin levels were decreased in obese mice on a HF diet. Lower serum adiponectin levels in obese mice on the HF diet were partially restored by replenishment of recombinant adiponectin. Importantly, replenishment of adiponectin significantly ameliorated HF diet-induced insulin resistance and hypertriglyceridaemia. These data raise the possibility that the replenishment of adiponectin may provide a treatment modality for insulin resistance and type 2 diabetes. Adiponectin structurally belongs to the complement 1q family and is known to form a characteristic homomultimer (Fig. 4.2). It has been demonstrated that simple SDS-PAGE under non-reducing and non-heat-denaturing conditions clearly separates multimeric species of adiponectin. Adiponectin in human or mouse serum and adiponectin expressed in NIH-3T3 cells or *Escherichia coli* forms a wide range of multimers from trimers and hexamers to high molecular weight multimers [9, 10]. Adiponectin can exist as full-length or a smaller, globular fragment; however, almost all adiponectin appears to exist as full-length adiponectin in plasma (Fig. 4.2).

The relationship between Ile and adipocytokine adiponectin is described in obese mice [7]. The hepatic and skeletal muscle triglyceride concentrations and degree of hyperinsulinaemia in the Ile group mice were lower than the control group. The levels of adiponectin in the adipose tissue of the Ile-treated obese mice were higher than in the controls (Fig. 4.3). This implies that a mechanism through the regulation of adiponectin contributes to the Ile-effects on insulin resistance and type 2 diabetes.

Ile and Glucose Homeostasis

BCAA Ile is dietary essential amino acid which plays a role in glucose and lipid metabolism. Ile suppresses the plasma glucose concentration with an activation of the glucose transporter in rats. In muscle cells, Ile stimulate glucose uptake in an insulin-independent manner [11]. Phosphatidylinositol

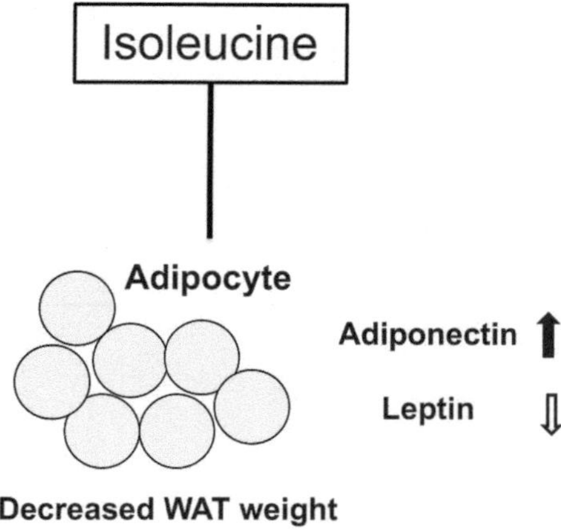

Fig. 4.3 Effects of Ile on leptin and adiponectin. Ile decreased leptin and increased adiponectin

3-kinase and protein kinase C are involved in the enhancement of glucose uptake by Ile. The hypoglycemic effect of Ile was confirmed in a human study of administration of Ile [12].

The maintenance of blood glucose levels is due to an optimal balance between glucose uptake by peripheral tissues and glucose production occurring mainly in the liver. Glucose uptake was significantly increased in the muscle of Ile-administered rats when compared with that in controls. A previous study revealed that glucose uptake by Ile is involved in increased glucose transporter GLUT1 and GLUT4 translocation in the skeletal muscle of rats with liver cirrhosis [13]. Hepatic gluconeogenesis, as well as glucose utilization by peripheral and hepatic tissues, may be a possible mechanism by which amino acids lower blood glucose levels. The mRNA levels of phosphoenolpyruvatecarboxykinase (PEPCK), which is a well-researched key gluconeogenic enzyme, parallel both PEPCK activity and the rate of gluconeogenesis [14]. Expression levels of hepatic PEPCK mRNA were lower in Ile-administered rats than in controls, thus suggesting that PEPCK activity was lower in Ile-administered rats and that the inhibitory effects of Ile on PEPCK are regulated at the transcriptional level. Furthermore, the expression levels of hepatic glucose-6-phosphatase (G6Pase) mRNA and G6Pase activity were also lower in Ile-administered rats.

In conclusion, Ile administration stimulates both glucose uptake in the muscle and whole body glucose oxidation, in addition to depressing gluconeogenesis in the liver, thereby leading to a hypoglycemic effect.

UCP1 in Brown Adipose Tissue

UCPs are mitochondrial transporters present in the inner mitochondrial membrane. The first member of the family, UCP1, is expressed in brown adipocytes and it confers on brown adipose tissue its thermogenic capacity. UCP1 confers to the mitochondrial inner membrane an enhanced conductivity to protons, thus resulting in the uncoupling of the respiratory chain and heat production. This action of UCP1 in brown adipose tissue constitutes the main molecular basis for nonshivering thermogenesisin rodents in response to cold exposure and diet [15]. The thermogenic activity of brown fat is mainly regulated by norepinephrine released from the sympathetic nervous system innervating the tissue, acting through beta-adrenergic, cAMP-dependent pathways.

Accumulating pieces of evidence over more than two decades have indicated that energy expenditure processes elicited by UCP1 are involved in the control of energy balance, and that UCP1 activity in brown adipose tissue may provide the basis for diet-induced thermogenesis. In fact, obesity models in rodents are in most cases associated with low levels and activity of UCP1 in brown fat.

UCPs and PPARs Are Involved in Fatty Acid Oxidation

Fatty acids oxidation was performed in liver and skeletal muscle (Fig. 4.4). The levels of FAT/CD36, UCP2, UCP3 and PPAR-alpha are key regulators of FFA uptake in the tissue (Fig. 4.5). PPAR-alpha expression increases in the liver in several models of murine obesity [16]. Chronic treatment of rodents with PPAR-alpha agonists increases hepatic UCP2mRNA expression. UCP2 mRNA levels are also up-regulated in cultured hepatocytes in response to polyunsaturated fatty acids, PPAR-alpha agonists.

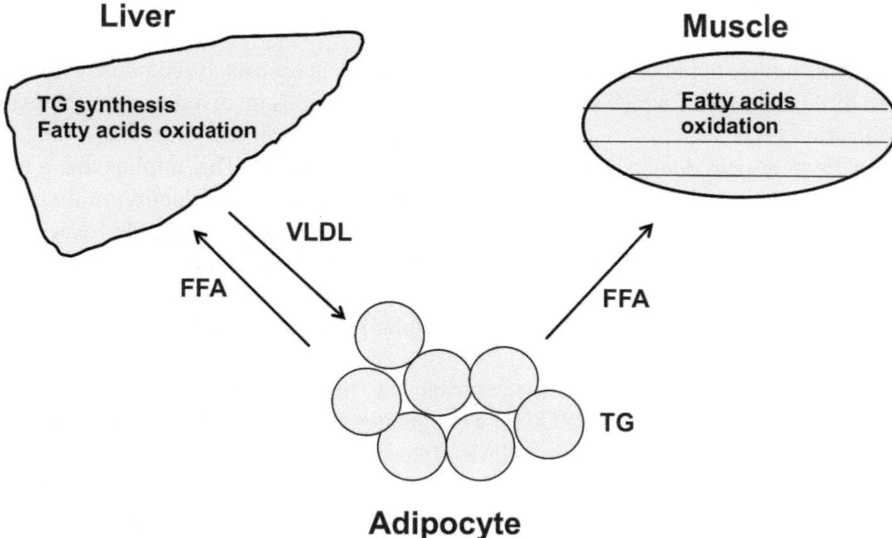

Fig. 4.4 Free fatty acid oxidation. Free fatty acids are oxidized in liver and skeletal muscle

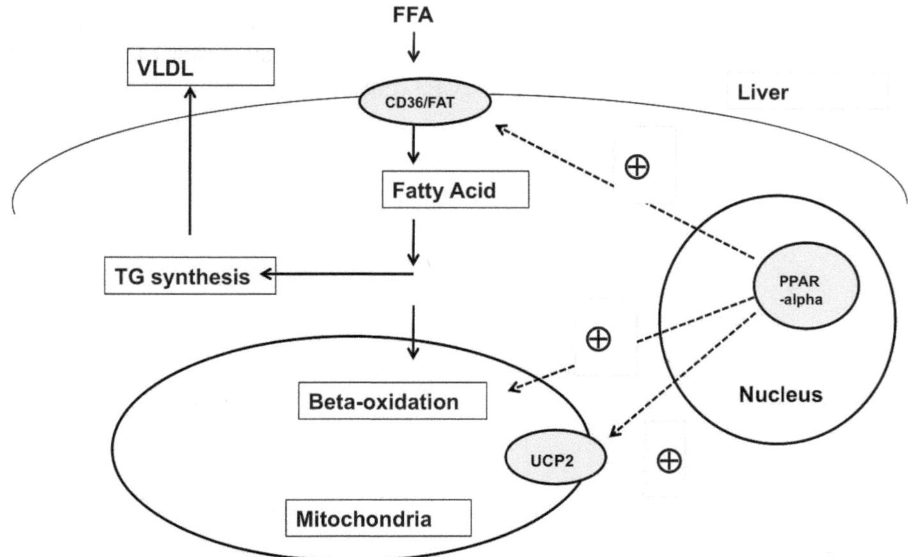

Fig. 4.5 Free fatty acid metabolism in the liver. Schematic representation of the regulation of UCP2 and CD36 expression by ligand-dependent activation of PPAR-alpha

Multiple lines of evidence have shown that PPAR-alpha plays a role in the induction of the UCP3 gene in response to fatty acids in skeletal muscle. Acute treatment of mice pups with the specific activator of PPAR-alpha mimics the postnatal skeletal muscle UCP3 gene. Moreover, PPAR-alpha-null mice show reduced levels of UCP3 gene expression. This occurs in concert with induction of many other genes involved in fatty acid oxidation. Thus, these observations strongly suggest that UCP3 function in skeletal muscle is likely to be related to fatty acid metabolism.

Ile Is Involved in Tissue Fatty Acid Oxidation

Ile treatment was further found to reduce the levels of adiposity in each analyzed tissue without affecting food intake of the DIO mice, which suggests that this amino acid is involved in energy metabolism or lipid mobilization [7].

However, Ile treatment does not affect BAT UCP1 expression [7]. This implies that a mechanism other than the regulation of BAT energy expenditure contributes to the reduction in tissue adiposity. The levels of PPAR-alpha expression in the liver and skeletal muscle of the Ile-treated mice were higher than in the controls. Given that the activation of PPAR-alpha in these tissues accelerates FFA oxidation. A previous study has demonstrated that PPAR-alpha activation prevents the development of DIO and insulin resistance. In contrast, PPAR-alpha-knockout mice become obese when fed a high-fat diet.

Thus, Ile-induced PPAR-alpha activation contributes to the reduction in tissue adiposity.

The levels of liver UCP2 and muscle UCP3 were greater after Ile treatment than in the controls [7]. UCP2 and UCP3 genes are regulated by PPAR-alpha in liver and muscle. The Ile-induced UCPs expression in both tissues is mediated by PPAR-alpha. UCP2 and UCP3 were reported to play a role in the control of FFA oxidation rather than in the regulation of energy expenditure. The up-regulation of muscle UCP3 and hepaticUCP2 may also contribute to the activation of FFA oxidation in the tissues. The levels of FAT/CD36, a key protein regulator of FFA uptake in skeletal muscle and liver, were greater in the Ile treatment group than in the control mice. Although FFA transporters other than FAT/CD36 have been shown to function in the liver, the Ile-induced differences in skeletal muscle and liver FAT/CD36 levels that we observed indicate that FFA uptake may be accelerated in muscle and liver. Taken together, Ile simultaneously activates liver and skeletal muscle FFA uptake and oxidation through PPAR-alpha, FAT/CD36, UCP2 or UCP3 (Fig. 4.6).

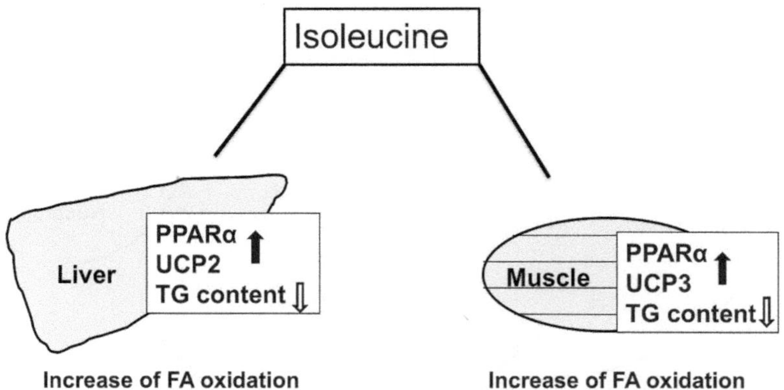

Fig. 4.6 Distinctive molecular pathway for ile-induced fatty acids oxidation. Ile increase PPAR-alpha expression and subsequent UCP-2 in liver and UCP-3 in muscle

Conclusions

PPAR-alpha is a major controller of UCPs gene expression in liver and skeletal muscle. Ile simultaneously activates liver and skeletal muscle FFA uptake and oxidation through PPAR-alpha and UCPs. The studies have revealed distinct cross-talk mechanisms between Ile, FFA oxidation and obesity.

References

1. Kern PA, Saghizadeh M, Ong JM, et al. The expression of tumor necrosis factor in human adipose tissue. Regulation by obesity, weight loss, and relationship to lipoprotein lipase. J Clin Invest. 1995;95:2111–9.
2. Xu H, Barnes GT, Yang Q, et al. Chronic inflammation in fat plays a crucial role in the development of obesity-related insulin resistance. J Clin Invest. 2003;112:1821–30.
3. Weisberg SP, McCann D, Desai M, et al. Obesity is associated with macrophage accumulation in adipose tissue. J Clin Invest. 2003;112:1796–808.
4. Trayhurn P, Wood IS. Adipokines: inflammation and the pleiotropic role of white adipose tissue. Br J Nutr. 2004;92:347–55.
5. Gautron L, Elmquist JK. Sixteen years and counting: an update on leptin in energy balance. J Clin Invest. 2011; 12:2087–93.
6. Wauman J, Tavernier J. Leptin receptor signaling: pathways to leptin resistance. Front Biosci. 2011;16:2771–93.
7. Nishimura J, Masaki T, Arakawa M, et al. Isoleucine prevents the accumulation of tissue triglycerides and upregulates the expression of PPARalpha and uncoupling protein in diet-induced obese mice. J Nutr. 2010;140:496–500.
8. Kadowaki T, Yamauchi T. Adiponectin and adiponectin receptors. Endocr Rev. 2005;26:439–51.
9. Waki H, Yamauchi T, Kamon J, et al. Impaired multimerization of human adiponectin mutants associated with diabetes: molecular structure and multimer formation of adiponectin. J Biol Chem. 2003;278:40352–63.
10. Tsao T-S, Murrey HE, Hug C, et al. Oligomerization state-dependent activation of NF-kB signaling pathway by adipocyte complement-related protein of 30 kDa (Acrp30). J Biol Chem. 2002;277:29359–62.
11. Doi M, Yamaoka I, Fukunaga T, et al. Isoleucine, a potent plasma glucose-lowering amino acid, stimulates glucose uptake in C2C12 myotubes. Biochem Biophys Res Commun. 2003;312:1111–7.
12. Nuttall FQ, Schweim K, Gannon MC. Effect of orally administered isoleucine with and without glucose on insulin, glucagon and glucose concentrations in non-diabetic subjects. E Spen Eur E J Clin Nutr Metab. 2008;3:e152–8.
13. Nishitani S, Takehana K, Fujitani S, et al. Branched-chain amino acids improve glucose metabolism in rats with liver cirrhosis. Am J Physiol Gastrointest Liver Physiol. 2005;288:G1292–300.
14. Xu H, Yang Q, Shen M, et al. Dual specificity MAPK phosphatase 3 activates PEPCK gene transcription and increases gluconeogenesis in rat hepatoma cells. J Biol Chem. 2005;280:36013–8.
15. Kozak LP. Brown fat and the myth of diet-induced thermogenesis. Cell Metab. 2011;11:263–7.
16. Villarroya F, Iglesias R, Giralt M. PPARs in the control of uncoupling proteins gene expression. PPAR Res. 2007;2007:74364.

Chapter 5
Leucine as a Stimulant of Insulin

Jun Yang, Michael Dolinger, Gabrielle Ritaccio, David Conti, Xinjun Zhu, and Yunfei Huang

Key Points

- Leucine regulates a large array of cellular processes in pancreatic β-cells, including metabolism, growth, proliferation and insulin secretion.
- Leucine stimulates insulin release by serving as a fuel source for ATP production and an allosteric regulator of glutamate dehydrogenase (GDH).
- Leucine also exerts its secretagogue effects by triggering calcium oscillations.
- Leucine regulates the expression of some key genes those are critical for insulin secretion in pancreatic cells.
- Leucine was found to regulate insulin release by influencing adrenergic activity via the mTOR pathway.

Keywords Leucine • ATP-sensitive K^+ channels • Glutamate dehydrogenase • mTOR • Insulin

Abbreviations

mTOR	The mammalian target of rapamycin
K_{ATP} channels	ATP-sensitive K^+ channels
KIC	α-ketoisocaproate

J. Yang, M.D., Ph.D.
Center for Neuropharmacology and Neuroscience; Center for Cardiovascular Science; Division of Gastroenterology, Department of Internal Medicine, Albany Medical College, Albany, NY 12208, USA

M. Dolinger • G. Ritaccio
Division of Gastroenterology, Department of Internal Medicine, Albany Medical College, Albany, NY 12208, USA

D. Conti
Department of Transplant Surgery, Albany Medical College, Albany, NY 12208, USA

X. Zhu
Center for Cardiovascular Science; Division of Gastroenterology, Department of Internal Medicine, Albany Medical College, Albany, NY 12208, USA

Y. Huang, M.D., Ph.D. (✉)
Center for Neuropharmacology and Neuroscience, Albany Medical College, Albany, NY 12208, USA
e-mail: huangy@mail.amc.edu

R. Rajendram et al. (eds.), *Branched Chain Amino Acids in Clinical Nutrition: Volume 1*, Nutrition and Health, DOI 10.1007/978-1-4939-1923-9_5,

GDH	Glutamate dehydrogenase
SUR	Sulfonylurea receptor
BCATs	Branched chain aminotransferase isozymes
BCKDH	Branched chain α-keto acid dehydrogenase
DES	Diethylstilbestrol
BCH	β-2-aminobicyclo [2.2.1]-heptane-2-carboxylic acid
PKB	Protein kinase B
PKCs	Protein kinase Cs
SGK1	Serum/glucocorticoid-induced kinase 1
TSC	Tuberous sclerosis complex
AMPK	AMP-activated protein kinase
M3R	Muscarinic receptor M3
Caβ2	β2 subunit of the L-type calcium channel
ATPSβ	ATP synthase β-subunit
cAMP	Cyclic adenosine monophosphate
GLP-1	Glucagon-like peptide-1
GK rat	Goto-Kakizaki rat
SNPs	Single-nucleotide polymorphisms
NODAT	New-onset diabetes after transplantation

Introduction

Leucine is one of the three branched chain amino acids and also an essential amino acid that cannot be synthesized de novo in mammals and other vertebrates. Therefore, it has to be acquired from the diet. Besides serving as a building block for protein synthesis, leucine also can be catabolically converted to generate energy or to produce intermediates for biosynthesis of non-essential amino acids, carbohydrates, nucleotides and fatty acids. Leucine has recently gained particular attention because it is a potent activator of the mammalian target of rapamycin (mTOR) pathway, a critical nutrient-sensing pathway that governs cell metabolism, cell growth and proliferation. This newly recognized role thus expands the known modes of biological action of leucine. In particular, we are now beginning to appreciate how leucine regulates insulin secretion and glucose homeostasis by its influence on mTOR activity. The mTOR pathway becomes complementary to canonical metabolic processes such as those that provide fuel for energy production or intermediates to regulate insulin secretion.

Historically, glucose was considered the major insulinotropic factor and it dominated research into the physiology of insulin secretion for nearly five decades before the role of amino acids in insulin secretion was recognized. The general consensus is that, after absorption into pancreatic cells, glucose is metabolized via glycolysis to generate intermediates and ATP, which in turn impinge on ATP-sensitive K^+ channels (K_{ATP} channels) in pancreatic β-cells, leading to membrane depolarization and intracellular Ca^{2+} mobilization and subsequently the triggering of insulin release (Fig. 5.1). However, it was not until the late 1950s and early 1960s that Cochrane et al., Yalow et al. and Floyd et al. were able to show that proteins and the branched-amino acid L-leucine also possess hypoglycemia-inducing activities [1–3]. The effect of leucine on blood glucose apparently does not solely reflect enhanced glucose uptake by tissues, because Floyd et al. were also able to demonstrate that proteins and leucine promote insulin secretion. These initial findings indicated that leucine is catabolized in pancreatic β-cells to produce metabolic intermediates and ATP so that it acts similarly to glucose to regulate insulin release. Consistent with this notion, the two other branched chain amino acids, valine and isoleucine, which share a similar metabolic pathway to leucine, also have insulinotropic effects; however, the effect of leucine is much more potent. For example, the insulinotropic effect of L-valine is less than 20 % of that of L-leucine. Moreover, leucine can act synergistically to facilitate insulin

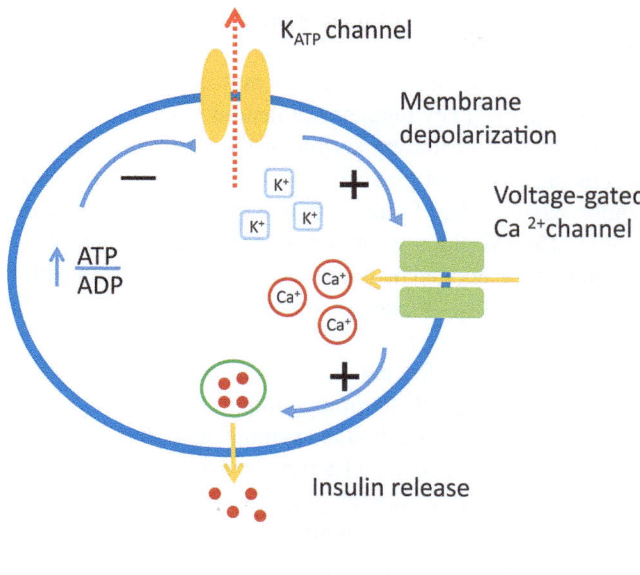

Fig. 5.1 Regulation of insulin secretion by leucine in pancreatic β-cell

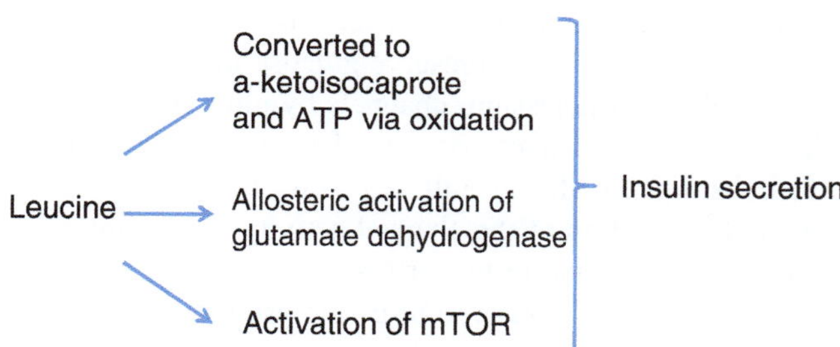

Fig. 5.2 Mechanism of insulin release in pancreatic β-cell

release elicited by glucose. These observations raise the possibility that the regulatory effects of leucine on insulin secretion may involve additional mechanisms besides those shared with glucose.

Indeed, over the past decade, leucine was found to exert its secretagogue effects via multiple mechanisms [4] (Fig. 5.2). Leucine can serve as a fuel source for ATP production, which inhibits K_{ATP} channel activity, leading to membrane depolarization and triggering of insulin secretion. Leucine can also be converted to α-ketoisocaproate (KIC), a metabolic intermediate which exhibits an insulinotropic effect via direct inhibition of K_{ATP} channel activity. Moreover, leucine was found to be an allosteric regulator of glutamate dehydrogenase (GDH) [5], a key enzyme that fuels amino acids into the tricarboxylic acid cycle, which is an alternate pathway able to influence insulin secretion. Additional routes of action include triggering calcium oscillations in pancreatic β-cells [6] and regulating the expression of some key genes that are critical for insulin secretion in pancreatic islets. Most recently, leucine was found to regulate insulin release by influencing adrenergic activity via the mTOR pathway. In this review, we will summarize the studies on the mechanisms of insulin secretion elicited by leucine.

K_{ATP} Channel in Insulin Secretion

The general consensus for the metabolic regulation of insulin release is that mitochondrial oxidation of glucose or amino acids in pancreatic β-cells leads to accumulation of high-energy triphosphate nucleotides (ATP and GTP). The rise in the ATP/ADP ratio triggers closing of the K_{ATP} channels, which in turn depolarizes membrane potential and subsequently opens voltage-gated Ca^{2+} channels. As a consequence, intracellular Ca^{2+} is mobilized, as reflected in the rise of free cytoplasmic Ca^{2+}, which triggers a series of membrane events, including insulin vesicle membrane docking, fusion and insulin secretion.

A wealth of studies using various molecular, cellular and electrophysiological tools revealed that the K_{ATP} channel is a key regulator of insulin release in pancreatic β-cells. In terms of glucose, which is a major physiological stimulus for insulin secretion in pancreatic β-cells, initial studies of pancreatic β-cells observed that glucose-induced insulin secretion proceeds through a slow membrane depolarization which is then followed by either oscillatory bursts of action potential at low glucose concentration (5–15 mM glucose) or continuous spiking when glucose is high (>16 mM) [7]. The membrane depolarization elicited by glucose is closely associated with reduced potassium permeability. By blocking the metabolism of glucose, the electrical and secretory responses to glucose are largely diminished, suggesting metabolic products from glucose regulate ion channel activity and insulin release.

In the early 1980s, patch-clamp, a revolutionary electrophysiological technique, was invented, which made it possible to directly monitor ion channel activity at the single-molecule level. Cook and Hales employed this method to examine potassium channels in pancreatic β-cells [8]. They identified a particular potassium channel activity sensitive to intracellular ATP. The potassium channel is spontaneously active in the membrane patch excised from β-cells in the recording bath solution, but was found to be rapidly and reversibly inhibited when the cytoplasmic side of the membrane was exposed to solution containing ATP. These experiments led to the discovery of the K_{ATP} channel, a key player in the coupling of cellular metabolism and membrane electrical activity. In the meantime, using a cell-attached configuration to preserve glucose metabolism in intact β-cells, Ashcroft et al. demonstrated that inhibition of potassium channels by glucose requires the metabolism of glucose in those β-cells [7]. These two studies represent a milestone in understanding how insulin release is coupled to glucose metabolism in β-cells. Overall, in β-cells, the K_{ATP} channel is the main regulator of resting membrane potential. Alteration of the channel activity, for example by inhibition by ATP, will cause membrane depolarization, which in turn triggers intracellular mobilization and insulin release. The intracellular ATP/ADP ratio is considered to be the primary factor regulating channel activity. Molecular cloning revealed that the K_{ATP} channel is a large heteromultimeric macromolecular complex in which four inwardly rectifying potassium channels (Kir6.2) form a central pore surrounded by four regulatory sulfonylurea receptor (SUR) subunits [9]. SUR1 is expressed in pancreatic β-cells and neurons. It has been shown that loss-of-function mutations in Kir6.2 (KCNJ11) and SUR1 (ABCC8) cause over-secretion of insulin and result in hyperinsulinism of infancy [10, 11], confirming a critical role of the K_{ATP} channel in the regulation of insulin secretion. Notably, leucine-sensitive hyperinsulinemia and hypoglycemia is a genetic disorder that was first reported by Cochrane in 1956 [1] and soon after by DiGeorge [12]. Patients tend to display hypoglycemia following high-protein feedings or administration of oral or IV infusions of leucine. Genomic sequencing revealed several mutations in SUR1 in pediatric patients. While these studies further confirm the role of the K_{ATP} channel in the regulation of insulin release, they also provide genetic evidence supporting a role for leucine in regulating insulin secretion and glucose homeostasis. Leucine can stimulate insulin release through at least two major metabolic routes. The first route involves a transamination reaction in which leucine is converted to KIC. The second route is to activate GDH allosterically, which in turn promotes oxidation of glutamate and catabolism of intermediates of α-ketoglutarate via the Krebs cycle.

Transamination Reaction of Leucine and Its Role in Regulating Insulin Secretion

In the first route, leucine is catabolized to produce KIC, which can be fed into the TCA cycle for ATP production (Fig. 5.3). The two initial enzymatic steps that mediate catabolism of leucine are shared by all three branched chain amino acids, and thus are critical in terms of maintaining the balance of the levels of individual branched chain amino acids in the cells. Leucine is first converted to KIC via a

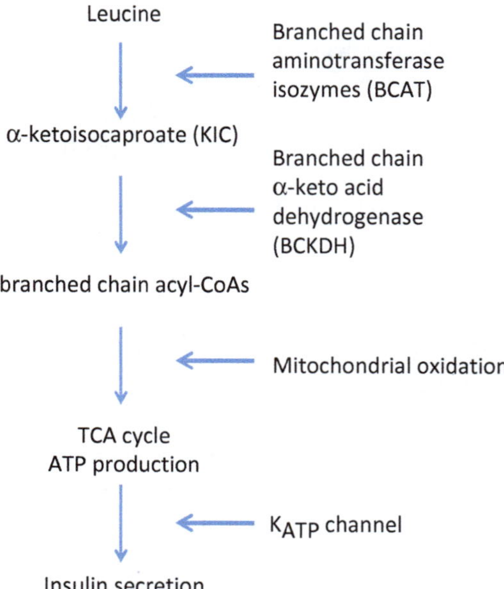

Fig. 5.3 A schematic view of the leucine catabolic pathway

transamination reaction catalyzed by branched chain aminotransferase isozymes (BCATs) (including mitochondrial BCAT and cytosolic BCAT). In this reaction, the α-amino group of leucine is reversibly transferred to α-ketoglutarate. In the second reaction, KIC is decarboxylated to branched chain acyl-CoAs by branched chain α-keto acid dehydrogenase (BCKDH). This enzyme is a complex formed by three enzymes, a branched chain α-keto acid decarboxylase (E1), a dihydrolipoyl transacylase (E2) and a dihydrolipoyl dehydrogenase (E3). The enzymatic activity of the BCKDH complex is entirely dependent on the status of phosphorylation of the E1 subunit. The BCKDH kinase phosphorylates the E1subunit, resulting in activation of BCKDH, whereas dephosphorylation of the E1subunit by BCKDH phosphatase inactivates BCKDH. The branched chain acyl-CoA can be further converted to intermediates and then fed into the TCA cycle for production of ATP.

Although ATP generated from oxidation of leucine or KIC regulates K_{ATP} channel activity [13], interestingly, KIC was also found to regulate K_{ATP} channel activity directly. In 1972, Panten et al. examined a few metabolites of leucine to determine if they have any effect on insulin secretion in pancreatic islets. They found that KIC at 10 mM can trigger a marked release of insulin, comparable to the effect elicited by leucine at 10 mM, but that other intermediates, including isovalerate, pyruvate, d,l mevalonate and acetoacetate which are also relevant to leucine metabolism even at similar concentrations, had either no or a very slight effect. Using the patch–clamp technique, Branstrom et al. proved that KIC can directly inhibit the K_{ATP} channel in isolated patches and intact pancreatic β-cells, with a half-maximal concentration of approximately 8 mM [14]. In contrast, leucine, the precursor of KIC, has no direct effect on K_{ATP} channel activity. However, other interesting observations complicate the interpretation of how KIC regulates insulin release. For example, it appears that KIC is not an effective resource for production of ATP in rat pancreatic β-cells [15]. Moreover, KIC can be reversibly converted to leucine through transamination in the pancreatic islet homogenate. This reverse route was reported to regulate Ca^{2+} mobilization and insulin release [16]. Macdonald et al. and Wollheim et al. reported that pre-exposure of pancreatic islets to high glucose for overnight diminishes the response to leucine-induced insulin release [17, 18]. It was thus postulated that BCKDH, the enzyme that mediates metabolism of leucine and KIC, is the rate-limiting step for the action of leucine and KIC in β-cells. However, Matschinsky and his colleagues found that leucine and KIC show distinct effects on stimulation of insulin secretion from pancreatic islets [19]. That group also found that, although pre-incubation of islets in high glucose for overnight completely blocks the effects of leucine-stimulated insulin release which is accompanied by slow mobilization of Ca^{2+}, high glucose does not attenuate the effect of KIC [19]. This observation suggests that breakdown of leucine and KIC via oxidation is not the essential step in the regulation of insulin release. The authors thus raised the argument that BCKDH perhaps is not the key regulator in leucine-induced insulin release. A further observation in support of this argument is that insulin secretion in patients with mutations in BCKDH appears to be normal, although mitochondrial oxidation of leucine or KIC is not compromised [20]. Therefore, these observations suggest that an additional target likely mediates leucine-induced insulin release.

Allosteric Regulation of GDH by Leucine and Its Role in Regulating Insulin Secretion

GDH is a mitochondrial matrix enzyme that catalyzes the reversible oxidative deamination of glutamate to α-ketoglutarate using NAD(H) or NADPH(H) as a co-factor. α-Ketoglutarate is a component of the TCA cycle. Therefore, when amino acids are abundant, GDH is a critical enzyme in the catabolism of amino acids and in driving the reaction toward the TCA cycle for production of energy and other important metabolic intermediates (Fig. 5.4). GDH is broadly expressed in pancreatic β-cells and several other organs, including the brain, heart, kidney, liver and lung. Although GDH could theoretically drive the reaction in the reverse direction (involving a reductive amination reaction) toward glutamate, under normal conditions the forward oxidative deamination reaction is more favorable [21]. This enzymatic reaction is very sensitive to the fuel status of the cell, thereby dictating the rate at which amino acids undergo catabolism when conditions become favorable, for example when cells experience low glucose levels. Among the triphosphate nucleotides, GTP is a potent inhibitor of the GDH-mediated deamination reaction. GDH has a 100-fold higher affinity for GTP over ATP [22]. GTP exerts its inhibitory effect through stabilizing the binding of the enzymatic product to GDH, thereby slowing the release of the product from GDH and consequently reducing the overall turnover of the enzymatic reaction [23]. The low-energy form of the nucleotide, GDP, is an activator of GDH. Binding of GDP to GDH promotes the release of the enzymatic products, thereby facilitating the forward reaction. Besides glutamate, l-leucine can also be catabolized by GDH, but it is a very weak substrate. Struck and Sizer reported that the rate of deamination of leucine is about 1.7 % of that

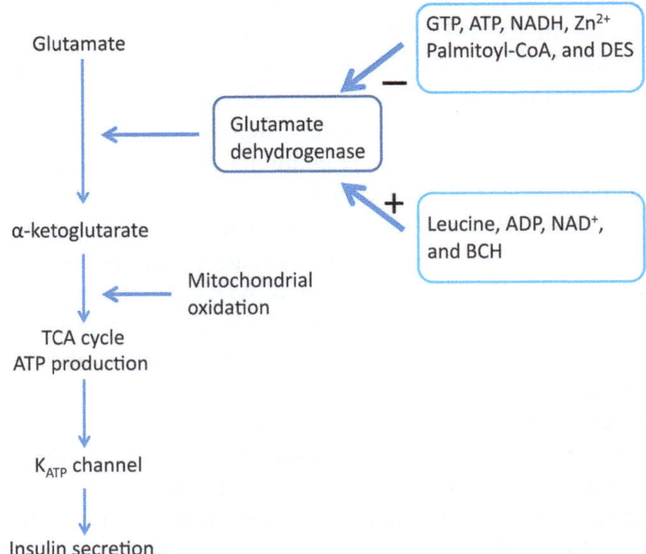

Fig. 5.4 Regulation of GDH and its role in regulating insulin secretion

of l-glutamate [24]. However, leucine can stimulate the oxidation of glutamate. Therefore, leucine is an activator of GDH. The branched chain amino acid isoleucine along with methionine and the unnatural amino acid norvaline are also able to stimulate GDH activity, but are considerably less effective than leucine. All other amino acids and the intermediate metabolites from leucine, including KIC or isocaproate, have no noticeable effect on the rate of the GDH-mediated reaction. Interestingly, several compounds that are relatively hydrophobic such as palmitoyl coenzyme A (CoA) [25], steroid hormones [26] and the hormone analogue diethylstilbestrol (DES) [27] have a negative impact on GDH enzymatic activity. Zn^{2+} also inhibits GDH activity, by stabilizing ineffective enzymatic multimeric protein complexes [5]. At a high concentration, NADH inhibits GDH activity, which represents feedback inhibition of the catabolic reactions [28]. A non-hydrolyzable form of leucine is a potent activator of GDH [29]. Together, ADP, NAD^+, leucine and BCH are activators of GDH, whereas ATP, GTP, NADH, palmitoyl-CoA, DES and Zn^{2+} are the inhibitors of GDH (Fig. 5.4). In a search for the mechanisms of regulation of GDH activity, Yielding et al. observed that DES disrupts the GDH protein complex into subunits, resulting in inhibition of GDH activity. The dissociation effect of DES can be blocked by leucine. This observation suggests that leucine could stimulate GDH activity by physically stabilizing the protein complex.

Structurally, GDH is an oligomeric protein complex consisting of two trimers that directly stack on top of each other. Keeping the entire complex intact appears to be essential for enzymatic activity [5]. However, recent crystal structural studies revealed that prolonged occupation of the double-stack complex from one side by coenzyme NADH and the other side by substrate such as glutamate could transform an active form of GDH into a non-functional or "abortive" complex. This is thought to be the mechanism by which GTP and Zn^{2+} inhibits GDH. Although the structure of GDH is in general conserved in the entire kingdom of living organisms, GDHs in mammals and other vertebrates have an additional helix-loop-helix stretch, called the "antenna". Interestingly, GDHs with an antenna tend to be regulated by ligands, whereas those without are not. Therefore, the antenna motif in GDH appears to play a major role in the regulation of enzymatic activity by ligands. In an effort to identify the amino acid residues responsible for GTP binding and inhibition of GDH activity, Lee et al. performed mutagenesis studies. Human GDH mutants at Tyr-266 or Lys-450 were made based on the fact that those residues were found to be mutated in hyperinsulinism-hyperammonemia syndrome. Whereas wild-type GDH was completely inhibited by 30 μM GTP, GDH with a Lys-450 mutation in the antenna domain was immune from inhibition and could maintain 90 % of basal activity in the presence of GTP up to 300 μM [30]. Humans have two GDH isoforms, *GLUD1* (ubiquitously expressed) and *GLUD2* (brain-specific), with marked differences in terms of their basal activities and allosteric regulation. To elucidate the structural basis for the functional differences, Zaganas et al. mutagenized four residues that are unique to *GLUD1* [31]. They found that substitution of Ser-443 with Arg, but not three other substitutions (Ser-331 replaced with Thr, Met-370 with Leu and Met-415 with Leu), nearly completely eliminated the basal enzymatic activity and also conferred loss of regulation by leucine in the absence of other effectors. However, in the presence of both leucine and ADP, GDH activity surged by 20-fold relative to the control, indicating a synergistic effect of ADP and leucine and a critical role of Ser-443 in regulation of GDH activity. Interestingly, Ser-443 resides in the antenna domain. These observations further support the notion that the helical antenna domain perhaps mediates a conformational change as well as inter-subunit interactions in response to allosteric regulators and during the catalytic reaction. The largely exclusive presence of this regulation in animals may help to explain the essential role of GDH in regulating insulin secretion.

To identify the binding site for leucine in GDH, Tomita et al. used *Thermus thermophilus* GDH as a model molecule to elucidate the structural basis for leucine-induced allosteric activation of GDH. *T. thermophilus* GDH is composed of two different subunits, GdhA (regulatory subunit) and GdhB (catalytic subunit). Crystallography revealed that the GdhA-GdhB-Leu complex is a heterohexamer composed of four GdhA subunits and two GdhB subunits. Each complex binds six leucine molecules. Leucine binds to the interface formed by three subunits in various combinations. The crystal structures also suggest that the binding site for leucine may be shared for glutamate-binding. Interestingly, leucine mainly impacts the rate of turnover of the reaction, but not the affinity of the substrate such as glutamate binding. Nevertheless, alteration of amino acid residues of *T. thermophilus* GDH responsible for leucine binding at the interface formed by three subunits markedly decreased the activation of GDH elicited by leucine. Similarly, mutation of amino acid residues Arg-151 and Asp-185 of human GDH (*GLUD1*), which are equivalent to those mutated above in *T. thermophilus* GDH, caused a loss of allosteric regulation to leucine [32].

In pancreatic β-cells, the importance of GDH as a key enzyme in the regulation of insulin secretion was recognized more than three decades ago [33]. The non-metabolizable leucine analogue BCH (β-2-aminobicyclo [2.2.1]-heptane-2-carboxylic acid) that activates GDH was also found to effectively stimulate insulin secretion [29], providing a first clue that leucine may regulate insulin secretion by acting on GDH. Inhibition of GDH activity was shown to decrease insulin release [34]. Subsequent genetic studies of congenital hyperinsulinism/hypoglycemia cases revealed that GDH is a critical mediator of leucine-induced insulin secretion [35–38].

There are various types of genetic defects that can cause congenital hyperinsulinism/hypoglycemia. In a particular group of cases that patients presented with hyperammonemia following a protein meal, the levels of ammonia in patients were up to 3–8 times the normal range. However, this hyperammonemia is not influenced by changes in blood glucose concentration nor is it associated with any defects in enzymes of the urea cycle. Genetic studies revealed that these patients carry dominant, gain-of-function mutations in the allosteric domain of GDH. GDH purified from lymphoblasts displayed reduced sensitivity to inhibition by GTP [35]. Transgenic mice carrying a GTP-insensitive mutation of GDH found in patients also developed hypoglycemia. Moreover, islets isolated from these mice were more sensitive to leucine-stimulated insulin release. These studies have established that GDH is the key enzyme that mediates leucine sensing in insulin release. GDH has since been postulated as a therapeutic target, at least in some patients carrying mutations in GDH.

The Mammalian Target of Rapamycin (mTOR) Pathway and Its Role in Leucine-Elicited Insulin Secretion

The mTOR pathway regulates cell metabolism, cell growth, proliferation and survival in response to environmental cues such as growth factors, insulin, nutrients and stress. Activation of mTOR promotes anabolic cellular processes, such as translation and nucleotide biosynthesis. mTOR is a Ser/Thr kinase belonging to the phosphatidylinositol kinase-related family of proteins. It resides in two distinct multiprotein complexes, mTORC1 and mTORC2, which mediate different signaling cascades (Fig. 5.5). mTORC1 contains primarily mTOR, mammalian orthologue of lethal with sec 13 (mLST8), regulatory-associated protein of target of rapamycin (raptor), pro-rich Akt substrate of 40 kDa (PRAS40) and several others. This protein complex is sensitive to rapamycin. It phosphorylates S6K1, 4E-BP1 and ULK1 when activated, and influences cell growth by regulating protein translation, de novo pyrimidine synthesis, ribosome biogenesis, autophagy and others. Inhibition of mTORC1 by rapamycin, nutrient starvation or hypoxic stress stimulates autophagy and stress-responsive transcription. In contrast, mTORC2 contains mTOR, mLST8 and rapamycin-insensitive companion of TOR (rictor). This complex is relatively insensitive to rapamycin, although it can still be inhibited when exposed to a high concentration of rapamycin for a long period of time. mTORC2 regulates actin organization and is responsible for the phosphorylation and activation of several members of the AGC kinase subfamily, including protein kinase B (PKB, otherwise known as AKT), serum/glucocorticoid-induced kinase 1 (SGK1), conventional protein kinase Cs (PKCs) and PKCε [39].

Growth factors and nutrients activate mTOR via different signaling cascades. Growth factors bind to their receptors to activate PI3 kinase, which in turn recruits Akt to the membrane, resulting in Akt phosphorylation and activation. Akt phosphorylates and inhibits its downstream target, tuberous sclerosis complex (TSC1/TSC2). TSC1 is a tumor suppressor that associates with TSC2 to inactivate the

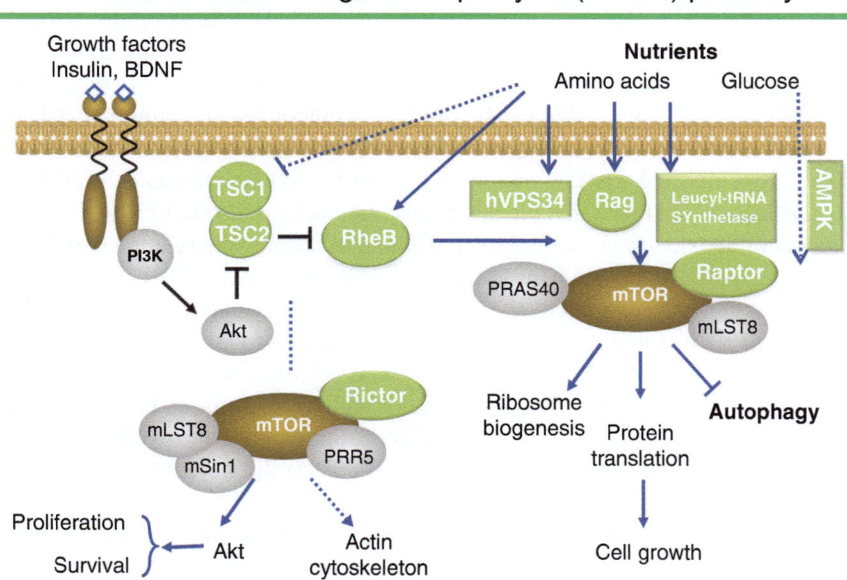

Fig. 5.5 The mTOR signaling pathway

small G-protein Rheb and inhibits mTORC1. Phosphorylation of TSC1 by Akt relieves the inhibitory effect of TSC1/TSC2 on Rheb, resulting in activation of mTOR. In contrast, amino acids activates mTOR through a PI3K-independent pathway, acting on TSC1/TSC2, Rheb, Rag family members and leucyl-tRNA-synthetase [40, 41].

It has become increasingly evident that the mTOR pathway plays a critical role in regulating glucose homeostasis, presumably through influencing pancreatic β-cell growth and proliferation, insulin secretion and glucose uptake. Excessive activation of mTOR increases β-cell size and promotes β-cell proliferation [42]. Mice with genetic deletion of TSC2, an upstream negative regulator of mTORC1, displayed increased β-cell mass largely due to increased cell size and proliferation [43]. Similarly, animals with restrictive deletion of TSC1 in β-cells exhibited larger β-cell size, produced more insulin and better handled blood glucose [44]. β-Cell-specific deletion of *Rictor* in mice, an essential component of the mTORC2 complex, impairs β-cell proliferation, resulting in mild hyperglycemia and glucose intolerance [45]. Mice with genetic deletion of S6K1, a downstream component of mTOR, become glucose intolerant. These animals show reduced β-cell mass and low blood insulin [46], further supporting a role for the mTOR pathway in regulation of insulin secretion.

Rapamycin is an mTOR inhibitor and an FDA-approved immunosuppressant for organ transplant. Chronic rapamycin treatment impaired glucose homeostasis, reduced the insulin content and β-cell mass of pancreatic islets and decreased glucose-stimulated insulin secretion in vivo and ex vivo [47]. Rapamycin reduces β-cell viability through inhibition of mTORC2 and inhibition of Akt. However, a recent study reported that knockdown of raptor, a component in the rapamycin-sensitive mTORC1 complex, in INS-1 cells using siRNA, surprisingly increased glucose-stimulated insulin secretion and intracellular insulin content. Possible explanations for this apparently contradictory result to the previous observations in pancreatic islets are the differences in cells used for the experiments and non-specific effects that could be associated with experimental approaches such as using siRNA. In pancreatic cells, rapamycin inhibits glucose-induced protein synthesis in pancreatic islets [48].

Besides growth factors, leucine is the most potent activator of the mTOR pathway among the amino acids. Depletion of amino acids, specifically leucine, is sufficient to inactivate mTOR and additionally attenuates growth factor signaling, indicating the importance of this nutrient-sensing pathway in regulating insulin secretion and glucose homeostasis. Leucine can effectively stimulate phosphorylation of PHAS1 and p70^{S6k}, important indicators of mTOR activation in RINm5F cells and rat pancreatic islets [48, 49]. Activation of mTOR appears to involve the intracellular metabolism of leucine in pancreatic islet cells and derived insulin secretion cell lines. First, two other branched chain amino acids (isoleucine and valine) that can be similarly metabolized also stimulate mTOR activation in RINm5F cells. Moreover, KIC, an intermediate from transamination of leucine, can effectively activate mTOR. Blocking the transamination of leucine by AOAA attenuates mTOR activation in RINm5F cells. A leucine analogue that blocks leucine metabolism also inhibits mTOR. Finally, inhibition of mitochondrial oxidation by azide largely eliminates the phosphorylation of p70^{S6K} in RINm5F cells. These data suggest that metabolism of leucine is what activates mTOR. This is consistent with the observation that glucose can effectively activate mTOR in pancreatic islets [50]. AMP-activated protein kinase (AMPK) is a key regulator of the energy status of cells and also regulates mTOR.It senses the intracellular levels of AMP and ATP. AMPK is activated when the ratio of AMP to ATP is high and inactivated when the ratio is low. Amino acids and glucose are catabolized in order to inhibit AMPK in MIN6 cells [51]. However, expression of a constitutive form of AMPK in INS-derived cells or pharmacological activation of AMPK in rat pancreatic islets did not affect insulin secretion, suggesting that the AMPK pathway is not involved in insulin release. Leucine-mediated activation of protein translation through mTOR may contribute to enhanced β-cell function by stimulating growth-related protein synthesis and enhancing the proliferation associated with the maintenance of β-cell mass. In malnourished mice, leucine supplementation increases the expression of key proteins involved in the mechanism of insulin secretion, such as the muscarinic receptor M3 (M3R), PKCα, PKAα, the β2 subunit of the L-type calcium channel (Caβ2) and phospho-AKT/AKT [52]. In addition, long-term treatment with leucine upregulates the levels of both glucokinase and ATP synthase β-subunit (ATPSβ) via mTOR-independent pathways [53].

G-protein-coupled receptors also influence insulin secretion through regulating the production of cyclic adenosine monophosphate (cAMP) via acting on G-protein Gs or Gi. This route appears to be involved in leucine-induced insulin secretion and is implicated in type-2 diabetes [54]. For example, glucagon-like peptide-1 (GLP-1), a hormone secreted from the intestine typically following a meal, regulates insulin release. Binding of GLP-1 to its receptor activates adenylyl cyclases to increase cAMP, which in turn mobilize intracellular Ca^{2+} stores, resulting in insulin release [55]. Pancreatic β-cells are also heavily innervated by the peripheral sympathetic nervous system. It is well known that the sympathetic system mediates acute regulation of blood glucose, for example under stress conditions. Adrenergic receptor α2A is the major subtype regulating insulin release. By coupling to Gi to inhibit cAMP production, activation of a2A adrenergic receptor negatively influences glucose-stimulated insulin secretion [56]. Previous studies revealed that activation of α2A by either electric stimulation or norepinephrine inhibits insulin secretion in islets, which is likely through inhibition of cAMP production [57]. Genetic studies revealed the importance of the α2A adrenergic receptor in regulating pancreatic function and glucose homeostasis. Genetic knockout of α2A increases insulin secretion [58]. Conversely, mice overexpressing α2A restrictively in β-cells displayed glucose intolerance [56]. The role of α2A in glucose homeostasis was further cemented by revealing that overexpression of α2A in the Goto-Kakizaki (GK) rat [59], a spontaneous diabetes model, causes impaired insulin granule membrane docking and secretion in β-cells. Pharmacological inhibition of α2A restores insulin secretion to normalcy. Genotyping of single-nucleotide polymorphisms (SNPs) in a large cohort of patients found that a SNP from GA to GG or AA in the 3′ untranslated region of the a2A gene is associated with type-2 diabetes. Islets from type-2 diabetes patients displayed increased expression of mRNA and protein of the a2A receptor, therefore establishing a pathogenic role of aberrant expression of α2A in type-2 diabetes. Interestingly, it has long been known that pancreatic islets prepared from GK rats exhibit a reduced response to leucine [60]. The mechanism underlying the impaired leucine sensing of pancreatic insulin release in GK rats was unclear. However, a recent study shed new light on how leucine regulates insulin secretion through the α2A receptor (Fig. 5.6). Leucine-induced insulin release was blocked by the mTOR inhibitor rapamycin and by an α2A receptor

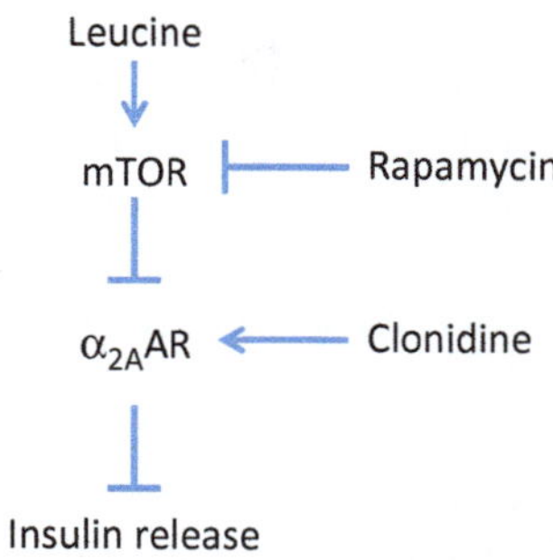

Fig. 5.6 Proposed model showing how rapamycin and clonidine synergistically increase the incidence of NODAT

agonist, clonidine, in pancreatic islets in vivo and in vitro, suggesting a convergence of the mTOR pathway and the adrenergic system in regulating leucine-induced insulin release. Leucine was found to increase the production of cAMP by stimulating the endocytosis of the α2A receptor. This process involves activation of mTOR, because rapamycin blocks both endocytosis of the α2A receptor and cAMP production. The impaired leucine-induced insulin secretion is associated with overexpression of the α2A receptor in GK rats, because blocking the α2A receptor at least partially restores leucine-induced insulin release. Together, by activating the mTOR pathway, leucine promotes the endocytosis of α2A receptors, thereby relieving the inhibitory effect on insulin secretion from the α2A receptor. The same study identified a novel regulatory path of insulin secretion by leucine in pancreatic islets. The detailed mechanism by which leucine regulates membrane trafficking of α2A remains to be elucidated. Leucine was reported to elicit a transient rise in intracellular calcium, resulting in activation of the class III lipid kinase Vps34 and mTORC1 [61]. Other studies revealed that leucine also induces slow calcium oscillation in pancreatic islets as well, which is relevant in regulating insulin granule docking and membrane fusion [13, 62]. Because Vps34 is known to regulate receptor vesicle recycling [63], it has been postulated that Vps34 may regulate α2A trafficking.

The role of the mTOR pathway and the α2A receptor in regulation of insulin secretion also led to a new understanding of new-onset diabetes after transplantation (NODAT). New-onset diabetes is a complication frequently associated with organ transplantation. NODAT creates additional challenges for patient care and likely generates significant negative impacts on long-term patient survival following organ transplantation [64]. NODAT is strongly associated with immunosuppressive medications [64]. Among the immunosuppressants, rapamycin is one of the major risk factors. Other previous studies have shown that rapamycin causes diabetes in obese sand rats (*Psammomys obesus*), lean rats and DIO mice. The α2A agonist clonidine was frequently administrated to patients with hypertension after renal transplantation. Interestingly, the incidence of NODAT was found to be significantly higher in patients treated with both rapamycin and clonidine than those just receiving rapamycin alone. Patients who received clonidine, but not rapamycin, experienced minimal NODAT, perhaps due to a general attenuation of adrenergic input into the pancreas because of clonidine's additional effect on sympathetic output [65]. Nevertheless, administration of both rapamycin and clonidine could have a synergistic action that contributes to the high incidence of NODAT in renal transplant patients. This finding has important clinical implications. For example, clonidine should be prescribed with caution in renal transplant patients who are undergoing treatment with rapamycin.

Conclusions

In summary, leucine regulates a large array of cellular processes in pancreatic β-cells, including metabolism, growth, proliferation and insulin secretion, which ultimately influence overall glucose homeostasis. Multiple routes are involved in acute regulation of insulin secretion. The catabolic courses of leucine, which generate energy by oxidation or intermediates that allosterically impinge on GDH, clearly play a role in insulin secretion. Leucine also influences insulin secretion through the mTOR pathway and adrenergic system. Of particular note, insulin binds its tyrosine kinase receptor, leading to phosphorylation of the insulin receptor substrate (IRS), which in turn activates the downstream PI3K/mTOR signaling pathway. This pathway mediates insulin actions, for example glucose uptake in the liver and skeletal muscles and perhaps the autocrine effect of insulin in pancreatic islet cells [48]. Leucine may exert its action on glucose homeostasis via impinging on the insulin/mTOR pathway. Therefore, the complexity of the regulatory pathways through which leucine acts on insulin secretion and glucose homeostasis helps to explain the huge diversity in the etiology of diabetes. Better understanding of the regulation of insulin release will ultimately guide development of more effective strategies for the management and prevention of diabetes.

References

1. Cochrane WA, Payne WW, Simpkiss MJ, Woolf LI. Familial hypoglycemia precipitated by amino acids. J Clin Invest. 1956;35(4):411–22.
2. Yalow RS, Berson SA. Immunoassay of endogenous plasma insulin in man. J Clin Invest. 1960;39:1157–75.
3. Floyd Jr JC, Fajans SS, Conn JW, Knopf RF, Rull J. Stimulation of insulin secretion by amino acids. J Clin Invest. 1966;45(9):1487–502.
4. Yang J, Chi Y, Burkhardt BR, Guan Y, Wolf BA. Leucine metabolism in regulation of insulin secretion from pancreatic beta cells. Nutr Rev. 2010;68(5):270–9.
5. Yielding KL, Tomkins GM. An effect of L-leucine and other essential amino acids on the structure and activity of glutamic dehydrogenase. Proc Natl Acad Sci U S A. 1961;47:983–9.
6. Jonkers FC, Henquin JC. Measurements of cytoplasmic Ca2+ in islet cell clusters show that glucose rapidly recruits beta-cells and gradually increases the individual cell response. Diabetes. 2001;50(3):540–50.
7. Ashcroft FM, Harrison DE, Ashcroft SJ. Glucose induces closure of single potassium channels in isolated rat pancreatic beta-cells. Nature. 1984;312(5993):446–8.
8. Cook DL, Hales CN. Intracellular ATP directly blocks K+channels in pancreatic B-cells. Nature. 1984;311(5983):271–3.
9. Inagaki N, Gonoi T, Clement JP, et al. Reconstitution of IKATP: an inward rectifier subunit plus the sulfonylurea receptor. Science. 1995;270(5239):1166–70.
10. Nestorowicz A, Inagaki N, Gonoi T, et al. A nonsense mutation in the inward rectifier potassium channel gene, Kir6.2, is associated with familial hyperinsulinism. Diabetes. 1997;46(11):1743–8.
11. Flanagan SE, Clauin S, Bellanne-Chantelot C, et al. Update of mutations in the genes encoding the pancreatic beta-cell K(ATP) channel subunits Kir6.2 (KCNJ11) and sulfonylurea receptor 1 (ABCC8) in diabetes mellitus and hyperinsulinism. Hum Mutat. 2009;30(2):170–80.
12. Mabry CC, Digeorge AM, Auerbach VH. Leucine-induced hypoglycemia. I. Clinical observations and diagnostic considerations. J Pediatr. 1960;57:526–38.
13. Malaisse WJ, Hutton JC, Carpinelli AR, Herchuelz A, Sener A. The stimulus-secretion coupling of amino acid-induced insulin release: metabolism and cationic effects of leucine. Diabetes. 1980;29(6):431–7.
14. Branstrom R, Efendic S, Berggren PO, Larsson O. Direct inhibition of the pancreatic beta-cell ATP-regulated potassium channel by alpha-ketoisocaproate. J Biol Chem. 1998;273(23):14113–8.
15. Lembert N, Idahl LA. Alpha-ketoisocaproate is not a true substrate for ATP production by pancreatic beta-cell mitochondria. Diabetes. 1998;47(3):339–44.
16. Malaisse WJ, Sener A, Malaisse-Legae F, Hutton JC, Christophe J. The stimulus-secretion coupling of amino acid-induced insulin release. Metabolic interaction of L-glutamine and 2-ketoisocaproate in pancreatic islets. Biochim Biophys Acta. 1981;677(1):39–49.
17. MacDonald MJ, McKenzie DI, Kaysen JH, et al. Glucose regulates leucine-induced insulin release and the expression of the branched chain ketoacid dehydrogenase E1 alpha subunit gene in pancreatic islets. J Biol Chem. 1991;266(2):1335–40.
18. Wollheim CB. Beta-cell mitochondria in the regulation of insulin secretion: a new culprit in type II diabetes. Diabetologia. 2000;43(3):265–77.
19. Gao Z, Young RA, Li G, et al. Distinguishing features of leucine and alpha-ketoisocaproate sensing in pancreatic beta-cells. Endocrinology. 2003;144(5):1949–57.
20. Morris AA, Leonard JV. Early recognition of metabolic decompensation. Arch Dis Child. 1997;76(6):555–6.

21. MacDonald MJ, Fahien LA. Glutamate is not a messenger in insulin secretion. J Biol Chem. 2000;275(44):34025–7.
22. Frieden C. Glutamate dehydrogenase. VI. Survey of purine nucleotide and other effects on the enzyme from various sources. J Biol Chem. 1965;240:2028–35.
23. Koberstein R, Sund H. Studies of glutamate dehydrogenase. The influence of ADP, GTP, and L-glutamate on the binding of the reduced coenzyme to beef-liver glutamate dehydrogenase. Eur J Biochem. 1973;36(2):545–52.
24. Struck Jr J, Sizer IW. The substrate specificity of glutamic acid dehydrogenase. Arch Biochem Biophys. 1960;86:260–6.
25. Fahien LA, Kmiotek E. Regulation of glutamate dehydrogenase by palmitoyl-coenzyme A. Arch Biochem Biophys. 1981;212(1):247–53.
26. Yielding KL, Tomkins GM. Structural Alterations in Crystalline Glutamic Dehydrogenase Induced by Steroid Hormones. Proc Natl Acad Sci U S A. 1960;46(11):1483–8.
27. Tomkins GM, Yielding KL, Curran JF. The influence of diethylstilbestrol and adenosine diphosphate on pyridine nucleotide coenzyme binding by glutamic dehydrogenase. J Biol Chem. 1962;237:1704–8.
28. Li M, Allen A, Smith TJ. High throughput screening reveals several new classes of glutamate dehydrogenase inhibitors. Biochemistry. 2007;46(51):15089–102.
29. Sener A, Malaisse WJ. L-leucine and a nonmetabolized analogue activate pancreatic islet glutamate dehydrogenase. Nature. 1980;288(5787):187–9.
30. Lee EY, Yoon HY, Ahn JY, Choi SY, Cho SW. Identification of the GTP binding site of human glutamate dehydrogenase by cassette mutagenesis and photoaffinity labeling. J Biol Chem. 2001;276(51):47930–6.
31. Zaganas I, Spanaki C, Karpusas M, Plaitakis A. Substitution of Ser for Arg-443 in the regulatory domain of human housekeeping (GLUD1) glutamate dehydrogenase virtually abolishes basal activity and markedly alters the activation of the enzyme by ADP and L-leucine. J Biol Chem. 2002;277(48):46552–8.
32. Tomita T, Kuzuyama T, Nishiyama M. Structural basis for leucine-induced allosteric activation of glutamate dehydrogenase. J Biol Chem. 2011;286(43):37406–13.
33. Sener A, Owen A, Malaisse-Lagae F, Malaisse WJ. The stimulus-secretion coupling of amino acid-induced insulin release. XI. Kinetics of deamination and transamination reactions. Horm Metab Res. 1982;14(8):405–9.
34. Bryla J, Michalik M, Nelson J, Erecinska M. Regulation of the glutamate dehydrogenase activity in rat islets of Langerhans and its consequence on insulin release. Metabolism. 1994;43(9):1187–95.
35. Stanley CA, Lieu YK, Hsu BY, et al. Hyperinsulinism and hyperammonemia in infants with regulatory mutations of the glutamate dehydrogenase gene. N Engl J Med. 1998;338(19):1352–7.
36. Stanley CA, Fang J, Kutyna K, et al. Molecular basis and characterization of the hyperinsulinism/hyperammonemia syndrome: predominance of mutations in exons 11 and 12 of the glutamate dehydrogenase gene. HI/HA Contributing Investigators. Diabetes. 2000;49(4):667–73.
37. Li C, Matter A, Kelly A, et al. Effects of a GTP-insensitive mutation of glutamate dehydrogenase on insulin secretion in transgenic mice. J Biol Chem. 2006;281(22):15064–72.
38. MacMullen C, Fang J, Hsu BY, et al. Hyperinsulinism/hyperammonemia syndrome in children with regulatory mutations in the inhibitory guanosine triphosphate-binding domain of glutamate dehydrogenase. J Clin Endocrinol Metab. 2001;86(4):1782–7.
39. Jacinto E, Loewith R, Schmidt A, et al. Mammalian TOR complex 2 controls the actin cytoskeleton and is rapamycin insensitive. Nat Cell Biol. 2004;6(11):1122–8.
40. Duran RV, Hall MN. Leucyl-tRNA synthetase: double duty in amino acid sensing. Cell Res. 2012;22(8):1207–9.
41. Han JM, Jeong SJ, Park MC, et al. Leucyl-tRNA synthetase is an intracellular leucine sensor for the mTORC1-signaling pathway. Cell. 2012;149(2):410–24.
42. Xie J, Herbert TP. The role of mammalian target of rapamycin (mTOR) in the regulation of pancreatic beta-cell mass: implications in the development of type-2 diabetes. Cell Mol Life Sci. 2011;69(8):1289–304.
43. Rachdi L, Balcazar N, Osorio-Duque F, et al. Disruption of Tsc2 in pancreatic beta cells induces beta cell mass expansion and improved glucose tolerance in a TORC1-dependent manner. Proc Natl Acad Sci U S A. 2008;105(27):9250–5.
44. Blandino-Rosano M, Chen AY, Scheys JO, et al. mTORC1 signaling and regulation of pancreatic beta-cell mass. Cell Cycle. 2012;11(10):1892–902.
45. Gu Y, Lindner J, Kumar A, Yuan W, Magnuson MA. Rictor/mTORC2 is essential for maintaining a balance between beta-cell proliferation and cell size. Diabetes. 2011;60(3):827–37.
46. Pende M, Kozma SC, Jaquet M, et al. Hypoinsulinaemia, glucose intolerance and diminished beta-cell size in S6K1-deficient mice. Nature. 2000;408(6815):994–7.
47. Lamming DW, Ye L, Katajisto P, et al. Rapamycin-induced insulin resistance is mediated by mTORC2 loss and uncoupled from longevity. Science. 2012;335(6076):1638–43.
48. Xu G, Kwon G, Marshall CA, Lin TA, Lawrence Jr JC, McDaniel ML. Branched-chain amino acids are essential in the regulation of PHAS-I and p70 S6 kinase by pancreatic beta-cells. A possible role in protein translation and mitogenic signaling. J Biol Chem. 1998;273(43):28178–84.

49. Xu G, Kwon G, Cruz WS, Marshall CA, McDaniel ML. Metabolic regulation by leucine of translation initiation through the mTOR-signaling pathway by pancreatic beta-cells. Diabetes. 2001;50(2):353–60.
50. Kwon G, Marshall CA, Pappan KL, Remedi MS, McDaniel ML. Signaling elements involved in the metabolic regulation of mTOR by nutrients, incretins, and growth factors in islets. Diabetes. 2004;53 Suppl 3:S225–32.
51. Gleason CE, Lu D, Witters LA, Newgard CB, Birnbaum MJ. The role of AMPK and mTOR in nutrient sensing in pancreatic beta-cells. J Biol Chem. 2007;282(14):10341–51.
52. Amaral AG, Rafacho A, Machado de Oliveira CA, et al. Leucine supplementation augments insulin secretion in pancreatic islets of malnourished mice. Pancreas. 2010;39(6):847–55.
53. Yang J, Wong RK, Park M, et al. Leucine regulation of glucokinase and ATP synthase sensitizes glucose-induced insulin secretion in pancreatic beta-cells. Diabetes. 2006;55(1):193–201.
54. Yang J, Dolinger M, Ritaccio G, et al. Leucine stimulates insulin secretion via down-regulation of surface expression of adrenergic alpha2A receptor through the mTOR (mammalian target of rapamycin) pathway: implication in new-onset diabetes in renal transplantation. J Biol Chem. 2012;287(29):24795–806.
55. Holz GG, Leech CA, Heller RS, Castonguay M, Habener JF. cAMP-dependent mobilization of intracellular Ca2+ stores by activation of ryanodine receptors in pancreatic beta-cells. A Ca2+ signaling system stimulated by the insulinotropic hormone glucagon-like peptide-1-(7-37). J Biol Chem. 1999;274(20):14147–56.
56. Devedjian JC, Pujol A, Cayla C, et al. Transgenic mice overexpressing alpha2A-adrenoceptors in pancreatic beta-cells show altered regulation of glucose homeostasis. Diabetologia. 2000;43(7):899–906.
57. Ahren B. Autonomic regulation of islet hormone secretion–implications for health and disease. Diabetologia. 2000;43(4):393–410.
58. Fagerholm V, Gronroos T, Marjamaki P, Viljanen T, Scheinin M, Haaparanta M. Altered glucose homeostasis in alpha2A-adrenoceptor knockout mice. Eur J Pharmacol. 2004;505(1–3):243–52.
59. Rosengren AH, Jokubka R, Tojjar D, et al. Overexpression of alpha2A-adrenergic receptors contributes to type 2 diabetes. Science. 2010;327(5962):217–20.
60. Giroix MH, Saulnier C, Portha B. Decreased pancreatic islet response to L-leucine in the spontaneously diabetic GK rat: enzymatic, metabolic and secretory data. Diabetologia. 1999;42(8):965–77.
61. Cuevas-Casado I, Romero-Fernandez MM, Royo-Bordonada MA. Use of nutrition marketing in products advertised on TV in Spain. Nutr Hosp. 2012;27(5):1569–75.
62. Martin F, Soria B. Amino acid-induced [Ca2+]i oscillations in single mouse pancreatic islets of Langerhans. J Physiol. 1995;486(Pt 2):361–71.
63. Stack JH, DeWald DB, Takegawa K, Emr SD. Vesicle-mediated protein transport: regulatory interactions between the Vps15 protein kinase and the Vps34 PtdIns 3-kinase essential for protein sorting to the vacuole in yeast. J Cell Biol. 1995;129(2):321–34.
64. Kaposztas Z, Gyurus E, Kahan BD. New-onset diabetes after renal transplantation: diagnosis, incidence, risk factors, impact on outcomes, and novel implications. Transplant Proc. 2011;43(5):1375–94.
65. Fagerholm V, Haaparanta M, Scheinin M. Alpha2-adrenoceptor regulation of blood glucose homeostasis. Basic Clin Pharmacol Toxicol. 2011;108(6):365–70.

Chapter 6
Effects of Leucine and Isoleucine on Glucose Metabolism

Fumiaki Yoshizawa

Key Points

- BCAAs have recently been recognized as having functions other than simple nutrition.
- The signaling action of leucine in protein synthesis has been well studied, but the pharmacological effects of isoleucine and valine have not been clarified.
- It has recently been reported that, among the BCAAs, leucine and isoleucine act as signals in glucose metabolism.
- We revealed that isoleucine stimulates both glucose uptake in the muscle and whole body glucose oxidation, in addition to depressing gluconeogenesis in the liver, thereby leading to a hypoglycemic effect in rats.
- We speculate that isoleucine signaling accelerates catabolism of incorporated glucose for energy production and consumption.

Keywords Isoleucine • Leucine • Skeletal muscle • Liver • Hypoglycemic effect • Glucose uptake • Glucose production

Abbreviations

AMPK	5′-AMP-activated protein kinase
ChREBP	Carbohydrate responsive element binding protein
DPP-4	Dipeptidyl peptidase-4
FAS	Fatty acid synthase
G6Pase	Glucose-6-phosphatase
GLUT	Glucose transporter
GSK-3	Glycogen synthase kinase-3
HbA1c	Hemoglobin A1c
HOMA-IR	Homeostasis model assessment of insulin resistance
IRS	Insulin receptor substrates

F. Yoshizawa, Ph.D. (✉)
Department of Agrobiology and Bioresources, Faculty of Agriculture, Utsunomiya University, 350 Mine-machi, Utsunomiya, Tochigi 321-8505, Japan
e-mail: fumiaki@cc.utsunomiya-u.ac.jp

R. Rajendram et al. (eds.), *Branched Chain Amino Acids in Clinical Nutrition: Volume 1*, Nutrition and Health, DOI 10.1007/978-1-4939-1923-9_6,

L-GK	Liver-type glucokinase
LXR	Liver X receptor
mTOR	Mammalian target of rapamycin
PEPCK	Phosphoenolpyruvate carboxykinase
PI3K	Phosphoinositide 3-kinase
PKB/Akt	Protein kinase B
PKC	Protein kinase C
PPAR	Peroxisome proliferator-activated receptor
S6K1	70-kDa ribosomal protein S6 kinase
SREBP	Sterol regulatory element binding protein

Introduction

The branched chain amino acids (BCAAs) leucine, isoleucine, and valine are the most abundant of the essential amino acids. In addition to their critical role as substrates for protein synthesis, these amino acids play a variety of roles in the body. It is believed that BCAAs contribute to energy metabolism during exercise as energy sources and substrates to expand the pool of citric acid-cycle intermediates (anaplerosis) and for gluconeogenesis. Moreover, BCAAs serve as regulatory (signaling) molecules that modulate numerous cellular functions (Table 6.1). BCAAs, acting as nutrient signals, regulate protein synthesis and degradation, and insulin secretion, and have been implicated in central nervous system control of food intake and energy balance. Of the BCAAs, leucine has been the most thoroughly investigated as a signaling molecule. In particular, the signaling action of leucine in protein synthesis has been well studied, and the mechanism is currently under investigation [1]. Leucine appears to be the specific effector of protein synthesis in several tissues including skeletal muscle [2], liver [3], and adipose tissue [4]. However, the pharmacological effects of the other BCAAs, isoleucine and valine, have not been well clarified. We have focused on the pharmacological effects of BCAAs for the last 10 years, and found that among the BCAAs, isoleucine acts as a nutrient regulator of glucose metabolism [5, 6]. Another group has also showed that leucine plays a key role in regulating glucose homeostasis [7]. Today, it is generally accepted that one of the features of BCAA administration is modification of glucose metabolism.

The major focus of this chapter is on the role of BCAAs in regulating glucose metabolism in rats.

Effects of Amino Acids on Glucose Metabolism

BCAAs, particularly leucine, play essential roles in hormonal secretion and action, as well as in intracellular signaling. In glucose metabolism, despite the fact that amino acids can stimulate the release of insulin, it has been shown that amino acid infusion actually inhibits glucose utilization. Previous studies have also shown that amino acids, particularly leucine, inhibit insulin-stimulated

Table 6.1 The functions of BCAAs

			Action of regulatory molecules			
			Protein metabolism			
	Substrates for protein synthesis	Sources of energy	Synthesis	Degradation	Glucose metabolism	Lipid metabolism
Leu	◎	◎	◎	◎	○	Δ
Ile	◎	◎	Δ	–	○	Δ
Val	◎	◎	–	–	–	Δ

◎: well known function; ○: there are some studies that demonstrate function; Δ: there are few studies that demonstrate function; –: at present unknown function

glucose uptake [8–11]. One mechanism through which this could occur is the preferential oxidation of amino acids leading to glucose sparing. As an alternative to glucose oxidation, amino acids may serve as fuel, and amino acids, including the glucogenic amino acids (alanine, valine, or glutamine), are considered to be able to increase glucose production and blood glucose levels. More recent work has begun to identify intracellular mechanisms through which amino acids appear to control glucose metabolism. This inhibitory action is mediated through the attenuated tyrosine phosphorylation of insulin receptor substrates (IRS)-1 and IRS-2 and subsequent interaction with the regulatory subunit of phosphoinositide 3-kinase (PI3K), leading to decreased activity of PI3K, protein kinase B (PKB/Akt), and mammalian target of rapamycin (mTOR) [11–13].

Of the amino acids, leucine is involved in glucose uptake in isolated muscle [7], in glycogen synthesis via the inactivation of glycogen synthase kinase-3 (GSK-3) [14], and in the insulin-secretion effect in the pancreas [15]. On the other hand, leucine, but not isoleucine or valine, also inhibits insulin-stimulated glucose uptake in L6 cells by degrading IRS-1 via activation of the mTOR/S6K1 signaling pathway, leading to desensitization of insulin signaling [11, 13, 16]. In addition, leucine reduces the duration of insulin-induced IRS-1-associated PI3K activity in rat skeletal muscle [17]. Given these results, it is to be expected that amino acids will decrease glucose oxidation and lead to amino acid-induced insulin resistance. However, it has been reported that amino acid infusion causes a decrease in blood glucose levels and an increase in glucose oxidation in humans [18, 19], although there have been few investigations of this hypoglycemic effect to date. These changes appear to occur via the action of insulin, as leucine, but not isoleucine or valine, stimulates insulin release from the pancreas, thereby decreasing blood glucose [20, 21]. Thus, this contradicts the amino acid-induced insulin resistance described above, and this issue therefore remains controversial.

Hypoglycemic Effect of Isoleucine

Some studies have demonstrated that a BCAAs mixture decreases plasma glucose levels in vivo. Oral administration of a BCAAs mixture has been shown to ameliorate hyperglycemia in a virus-induced noninsulin-dependent diabetes mellitus mouse model [22]. In streptozotocin-induced rats, oral administration of a BCAAs mixture (0.75–1.5 g/kg body weight) significantly decreased plasma glucose levels [23]. However, it was unknown whether this reflected glucose metabolism caused by leucine, isoleucine, or valine, and the mechanism of action of the individual BCAAs was not understood in vivo or in vitro.

Our collaborators reported that isoleucine prevents a rise in the plasma glucose concentration and that the effect of isoleucine is greater than that of leucine or valine in oral glucose tolerance tests in normal 7-week-old rats [24]. Oral administration of isoleucine (0.30 g/kg body weight) significantly alleviated the increase of plasma glucose at 30 and 60 min after glucose infusion in these young rats compared to saline-administered control rats. In contrast, the administration of valine (0.30 g/kg body weight) significantly increased the glucose level at 30 min after glucose administration, which suggests that valine, a glycogenic amino acid, is used as a substrate for gluconeogenesis in the liver. Leucine had a similar effect at 120 min compared to saline-administered control rats. Oral administration of leucine, isoleucine, and valine (0.30 g/kg body weight) in these rats did not alter plasma insulin at 30, 60, or 120 min after oral glucose administration compared to the control. Moreover, a dose-dependent effect that lowered plasma glucose levels by isoleucine was observed in 18-week-old rats. In these older rats, oral administration of isoleucine (0.30 g/kg body weight) significantly lowered plasma glucose levels at 30, 60, and 120 min after glucose infusion, whereas oral administration of lower amounts of isoleucine (0.05 or 0.10 g/kg body weight) caused no significant changes in plasma glucose levels. The plasma insulin levels in the isoleucine-administered 18-week-old rats were below those in the controls at 30 and 60 min after glucose infusion. The hypoglycemic effect of isoleucine was recently confirmed in a human study using oral administration of isoleucine [25]. In C_2C_{12} myotubes, leucine and isoleucine stimulate glucose uptake in an insulin-independent manner, and the effect of isoleucine is greater than that of leucine [24]. In such cells, signaling pathway analysis using a PI3K inhibitor (LY294002), a protein kinase C (PKC) inhibitor (GF109203X), and an mTOR inhibitor (rapamycin) suggests that PI3K and PKC, but not mTOR, are involved in the enhancement of glucose uptake by isoleucine. These data suggest that isoleucine assumes the role of a signal for glucose metabolism, thereby stimulating insulin-independent and mTOR-independent glucose transport in cultured skeletal muscle cells.

We focused on the blood glucose-lowering effects of isoleucine and examined whether isoleucine decreased the plasma glucose concentration in food-deprived rats, and whether isoleucine increased glucose uptake in skeletal muscles in vivo. Valine was excluded from the scope of this research, as valine caused an increase in plasma glucose levels. Oral administration of isoleucine, but not leucine, significantly decreased plasma glucose concentrations in food-deprived rats [5]. Glucose uptake in the skeletal muscle did not differ after leucine administration, but glucose uptake in the muscles of rats given isoleucine was 73 % greater than that in food-deprived controls, suggesting that isoleucine increases skeletal muscle glucose uptake in vivo. These results indicate a relationship between the reduction in blood glucose and the increase in skeletal muscle glucose uptake that occur with isoleucine administration in rats (Table 6.2).

Furthermore, we investigated the possible involvement of the energy sensor 5′-AMP-activated protein kinase (AMPK) in the modulation of glucose uptake in skeletal muscle, which is independent of insulin, and also in isoleucine-stimulated glucose uptake [5]. AMPK is a serine/threonine kinase consisting of a catalytic subunit (α) and two regulatory subunits (β and γ). The catalytic α subunit occurs in two distinct isoforms in mammals. AMPK α1 is widely expressed, whereas the α2 isoform is expressed predominantly in the liver, heart, and skeletal muscle [26]. AMPK α1 activity in skeletal muscle was not affected by leucine or isoleucine administration (Table 6.3). However, isoleucine,

Table 6.2 Plasma concentrations of glucose, insulin, leucine, and isoleucine in rats 1 h after administration of saline (control), leucine, or isoleucine, and the effect of leucine or isoleucine on glucose uptake and [U-^{14}C]-glucose incorporation into glycogen in gastrocnemius muscles in rats 1 h after administration of saline (control), leucine, or isoleucine

Treatment group	Glucose	Insulin	Leucine	Isoleucine	Glucose uptake	[U-^{14}C]-Glucose incorporation into glycogen
% of control						
Control	100.0±2.6[a]	100.0±22.9	100.0±9.7[a]	100.0±4.6[a]	100.0±18.9[a]	100.0±16.6[a]
Leucine	91.7±5.8[a,b]	105.7±13.8	1,280.6±158.3[b]	47.2±3.7[a]	118.4±17.1[a,b]	174.5±8.5[b]
Isoleucine	75.9±4.4[a]	110.0±11.1	97.2±6.3[a]	4,029.6±148.1[b]	172.8±15.7[b]	119.3±10.4[a,b]

[a]Rats were food-deprived for 18 h and then administered saline (0.155 mol/L), 1.35 g L-leucine/kg body weight (prepared as 54.0 g/L of the L-amino acid in distilled water), or 1.35 g L-isoleucine/kg body weight by oral gavage
[b]Value are means±SEM, $n=5$ for plasma concentrations of glucose and insulin; $n=6$ for plasma concentrations of leucine and isoleucine and glucose uptake; $n=4$ for [U-^{14}C]-glucose. Means in a column with superscripts without a common letter differ, $P<0.05$. This table is partially modified from the original reported by Doi et al. [5]

Table 6.3 Effects of leucine or isoleucine on adenine nucleotide contents, the AMP:ATP ratio, and activity of the AMPK α1 and α2 isoforms in gastrocnemius muscles of rats 1 h after administration of saline (control), leucine, or isoleucine

Treatment group	AMP	ADP	ATP	AMP:ATP	AMPK α1	AMPK α2
% of control						
Control	100.0±3.7[a]	100.0±3.0	100.0±2.6	100.0±3.8[a]	100.0±10.3	100.0±9.5[a]
Leucine	100.0±6.5[a]	106.1±3.0	108.9±3.2	91.0±4.5[a,b]	97.4±2.5	97.0±7.6[a,b]
Isoleucine	78.5±4.7[b]	96.0±3.0	99.4±3.4	79.7±6.0[b]	91.3±5.1	71.9±5.1[b]

[a]Rats were food-deprived for 18 h and then administered saline (0.155 mol/L), 1.35 g L-leucine/kg body weight (prepared as 54.0 g/L of the L-amino acid in distilled water), or 1.35 g L-isoleucine/kg body weight by oral gavage
[b]Value are means±SEM, $n=5$ for adenine nucleotide contents and the AMP:ATP ratio; $n=6$ for activity of the AMPK α1 and α2 isoforms. Means in a column with superscripts without a common letter differ, $P<0.05$. This table is partially modified from the original reported by Doi et al. [5]

but not leucine, significantly decreased AMPK α2 activity (Table 6.3). These results indicate that isoleucine-stimulated glucose uptake increases in the absence of increased AMPK α1 and α2 activity in skeletal muscle in food-deprived rats.

Effects of Isoleucine on Glycogen Synthesis and Glucose Oxidation

As mentioned above, the administration of isoleucine leads to an increase in glucose uptake in skeletal muscle in vivo [5]. However, it remains unknown how the glucose incorporated by isoleucine is metabolized. Leucine stimulates glycogen synthesis through the inactivation of GSK-3 in L6 muscle cells in a manner that is dependent on mTOR and independent of insulin [14]. Our collaborators reported that leucine causes a significant increase in D-[U-^{14}C] glucose incorporation into intracellular glycogen in myotube cells in vitro, whereas isoleucine does not affect glycogen synthesis when compared to controls [24]. Therefore, we first examined the effects of isoleucine on glycogen synthesis in skeletal muscle.

Muscle glycogen synthesis, as determined by [U-^{14}C] glucose incorporation into glycogen, was significantly increased by administration of leucine, but not isoleucine, in rat skeletal muscles in vivo when compared with food-deprived control rats (Table 6.2). Although leucine has less of an effect on glucose uptake in skeletal muscle, it stimulates glycogen synthesis in skeletal muscle (Table 6.2). In contrast, isoleucine stimulates glucose uptake, although it has less of an effect on glycogen synthesis (Table 6.2) [5]. We also measured the contents of high-energy phosphate metabolites (AMP, ADP, and ATP) in the skeletal muscles of rats administered with leucine and isoleucine to evaluate the cellular energy state [5]. Oral administration of leucine or isoleucine did not alter the ADP or ATP contents in the skeletal muscle when compared with control rats (Table 6.3). Although isoleucine caused a decrease in AMP content in skeletal muscle when compared with the control and leucine groups, the AMP content was not affected after administration of leucine when compared with the control group (Table 6.3). Furthermore, although leucine did not change the AMP:ATP ratio, isoleucine caused a significant decrease in this ratio in the skeletal muscle when compared with the control group (Table 6.3). We assume that a depletion of cellular AMP would result in a decrease in the AMP:ATP ratio and improve the availability of ATP in the skeletal muscle without any marked increase in ATP concentration, thereby resulting in an improvement in the cellular energy state.

In order to determine whether the hypoglycemic effect of isoleucine affects whole body glucose oxidation, we examined the effects of isoleucine on the expiratory excretion of $^{14}CO_2$ from [U-^{14}C] glucose in vivo at a dose where the hypoglycemic effect was greatest [6]. The expiratory excretion of $^{14}CO_2$ of rats treated with isoleucine was significantly elevated between 60 and 90 min after administration when compared with controls. Based on the above results, muscle glucose uptake in isoleucine-administered rats was elevated at 60 min. As the time to achieve maximum plasma isoleucine levels was 60 min [5, 24], this indicates a strong correlation among the increase in muscle glucose uptake, the decrease in blood glucose, and the subsequent increase in glucose oxidation. In contrast, it has been reported that isoleucine suppresses $^{14}CO_2$ production from [1-^{14}C] pyruvate in isolated skeletal muscle [8]. As a potential explanation, the suppression might have been caused by an increase in total pyruvate content in the skeletal muscle [27] due to increased glucose uptake by isoleucine [5, 24], which results in a diluting effect for [1-^{14}C] pyruvate and an increase in total CO_2 production. Meanwhile, glycogen synthesis in isoleucine-administered rats was not altered in skeletal muscle [5], thus suggesting that when there is a lowering of blood glucose levels, isoleucine increases glucose uptake in skeletal muscle with the incorporated glucose mainly oxidized in the muscle immediately after uptake (Table 6.4).

Table 6.4 The effects of leucine and isoleucine on blood glucose and insulin in rats and on glucose metabolism in skeletal muscle in vivo

Blood glucose lowering effect	:	Isoleucine	>	Leucine
Insulinotropic effect	:	Leucine	>	Isoleucine
Glucose uptake in vivo	:	Isoleucine	>	Leucine
Glycogen synthesis in vivo	:	Leucine	>	Isoleucine
Glucose oxidation in vivo	:	Isoleucine	>	Leucine

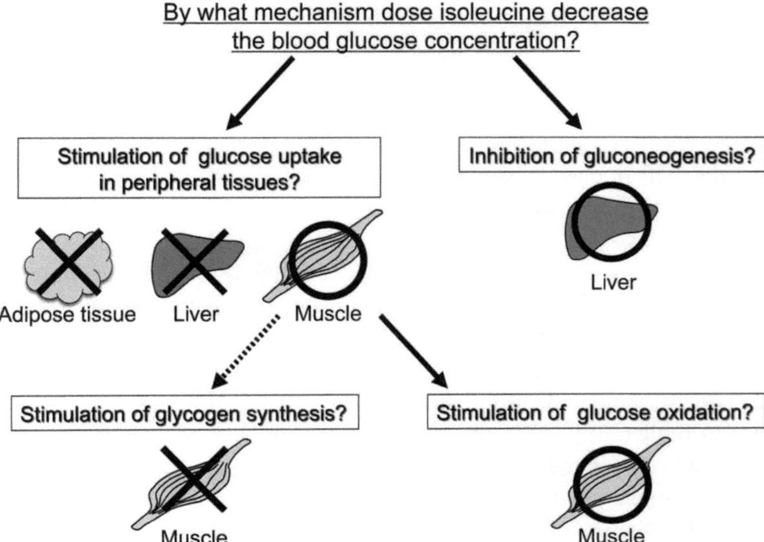

Fig. 6.1 A possible mechanism for the glucose-lowering effects of isoleucine in rats. Glucose uptake was significantly increased in the muscles of isoleucine-administered rats when compared with controls. In contrast, there were no significant differences in the glucose uptake in liver or adipose tissue in isoleucine-administered rats. Glycogen synthesis in isoleucine-administered rats was not altered in skeletal muscle, thus suggesting that when there is a lowering of blood glucose levels, isoleucine increases glucose uptake in skeletal muscle with the incorporated glucose mainly oxidized immediately after uptake. Isoleucine also reduces hepatic glucose production in vitro and the expression and activity of hepatic gluconeogenic enzymes both in vitro and in vivo

Effects of Isoleucine on Glucose Uptake in Peripheral Tissues and Hepatic Glucose Production

Generally, the maintenance of blood glucose levels is due to an optimal balance between glucose uptake by peripheral tissues and glucose production occurring mainly in the liver. Therefore, we examined the effects of isoleucine administration on glucose uptake in peripheral tissues and hepatic glucose production.

Glucose uptake was significantly increased in the muscles of isoleucine-administered rats when compared with controls at the most effective dose of isoleucine. In contrast, there were no significant differences in the glucose uptake in liver or adipose tissue in isoleucine-administered rats when compared with controls. These results suggest that skeletal muscle is the major organ contributing to the hypoglycemic effects of isoleucine on glucose uptake (Fig. 6.1). Leucine has a stimulatory effect on insulin secretion [9]. As a temporal increase in plasma insulin levels after oral administration of leucine was observed in this study, the effect of leucine on glucose metabolism may primarily be insulin dependent. On the other hand, the hypoglycemic effect of isoleucine was more potent than that of leucine, although significant changes in plasma insulin and glucagon levels were not observed after isoleucine administration. Furthermore, isoleucine had an additive effect on insulin-stimulated glucose uptake via PI3K [24], in contrast to leucine, which inhibited insulin-stimulated glucose uptake in skeletal muscle cells [11]. Although the molecular basis of increased glucose uptake by isoleucine remains unclear, a previous study by another group revealed that glucose uptake by isoleucine is involved in increased glucose transporter (GLUT)1 and GLUT4 translocation in the skeletal muscles of rats with liver cirrhosis [28]. These data suggest that isoleucine improves insulin sensitivity in skeletal muscle via an intracellular signaling pathway.

The liver plays a role in the uptake of blood glucose as with skeletal muscles. The primary transporter for glucose uptake in the liver is GLUT2. Glucose is taken up into the hepatocytes through GLUT2, and liver-type glucokinase (L-GK) traps glucose in the cytoplasm by phosphorylation. Therefore, GLUT2 and L-GK play important roles in the liver as a glucose-sensing apparatus [29]. We have not evaluated the effect of leucine or isoleucine on the glucose-sensing apparatus in the liver, because there were no significant differences in the glucose uptake in liver in leucine-administered or isoleucine-administered rats when compared with controls. Recently, it was demonstrated that BCAAs strongly accelerated GLUT2 and L-GK mRNA expression in HepG2 cells in a glucose-dependent manner, and dose-dependently enhanced the mRNA levels of L-GK in rat liver [30]. The glucose-sensing apparatus enables glucose to regulate the expression of glucose-responsive genes such as L-type pyruvate kinase, S14, fatty acid synthase (FAS), and GLUT2 [31]. Thus, the glucose-sensing apparatus exerts a strong influence on glucose utilization and glycogen synthesis. The biological activity of the glucose-sensing apparatus is tightly regulated by transcriptional mechanisms. The transcriptional factors sterol regulatory element binding protein (SREBP)-1c and peroxisome proliferator-activated receptor (PPAR)-γ are thought to be involved in the transcriptional regulation of the glucose-sensing apparatus, because the functional binding sites for SREBP-1c and PPAR-γ have been identified in the GLUT2 and L-GK promoter regions, and the mRNA expression of these genes is upregulated by SREBP-1c and PPAR-γ [29, 32–34]. BCAAs markedly upregulated the expression of SREBP-1c, carbohydrate responsive element binding protein (ChREBP), and liver X receptor (LXR) α in HepG2 cells, and upregulated SREBP-1c and LXRα in rat liver [30]. Because LXRα is known to transactivate SREBP-1c and ChREBP in the presence of glucose or glucose-6-phosphate, as direct agonists, the increased glucose or glucose-6-phosphate, through the activation of the glucose-sensing apparatus by BCAAs, may have a role in the activation of LXRα. From these results, it has been speculated that the LXRα-induced SREBP-1c-dependent mechanism may be the main signaling pathway for BCAAs-induced transactivation of the glucose-sensing apparatus [30].

Hepatic gluconeogenesis, as well as glucose utilization by peripheral and hepatic tissues, may be a possible mechanism by which amino acids lower blood glucose levels. During the fasting state, glucose production is largely a result of gluconeogenesis as opposed to hepatic glycogenolysis [35]. Therefore, we examined the effects of isoleucine on the gluconeogenic rate-limiting enzymes in vivo. The mRNA levels of phosphoenolpyruvate carboxykinase (PEPCK), which is a well-researched key gluconeogenic enzyme, parallel both PEPCK activity and the rate of gluconeogenesis [36, 37]. The data showed that expression levels of hepatic PEPCK mRNA were lower in isoleucine-administered rats than in controls, suggesting that PEPCK activity was lower in isoleucine-administered rats and that the inhibitory effects of isoleucine on PEPCK are regulated at the transcriptional level [6]. Furthermore, we demonstrated that expression levels of hepatic glucose-6-phosphatase (G6Pase) mRNA, and G6Pase activity were also lower in isoleucine-administered rats [6]. Enzyme activity is regulated by controlling both protein expression and existing enzymes. Although whether isoleucine regulates existing enzymes is unknown, we believe that isoleucine inhibits G6Pase activity by decreasing G6Pase mRNA expression. These findings suggest that isoleucine also downregulates G6Pase activity and associated mRNA, in addition to inhibiting gluconeogenesis in the liver in vivo (Fig. 6.1).

Under in vitro conditions, there is an inhibitory effect of isoleucine on the expression of PEPCK and G6Pase in isolated hepatocytes [6]. It has also been demonstrated that the G6Pase activity is lower in isoleucine-added cells when compared with controls [6]. These findings suggest that isoleucine downregulates the transcription of gluconeogenic enzymes and inhibits glucose production in the liver under insulin-free conditions. This suggests that the inhibitory effects of gluconeogenesis by isoleucine involve an insulin-independent signaling pathway as well as the insulinotropic effects of isoleucine in vivo.

Although we can calculate the glucose uptake and endogenous glucose production values at the same time through the use of a tracer, the value of endogenous glucose production may be underestimated in experiments in which glucose uptake values are markedly elevated. Therefore, we measured glucose production by using isolated hepatocytes in order to determine the mechanism underlying the hypoglycemic effects of isoleucine [6]. As a glucogenic substrate, we examined alanine, because plasma alanine levels were found to be significantly higher in the 0.45 g/kg body weight isoleucine group, which was also the most effective dose for decreasing plasma glucose when compared with controls [6]. Isoleucine significantly inhibited glucose production when alanine was used as a glucogenic substrate in isolated hepatocytes [6]. In addition, phenylalanine, which is a neutral amino acid transported via the same neutral amino acid transport system as alanine, leucine, and isoleucine [38, 39], also significantly reduced glucose production [6]. These results indicate that the inhibitory effects of isoleucine on glucose production with alanine may be due to a competitive inhibitory effect with alanine for transport via the neutral amino acid transporter.

In conclusion, isoleucine administration stimulates both glucose uptake in the muscles and whole body glucose oxidation, in addition to depressing gluconeogenesis in the liver, thereby leading to a hypoglycemic effect in rats (Fig. 6.2).

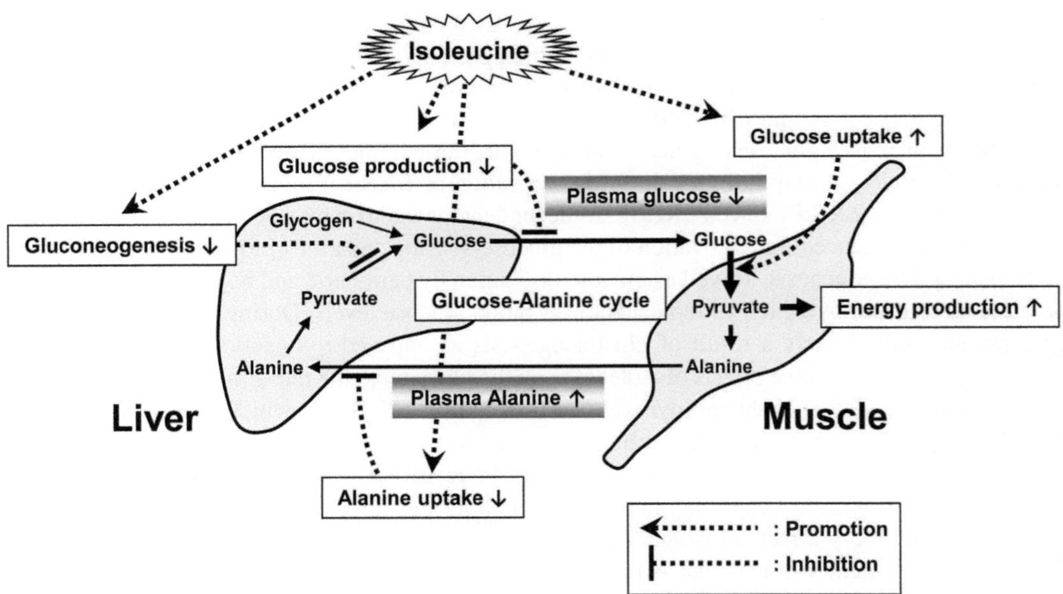

Fig. 6.2 Schematic diagram of the effects of isoleucine on glucose metabolism. Isoleucine stimulates glucose uptake in skeletal muscle and the incorporated glucose is oxidized without significant elevation of plasma insulin levels. In the liver, isoleucine decreases hepatic gluconeogenic enzyme activity and glucose production. These mechanisms are responsible for the hypoglycemic effect of isoleucine that improves the energy state of the muscle and liver, and that may improve insulin resistance in vivo. This figure is partially modified from an original reported by Doi et al. [6]

The Clinical Utility of BCAAs as Regulators of Glucose Metabolism

The clinical importance of BCAAs as regulators of glucose metabolism has been studied. Nuttall et al. have systematically evaluated the insulin and glucose responses to individual amino acids ingested with and without glucose in healthy nondiabetic volunteers. When ingested alone, isoleucine did not affect the insulin concentration but did decrease the glucose concentration [25]. It increased glucagon modestly. When isoleucine was ingested with glucose it resulted in a smaller increase in glucose concentration, but a similar increase in insulin. The data suggest that this was associated with an increased removal rate of peripheral circulating glucose. They also determined whether leucine stimulates insulin and/or glucagon secretion, and whether when ingested with glucose it modifies the glucose, insulin, or glucagon response [40]. Leucine at a dose equivalent to that present in a high-protein meal had little effect on serum glucose or insulin concentrations but did increase the glucagon concentration. When leucine was ingested with glucose, it attenuated the serum glucose response and strongly stimulated additional insulin secretion. Leucine also attenuated the decrease in glucagon expected when glucose alone is ingested. The data suggest that a rise in glucose concentration is necessary for leucine to stimulate significant insulin secretion. This in turn reduces the glucose response to ingested glucose.

The blood glucose-lowering effects of BCAAs in patients with chronic liver disease have been reported [41, 42]. BCAAs supplementation improved the homeostasis model assessment of insulin resistance (HOMA-IR) [41] and decreased plasma glucose levels in a glucose tolerance test [42]. Chronic liver disease, especially hepatitis C, is associated with insulin resistance and diabetes. The effects of BCAAs on glucose tolerance and insulin sensitivity in patients with chronic hepatitis C and insulin resistance have been reported [43]. BCAAs supplementation therapy did not have adverse effects on glucose tolerance or insulin sensitivity in patients with chronic hepatitis C and insulin resistance. BCAAs did not significantly improve overall glycemic control in this study. However, BCAAs therapy may exert a beneficial effect on hemoglobin A1c (HbA1c) values in patients with marked insulin resistance in skeletal muscles. Based on these preliminary observations, future clinical trials should also evaluate the effect of BCAAs in patients with type 2 diabetes.

Conclusions

Numerous hormones play important roles in the metabolic regulation of nutrients. Among these, insulin functions to regulate the metabolism of all macronutrients (e.g., proteins, carbohydrates, and fats), making it a crucial metabolic hormone. Insulin acts anabolically on protein metabolism by stimulating protein synthesis and inhibiting its breakdown. Similarly, leucine functions to stimulate protein synthesis and inhibit its breakdown. The role of isoleucine has been overshadowed by leucine, but it has also been demonstrated to have glucose metabolism-regulating functions similar to insulin, such as stimulation of glucose uptake into cells and inhibition of gluconeogenesis. These findings suggest that leucine and isoleucine share metabolic regulatory functions with insulin. Recent studies suggest a close relationship between BCAAs and insulin resistance and demonstrate that BCAAs may play a major role in the modulation of insulin action [44]. In addition, it has been reported that certain BCAA-containing bioactive peptides derived from whey protein can reduce postprandial glucose levels and stimulate insulin release in healthy subjects and in subjects with type 2 diabetes by reducing dipeptidyl peptidase-4 (DPP-4) activity in the proximal bowel, hence increasing intact incretin levels [45]. Although we have not been able to find reports on the effects of single BCAAs on DPP-4 activity, the possibility that a single BCAA can regulate glucose metabolism by affecting the activity of DPP4 cannot be excluded.

As BCAAs serve to regulate the metabolism of major nutrients, similarly to insulin, the value of their use as biological regulators cannot be overestimated. BCAAs are nutrients that should therefore be a focus of investigation as next-generation biological regulators.

References

1. Yoshizawa F. Regulation of protein synthesis by branched-chain amino acids in vivo. Biochem Biophys Res Commun. 2004;313:417–22.
2. Anthony JC, Yoshizawa F, Anthony TG, Vary TC, Jefferson LS, Kimball SR. Leucine stimulates translation initiation in skeletal muscle of postabsorptive rats via a rapamycin-sensitive pathway. J Nutr. 2000;130:2413–9.
3. Anthony TG, Anthony JC, Yoshizawa F, Kimball SR, Jefferson LS. Oral administration of leucine stimulates ribosomal protein mRNA translation but not global rates of protein synthesis in the liver of rats. J Nutr. 2001;131:1171–6.
4. Lynch CJ, Patson BJ, Anthony J, Vaval A, Jefferson LS, Vary TC. Leucine is a direct-acting nutrient signal that regulates protein synthesis in adipose tissue. Am J Physiol Endocrinol Metab. 2002;283:E503–13.
5. Doi M, Yamaoka I, Nakayama M, Mochizuki S, Sugahara K, Yoshizawa F. Isoleucine, a blood glucose-lowering amino acid, increases glucose uptake in rat skeletal muscle in the absence of increases in AMP-activated protein kinase activity. J Nutr. 2005;135:2103–8.
6. Doi M, Yamaoka I, Nakayama M, Sugahara K, Yoshizawa F. Hypoglycemic effect of isoleucine involves increased muscle glucose uptake and whole body glucose oxidation and decreased hepatic gluconeogenesis. Am J Physiol Endocrinol Metab. 2007;292:E1683–93.
7. Nishitani S, Matsumura T, Fujitani S, Sonaka I, Miura Y, Yagasaki K. Leucine promotes glucose uptake in skeletal muscles of rats. Biochem Biophys Res Commun. 2002;299:693–6.
8. Chang TW, Goldberg AL. Leucine inhibits oxidation of glucose and pyruvate in skeletal muscles during fasting. J Biol Chem. 1978;253:3696–701.
9. Flakoll PJ, Wentzel LS, Rice DE, Hill JO, Abumrad NN. Short-term regulation of insulin-mediated glucose utilization in four-day fasted human volunteers: role of amino acid availability. Diabetologia. 1992;35:357–66.
10. Tessari P, Inchiostro S, Biolo G, et al. Hyperaminoacidaemia reduces insulin-mediated glucose disposal in healthy man. Diabetologia. 1985;28:870–2.
11. Tremblay F, Marette A. Amino acid and insulin signaling via the mTOR/p70 S6 kinase pathway. A negative feedback mechanism leading to insulin resistance in skeletal muscle cells. J Biol Chem. 2001;276:38052–60.
12. Patti ME, Brambilla E, Luzi L, Landaker EJ, Kahn CR. Bidirectional modulation of insulin action by amino acids. J Clin Invest. 1998;101:1519–29.
13. Takano A, Usui I, Haruta T, et al. Mammalian target of rapamycin pathway regulates insulin signaling via subcellular redistribution of insulin receptor substrate 1 and integrates nutritional signals and metabolic signals of insulin. Mol Cell Biol. 2001;21:5050–62.
14. Peyrollier K, Hajduch E, Blair AS, Hyde R, Hundal HS. L-leucine availability regulates phosphatidylinositol 3-kinase, p70 S6 kinase and glycogen synthase kinase-3 activity in L6 muscle cells: evidence for the involvement of the mammalian target of rapamycin (mTOR) pathway in the L-leucine-induced up-regulation of system A amino acid transport. Biochem J. 2000;350:361–8.
15. Yang J, Chi Y, Burkhardt BR, Guan Y, Wolf BA. Leucine metabolism in regulation of insulin secretion from pancreatic beta cells. Nutr Rev. 2010;68:270–9.
16. Khamzina L, Veilleux A, Bergeron S, Marette A. Increased activation of the mammalian target of rapamycin pathway in liver and skeletal muscle of obese rats: possible involvement in obesity-linked insulin resistance. Endocrinology. 2005;146:1473–81.
17. Baum JI, O'Connor JC, Seyler JE, Anthony TG, Freund GG, Layman DK. Leucine reduces the duration of insulin-induced PI 3-kinase activity in rat skeletal muscle. Am J Physiol Endocrinol Metab. 2005;288:E86–91.
18. Tappy L, Acheson K, Normand S, Pachiaudi C, Jéquier E, Riou JP. Effects of glucose and amino acid infusion on glucose turnover in insulin-resistant obese and type II diabetic patients. Metabolism. 1994;43:428–34.
19. Tappy L, Acheson K, Normand S, et al. Effects of infused amino acids on glucose production and utilization in healthy human subjects. Am J Physiol. 1992;262:E826–33.
20. Fajans SS, Knopf RF, Floyd JC, Power L, Conn JW. The experimental induction in man of sensitivity of leucine hypoglycemia. J Clin Invest. 1963;42:216–29.
21. Milner RD. The stimulation of insulin release by essential amino acids from rabbit pancreas in vitro. J Endocrinol. 1970;47:347–56.
22. Utsugi T, Yoshida A, Kanda T, et al. Oral administration of branched chain amino acids improves virus-induced glucose intolerance in mice. Eur J Pharmacol. 2000;398:409–14.

23. Eizirik DL, Kettelhut IC, Migliorini RH. Administration of branched-chain amino acids reduces the diabetogenic effect of streptozotocin in rats. Braz J Med Biol Res. 1987;20:137–44.
24. Doi M, Yamaoka I, Fukunaga T, Nakayama M. Isoleucine, a potent plasma glucose-lowering amino acid, stimulates glucose uptake in C_2C_{12} myotubes. Biochem Biophys Res Commun. 2003;312:1111–7.
25. Nuttall FQ, Schweim K, Gannon MC. Effect of orally administered isoleucine with and without glucose on insulin, glucagon and glucose concentrations in non-diabetic subjects. E Spen Eur E J Clin Nutr Metab. 2008;3:e152–8.
26. Stapleton D, Mitchelhill KI, Gao G, et al. Mammalian AMP-activated protein kinase subfamily. J Biol Chem. 1996;271:611–4.
27. Goldstein L, Newsholme EA. The formation of alanine from amino acids in diaphragm muscle of the rat. Biochem J. 1976;154:555–8.
28. Nishitani S, Takehana K, Fujitani S, Sonaka I. Branched-chain amino acids improve glucose metabolism in rats with liver cirrhosis. Am J Physiol Gastrointest Liver Physiol. 2005;288:G1292–12300.
29. Kim HI, Ahn YH. Role of peroxisome proliferator-activated receptor-gamma in the glucose-sensing apparatus of liver and beta-cells. Diabetes. 2004;53 Suppl 1:S60–5.
30. Higuchi N, Kato M, Miyazaki M, et al. Potential role of branched-chain amino acids in glucose metabolism through the accelerated induction of the glucose-sensing apparatus in the liver. J Cell Biochem. 2011;112:30–8.
31. Leturque A, Brot-Laroche E, Le Gall M, Stolarczyk E, Tobin V. The role of GLUT2 in dietary sugar handling. J Physiol Biochem. 2005;61:529–37.
32. Iynedjian PB, Gjinovci A, Renold AE. Stimulation by insulin of glucokinase gene transcription in liver of diabetic rats. J Biol Chem. 1988;263:740–4.
33. Magnuson MA, Andreone TL, Printz RL, Koch S, Granner DK. Rat glucokinase gene: structure and regulation by insulin. Proc Natl Acad Sci U S A. 1989;86:4838–42.
34. Im SS, Kang SY, Kim SY, et al. Glucose-stimulated upregulation of GLUT2 gene is mediated by sterol response element-binding protein-1c in the hepatocytes. Diabetes. 2005;54:1684–91.
35. Landau BR, Wahren J, Chandramouli V, Schumann WC, Ekberg K, Kalhan SC. Contributions of gluconeogenesis to glucose production in the fasted state. J Clin Invest. 1996;98:378–85.
36. Iynedjian PB, Hanson RW. Increase in level of functional messenger RNA coding for phosphoenolpyruvate carboxykinase (GTP) during induction by cyclic adenosine 3′: 5′-monophosphate. J Biol Chem. 1977;252:655–62.
37. Xu H, Yang Q, Shen M, et al. Dual specificity MAPK phosphatase 3 activates PEPCK gene transcription and increases gluconeogenesis in rat hepatoma cells. J Biol Chem. 2005;280:36013–8.
38. Bröer A, Klingel K, Kowalczuk S, Rasko JE, Cavanaugh J, Bröer S. Molecular cloning of mouse amino acid transport system B^0, a neutral amino acid transporter related to Hartnup disorder. J Biol Chem. 2004;279:24467–76.
39. Bröer A, Tietze N, Kowalczuk S, et al. The orphan transporter v7-3 (slc6a15) is a Na^+-dependent neutral amino acid transporter (B^0AT2). Biochem J. 2006;393:421–30.
40. Kalogeropoulou D, LaFave L, Schweim K, Gannon MC, Nuttall FQ. Leucine, when ingested with glucose, synergistically stimulates insulin secretion and lowers blood glucose. Metabolism. 2008;57:1747–52.
41. Kawaguchi T, Nagao Y, Matsuoka H, Ide T, Sata M. Branched-chain amino acid-enriched supplementation improves insulin resistance in patients with chronic liver disease. Int J Mol Med. 2008;22:105–12.
42. Korenaga K, Korenaga M, Uchida K, Yamasaki T, Sakaida I. Effects of a late evening snack combined with alpha-glucosidase inhibitor on liver cirrhosis. Hepatol Res. 2008;38:1087–97.
43. Takeshita Y, Takamura T, Kita Y, et al. Beneficial effect of branched-chain amino acid supplementation on glycemic control in chronic hepatitis C patients with insulin resistance: implications for type 2 diabetes. Metabolism. 2012;61:1388–94.
44. Lu J, Xie G, Jia W, Jia W. Insulin resistance and the metabolism of branched-chain amino acids. Front Med. 2013;7:53–9.
45. Tulipano G, Sibilia V, Caroli AM, Cocchi D. Whey proteins as source of dipeptidyl dipeptidase IV (dipeptidyl peptidase-4) inhibitors. Peptides. 2011;32:835–8.

Chapter 7
Regulation of Liver Glucose Metabolism by the Metabolic Sensing of Leucine in the Hypothalamus

Roger Gutiérrez-Juárez

Key Points

- Leucine has profound effects on insulin action and food intake in mammals
- Direct as well as indirect actions of leucine in the liver and skeletal muscle account in part for the effects of leucine on insulin action
- The sensing of leucine levels in the hypothalamus controls glucose metabolism
- Circulating leucine modulates liver glucose production by a mechanism requiring leucine metabolism in the hypothalamus
- Faltering of the hypothalamic leucine sensing mechanism leads to disregulation of glycemic control
- The glucoregulatory action of leucine is a contributing factor in the maintenance of glucose homeostasis and is altered in diet-induced insulin resistance

Keywords Leucine • Mediobasal hypothalamus • Glucose homeostasis • Nutrient sensing • Liver glucose metabolism • Brain-liver circuit • Branched chain amino acids • Insulin action • Diabetes

Abbreviations

HYP	Hypothalamus
MBH	Mediobasal hypothalamus
ARC	Arcuate nucleus of the hypothalamus
NTS	Nucleus of the solitary tract
DMX	Dorsal motor nucleus of the vagus
BCAT	Branched chain amino acid aminotransferase
BCKDH	Branched chain ketoacid dehydrogenase
BCKDK	Branched chain ketoacid dehydrogenase kinase
ACC	Acetyl CoA carboxylase

R. Gutiérrez-Juárez, M.D., Ph.D. (✉)
Department of Medicine and Diabetes Research Center, Albert Einstein College of Medicine, Yeshiva University, Jack and Pearl Resnick Campus, Belfer 1300 Morris Park Ave, Bronx, NY 10461, USA
e-mail: roger.gutierrez@einstein.yu.edu; rgutierrezj@gmail.com

R. Rajendram et al. (eds.), *Branched Chain Amino Acids in Clinical Nutrition: Volume 1*, Nutrition and Health, DOI 10.1007/978-1-4939-1923-9_7,

MCD	Malonyl CoA decarboxylase
AMPK	AMP-activated kinase
FAS	Fatty acid synthase
K_{ATP} Ch	ATP-dependent potassium channel
Sur1	Sulfonylurea receptor 1
mTOR	Mammalian target of rapamycin
Leu	Leucine
KIC	α-keto-isocaproate
CoA	Coenzyme A
α-CIC	α-chloro-isocaproate
AICAR	5-aminoimidazole-4-carboxamide ribonucleotide
BCATi	BCAT inhibitor
AOAA	Amino oxyacetic acid
PEP	Phosphoenol pyruvate
GNG	Gluconeogenesis
GLS	Glycogenolysis

Introduction

Cellular sensing of branched chain amino acids constitutes an ancestral form of nutrient sensing that is preserved across multiple species including mammals [1–3]. Overwhelming evidence has shown that protein and amino acids exert pronounced effects on food intake, insulin action, and glucose metabolism in mammals [4–9] and the mechanisms responsible for these effects are beginning to emerge thanks to recent work by numerous groups. A loss of the ability to properly modulate insulin action and glucose metabolism are at the center of a number of metabolic diseases of pandemic dimensions such as diabetes and obesity. We will focus here on the effects of amino acids on insulin action and glucose metabolism and specifically examine the effects of leucine. Although, there has been controversy as to whether amino acids improve or worsen insulin action, recent in vivo studies point towards a positive (improving) effect [6, 7, 10, 11]. Early work identified a variety of direct and indirect effects of amino acids on insulin action mainly in the liver and skeletal muscle. These tissues were considered the principal candidates as mediators of the effects of amino acids. What are these direct and indirect effects? Postprandial elevations of plasma amino acids stimulate endogenous secretion of insulin and glucagon [12, 13] which in turn are proposed to modulate hepatic glucose metabolism (indirect effect) by changing the portal insulin/glucagon ratio [14]. Amino acids may act as substrates (direct effects) for gluconeogenesis [15], a metabolic pathway that contributes to glucose production in the liver. The direct and indirect effects of amino acids on glucose metabolism have been generally examined using complex mixtures of amino acids, an experimental approach that made hard to identify the individual actions of specific amino acids. Fortunately, further studies identified the branched chain amino acid (BCAA) leucine as a major contributor to the effects on insulin action [5]. In this work we will review recent evidence, supporting a role for the central nervous system in the effects of leucine on insulin action and glucose metabolism. We intend to advance the hypothesis that in the postabsorptive state, in response to elevations of circulating levels of amino acids, the sensing of leucine in the hypothalamus contributes to the maintenance of normal blood glucose levels by acutely inhibiting hepatic glucose production.

Nutrient Sensing, Overnutrition, and Diabetes

The maintenance of glucose homeostasis is a vital requirement for mammalian survival and failure to maintain glucose homeostasis is a key component of important metabolic diseases such as diabetes and obesity. Worldwide prevalence of obesity has been on the rise, along with its dreaded health consequences: diabetes, vascular diseases, and hypertension [16, 17]. Mammals have developed redundant neural mechanisms to control energy and glucose homeostasis [18–20]. Currently, the mediobasal hypothalamus (MBH) has been established as major center for the integration of multiple nutritional cues that inform the brain on the nutritional status of the body. This concept is often referred to as hypothalamic "nutrient sensing". Hypothalamic centers monitor the availability of circulating nutrients via nutrient-induced peripheral signals (e.g., leptin and insulin) as well as direct nutrient metabolic signaling (Fig. 7.1). As we will see later, the second mechanism relies on the key role played by malonyl-CoA in the regulation of metabolism [21, 22]. The activation of hypothalamic "sensors" networks exerts a negative feedback on food intake, and liver glucose and lipid output. Examples of this kind of regulatory loop include the modulation of hepatic glucose metabolism and triglyceride secretion by glucose, lactate, and oleic acid [23–26]. Until recently, only the third major group of macronutrients, amino acids, was absent from the list until recently. In fact, the regulation of hepatic glucose production by leucine was the first example of an amino acid acting in the brain to modulate glucose metabolism, soon followed by a second one, proline [27, 28]. Because the topic of this book is on BCAAs, in this review, we will focus on the role of leucine in the regulation of liver glucose metabolism. Leucine has all the metabolic features necessary to participate in hypothalamic nutrient sensing, mainly because the metabolism of leucine directly generates acetyl-CoA. This metabolite can be oxidized to carbon dioxide for energy production or used for the synthesis of malonyl-CoA, a precursor for fatty acid synthesis. Amino acids enter the central nervous system (CNS) through a facilitative transport system [29] and their concentration in the hypothalamus reflects the variations of circulating amino acid levels [30]. Moreover, leucine is a well-known activator of the mammalian

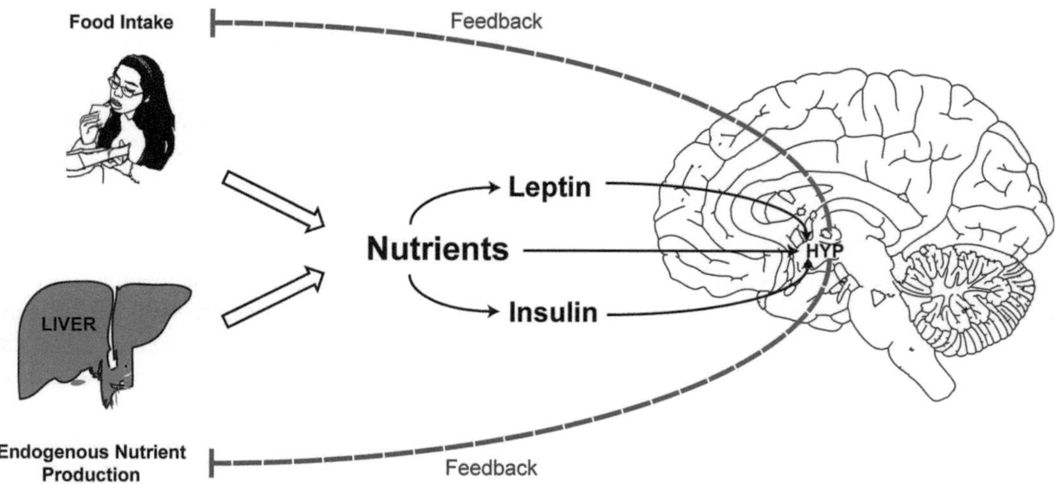

Fig. 7.1 Simplified schematic of the central regulation of food intake and endogenous nutrient production. The levels of circulating nutrients are informed to the brain via two major mechanisms: (a) indirectly through the hypothalamic action of nutrient-dependent hormones such as leptin and insulin; (b) directly, through metabolic sensing of circulating nutrients in the hypothalamus. *HYP* hypothalamus

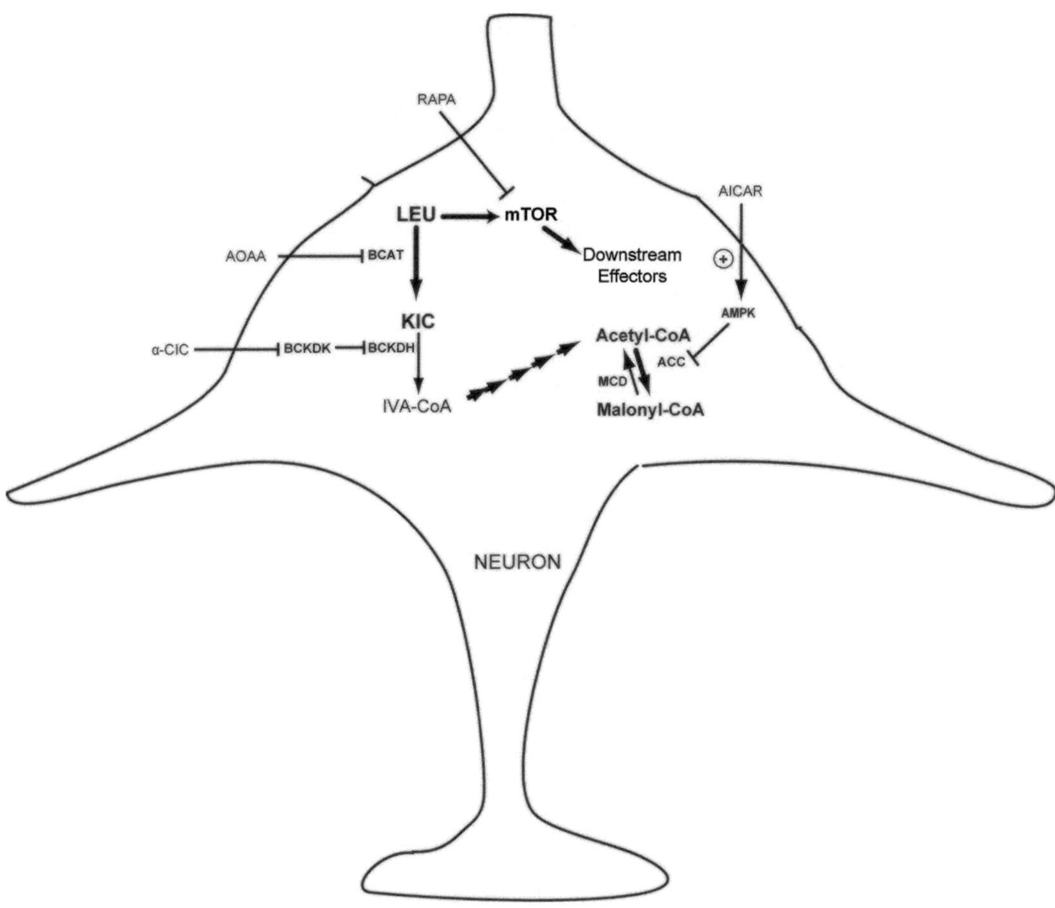

Fig. 7.2 Schematic of mTOR activation by leucine and leucine cellular metabolism in the brain. The ketogenic amino acid leucine is metabolized first to α-ketoisocaproic acid (KIC), then to isovaleryl-CoA (IVA-CoA), and eventually to acetyl-CoA. Leucine is also a well-known activator of the mTOR pathway via the mTOR complex 1. The schematic also shows pharmacological and molecular tools that have been used to delineate the mechanism of action of leucine in the hypothalamus. *BCAT* branched chain amino acid amino transferase, *BCKDH* branched chain ketoacid dehydrogenase, *ACC* acetyl CoA carboxylase, *MCD* malonyl CoA decarboxylase, *AMPK* AMP-dependent kinase, *α-CIC* α-chloroisocaproic acid, *AICAR* 5-Aminoimidazole-4-carboxyamide ribonucleoside

target of rapamycin (mTOR) pathway [1–3], and any effects of this amino acid activation of mTOR should be considered as a potential mechanism (Fig. 7.2). For example, it has been shown that leucine administered in the third cerebral ventricle decreases food intake via mTOR activation [8]. The focus of this article will be on the role of the CNS in the effects of leucine on insulin action and glucose metabolism. We will now briefly review the metabolism of leucine, graphically summarized in Fig. 7.2. Leucine is first step transaminated to α-ketoisocaproic acid (KIC) by the enzyme branched chain aminotransferase (BCAT) [31]. Next, KIC is oxidatively decarboxylated via the branched chain ketoacid dehydrogenase (BCKDH) [31, 32] to produce isovaleryl-CoA. KIC is further oxidized to acetyl-CoA through the sequential action of five enzymes. Finally, acetyl-CoA can be carboxylated to malonyl-CoA by the enzyme acetyl CoA carboxylase (ACC). ACC activity is tightly regulated (inhibited) by AMP kinase (AMPK). Here we will discuss our own evidence supporting the concept that the metabolism of leucine in the MBH couples the central sensing of leucine with the control of hepatic glucose production [27], and put this evidence in context with work by other groups.

Leucine Regulates Glucose Metabolism

Figure 7.3a, b shows a schematic of the experimental paradigm used in the studies described below. The strategy consisted of a combination of intraparenchymal infusions of leucine, KIC, pharmacological compounds, and recombinant vectors, in the MBH or rats combined with pancreatic insulin clamps to monitor hepatic glucose production. Figure 7.3c shows a simplified schematic of the metabolic fluxes that contribute to hepatic glucose production, mainly glucogenolysis and gluconeogenesis. First to determine if leucine modulates hepatic glucose metabolism, we delivered this amino acid into the MBH of conscious rats. Under basal conditions, leucine lowered the circulating levels of glucose. To better understand this glucose lowering effect, pancreatic clamps were used. As we know, changes in blood glucose levels produce adaptive regulatory changes of hormones that control glucose metabolism. Pancreatic insulin clamps are designed to control for changes in glucoregulatory hormones and eliminate their variations as confounding factors. Through measurements of whole body glucose kinetics using isotopic dilution, it was possible to determine whether the decrease of circulating glucose was due to increased glucose uptake (mainly in skeletal muscle and adipose tissue) or to a decrease of endogenous glucose production (mainly in the liver). The studies showed that central leucine inhibits hepatic glucose production without changing peripheral glucose disposal. The decrease of hepatic glucose production was the result of a marked inhibition in the in vivo rate of glycogenolysis and gluconeogenesis in the liver. These results are schematically summarized in Fig. 7.4. Because leucine is an activator of the mTOR complex 1 [2, 3], rapamicyn was used to prevent

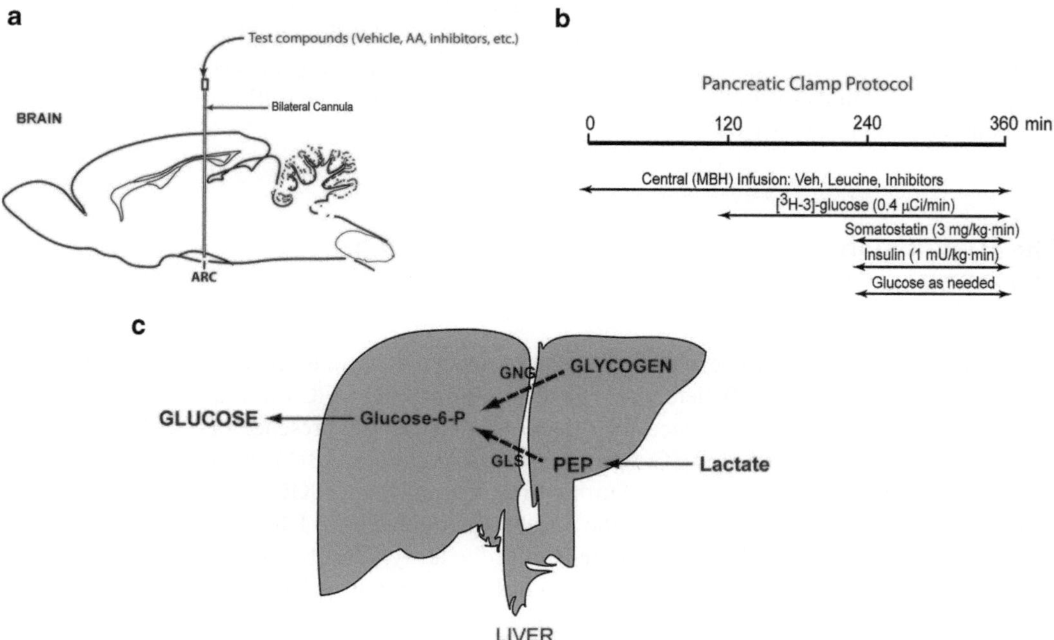

Fig. 7.3 Experimental paradigm for the in vivo study of hypothalamic leucine sensing and liver glucose metabolism. (**a**) A double cannula is implanted in the mediobasal hypothalamus (MBH) in the vicinity of the arcuate nucleus (ARC) for delivering of amino acids (AA) and other test compounds. (**b**) During the central infusions a pancreatic clamp study is performed in conscious unrestrained rodents according to the protocol shown. During the clamp, whole body glucose kinetics is assessed by isotopic tracer dilution methodology using tritiated glucose. (**c**) Simplified schematic of metabolic pathways contributing to glucose production in the liver: gluconeogenesis and glycogenolysis. *PEP* phospho-*enol* pyruvate, *GLS* glycogenolysis, *GNG* gluconeogenesis

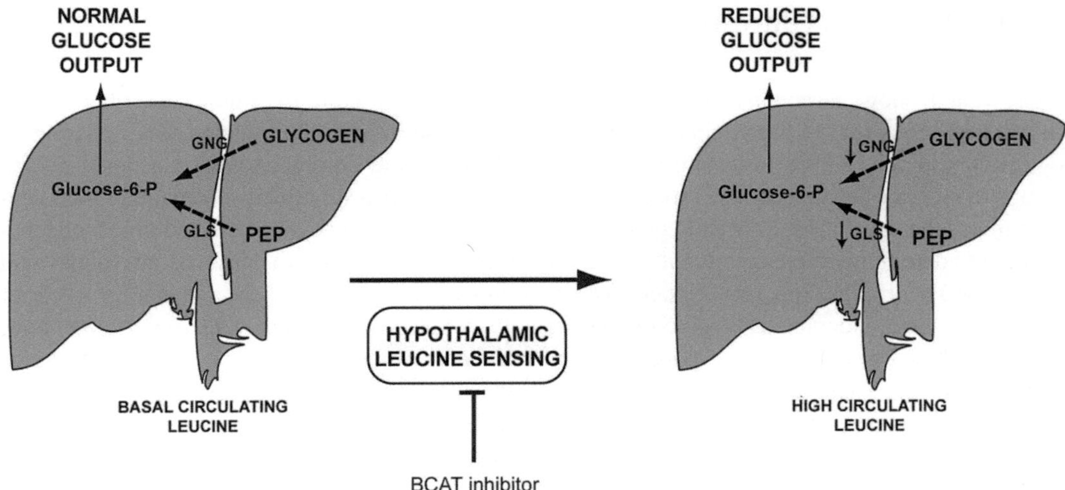

Fig. 7.4 Metabolic leucine sensing in the hypothalamus decreases glucose production by inhibiting hepatic glucose fluxes. Centrally acting leucine decreases the levels of circulating glucose and hepatic glucose production by a mechanism requiring its central metabolism. Inhibition of hypothalamic leucine metabolism with a BCAT inhibitor (BCATi) blunts the glucoregulatory effect. *BCAT* branched chain amino acid aminotransferase, *PEP* phospho-*enol* pyruvate, *GLS* glycogenolysis, *GNG* gluconeogenesis

this activation (Fig. 7.2). In the presence of rapamycin, the inhibitory effects of central leucine on plasma glucose, endogenous glucose production, liver glycogenolysis, and gluconeogenesis were all completely preserved, ruling out the participation of mTOR. Separate experiments showed that the dose of rapamycin used in these studies was sufficient to effectively block mTOR activation by leucine or insulin in the hypothalamus [27].

The Hypothalamic Metabolism of Leucine Modulates Liver Glucose Production

As mentioned earlier, the first step in the metabolism of leucine is its transamination to α-ketoisocaproic acid (KIC) (Fig. 7.2) [31, 32]. To determine if leucine conversion to KIC was required for the modulation of glucose metabolism, we tested the effect of leucine in the presence of the BCAT inhibitor amino-oxyacetic acid (AOAA) [33]. Coinfusion of AOAA into the MBH markedly blunted the effect of leucine on liver glucose metabolism. Furthermore, the infusion of KIC into the MBH completely recapitulated the effects of leucine. Thus, the metabolism of leucine to KIC is required for the inhibition of glucose production. KIC is further metabolized to acetyl-CoA by the sequential action of five enzymes [31, 32]. The oxidative decarboxylation of KIC to form isovaleryl-CoA is the first and rate-limiting step in this five-step process. This reaction is catalyzed by the BCKDH complex. The activity of BCKDH is robustly inhibited after a phosphorylation by the enzyme BCKDH kinase or BCKDK (Fig. 7.2). Pharmacologically, BCKDH can be activated by the compound α-chloro-isocaproic acid (α-CIC), a potent and specific inhibitor of BCKDK [34]. α-CIC relieves the inhibitory tone on BCKDH and can be used to stimulate endogenous leucine metabolism. The infusion of α-CIC into the MBH fully recapitulated the effect of leucine in the absence of either exogenous leucine or KIC, further supporting the concept that hypothalamic leucine metabolism can regulate glucose metabolism.

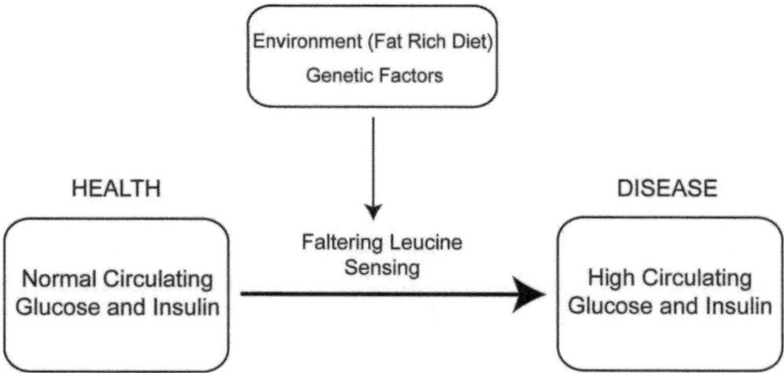

Fig. 7.5 Regulation of hepatic glucose metabolism by leucine sensing in the hypothalamus. Circulating leucine enters the hypothalamus where its metabolism generates a regulatory signal which, relayed via the vagus nerve, inhibits hepatic glucose production. *MBH* mediobasal hypothalamus, *Leu* leucine, K_{ATP} *Ch* ATP-dependent potassium channel

Redundant and independent confirmation of this finding was obtained through the overexpression of BCKDK in the MBH which was predicted to inhibit leucine metabolism. As expected, this intervention blunted the effect of leucine on hepatic glucose fluxes. Altogether, these studies demonstrate that the metabolism of leucine in the MBH is required for the inhibition of liver glucose production by leucine sensing.

Leucine can generate malonyl-CoA from acetyl-CoA by the action of acetyl-CoA carboxylase (ACC) (Fig. 7.2). Because our hypothesis postulates that malonyl-CoA is key intermediate required for generating the signal(s) that modify liver glucose metabolism we used AMPK activator 5-amino-imidazole-4-carboxamide ribonucleotide (AICAR) to antagonize the effect of KIC. The activation of AMPK by AICAR should result in inhibition of ACC and prevent the increase of malonyl-CoA in response to KIC (or leucine) infusion. The coinfusion of AICAR obliterated the effect of KIC on hepatic glucose metabolism thus indicating that malonyl-CoA was required for glucoregulation by leucine/KIC. The opposite approach was also tested. Malonyl-CoA is decarboxylated back to acetyl-CoA (Fig. 7.2) by the action of malonyl-CoA decarboxylase (MCD). We have previously shown that overexpression of this enzyme in the MBH reduces the local pool of malonyl-CoA [22]. We applied this strategy to the case of leucine. When we overexpressed MCD in the MBH of rats, the ability of leucine to modulate glucose metabolism was significantly diminished compared to animals overexpressing a control protein. Taken together, these results indicate that the metabolism of leucine to malonyl-CoA in the MBH is necessary for the regulation of glucose metabolism (Fig. 7.5).

We end our detailed description of the molecular characterization of hypothalamic leucine sensing with a discussion about the role of K_{ATP} channels. Studies have shown that hypothalamic ATP-sensitive potassium (K_{ATP}) channels are critical for the modulation of glucose metabolism by insulin [35] and oleic acid [19, 23, 24]. Of great interest, we found that the central administration of leucine or KIC increases the levels of oleyl-CoA in the MBH. Thus we investigated if functional K_{ATP} channels were required for leucine's effects. We first used the K_{ATP} channel blocker glibenclamide. In the presence of glibenclamide, the administration of KIC in the MBH failed to modulate hepatic glucose metabolism, indicating functional K_{ATP} channels were required. To obtain redundant but independent confirmation, we also examined amino acid sensing in mice lacking functional K_{ATP} channels in the brain, the *Sur1*$^{-/-}$ mice [36]. In these animals the central administration of KIC failed to modulate liver glucose metabolism, in contrast to wild type animals which were fully responsive [27]. These mice studies confirmed the pharmacological evidence and demonstrated that the action of leucine on glucose metabolism requires functional brain K_{ATP} channels (See Fig. 7.5).

Physiological and Pathophysiological Relevance of Hypothalamic Leucine Sensing

Having outlined the brain-liver circuit by which leucine modulates glucose production, the next most relevant issue was to determine the physiological significance of hypothalamic leucine sensing. In light of the various reports showing beneficial effects of protein-rich diets on insulin action, we wondered if central amino acid sensing could be a contributing factor. Some studies have shown that feeding protein-rich diets to rats produces an increase in the levels of BCAAs and specifically of leucine in plasma and brain [37, 38]. Thus, to examine the physiological relevance of leucine sensing, we performed systemic leucine infusions to generate moderate increases of its plasma concentration similar to those observed after consumption of diets enriched in protein [37–39]. In the course of pancreatic clamps with basal insulin and with vehicle infused in the MBH, gradual elevations of circulating leucine inhibited endogenous glucose production in a dose-independent manner, similarly to the centrally administered leucine. Furthermore, the sole infusion of a BCAT inhibitor into the MBH blunted the inhibitory effect of increasing circulating leucine. These results, summarized schematically in Fig. 7.4, indicate that the action of central leucine is not just a pharmacological artifact but that physiologically relevant variations of circulating leucine modulate hepatic glucose production. Interestingly, high circulating levels of leucine activate the mTOR-S6K pathway in liver and skeletal muscle and this has been suggested to induce insulin resistance [40] and oppose the effects of central leucine. In light of our findings we believe that when circulating leucine is high, its central actions predominate over the peripheral effects, overall resulting in enhanced insulin action. In this regard, it is of interest that mice, with a global deletion of mitochondrial BCAT, display a lean phenotype, high levels of circulating leucine, enhanced insulin sensitivity, and resistance to diet-induced obesity [41]. In these animals only the brain was still capable of metabolizing leucine due to the presence of an alternative isoform of BCAT that was not affected by the gene disruption. It is possible that central leucine sensing could at least in part explain the phenotype of these mice. Recent studies also showed that oral leucine supplementation improves glucose metabolism in both normal and diabetic mice [10, 11], in agreement with the effects of central leucine.

Studies in humans have shown similar effects of leucine on glucose metabolism. Subjects [42, 43] receiving systemic infusions of leucine showed a decrease in glucose production although, in contrast with rodents, a decrease in glucose utilization was also observed. Of interest, studies in humans have found elevated levels of circulating leucine linked to the development of insulin resistance and diabetes [44, 45]. Although it is currently unclear how these results relate to our findings on the central sensing of leucine, add to the concept that leucine plays a significant role in the control of glucose metabolism, insulin action, and energy metabolism.

The consumption of calorie dense (carbohydrate and fat rich) diets play a major role in the increasing incidence of obesity and diabetes in susceptible individuals. Studies have shown that rodents presented with a highly palatable, fat-rich diet promptly double their daily caloric intake and develop severe hepatic insulin resistance. Under these conditions, the animals develop a defect in the nutrient sensing mechanisms leading to a failure to decrease food intake and hepatic glucose or lipid production in response to macronutrients [26, 46]. This faltering of hypothalamic nutrient sensing (Fig. 7.6) is proposed to contribute to the susceptibility to obesity and insulin resistance in response to voluntary overfeeding. It was thus important to examine the impact of diet-induced insulin resistance on hypothalamic leucine sensing. Unpublished results from our laboratory showed that diet-induced insulin resistance obliterates the ability of animals to respond to leucine sensing. In these studies, the central administration of leucine to short-term overfed (3 days) rats failed to decrease liver glucose production when compared to fully responsive animals fed with a regular chow diet. This is a very rapid effect, long before any effects on body adiposity develop. Furthermore, genetic attenuation of hypothalamic leucine sensing through the overexpression of BCKDK in the MBH impaired the ability of animals to maintain glycemic control as indicated by the higher circulating glucose and insulin

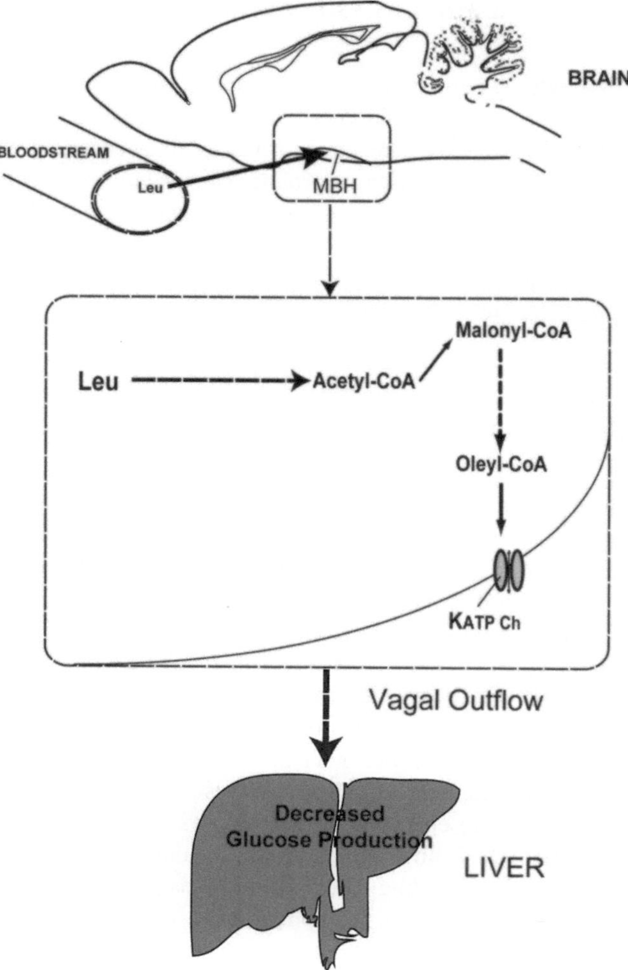

Fig. 7.6 Proposed role of hypothalamic leucine sensing in the maintenance of euglycemia. Under normal circumstances physiological elevations of circulating leucine inhibit hepatic glucose production by a central metabolic sensing mechanism. However, in susceptible individuals a combination of genetic and environmental factors leads to the faltering of the central sensing mechanism. Ultimately, sustained failure of leucine sensing makes these individuals prone to the development of hyperglycemia

observed during fasting and in response to a protein-rich meal in BCKDK overexpressing animals compared to control rats. These experiments strongly suggest that the hypothalamic sensing of leucine is required to maintain proper glycemic control. It is conceivable that a combination of environmental and genetic factors such as individual predisposition and exposure to fat-rich diets could lead to chronic faltering of hypothalamic nutrient sensing and, in the long run, hyperglycemia (Fig. 7.6).

Based on the evidence just presented, we postulate a novel brain-liver circuit that couples the metabolism of leucine in the MBH with the regulation of glucose metabolism in the liver (summarized in Fig. 7.5). In the postabsorptive state, this circuit is engaged by leucine in response to elevations of its circulating levels and contributes to maintain physiological levels of blood glucose by reducing hepatic glucose production. The sensing of leucine is completely dependent on its metabolism to acetyl-CoA and malonyl-CoA, in contrast with the regulation of food intake studies which is mTOR dependent [8]. Perhaps these distinct responses to leucine sensing represent mechanisms of the differential regulation of food intake and glucose production. Interestingly, a recent report suggesting that leucine metabolism may also play a role [9] in food intake regulation adds yet another layer of complexity. In this regard, considering that leucine can reduce food intake and stimulate glucose metabolism, it is tempting to speculate that the sensing of leucine derived from dietary protein, could be a contributing factor to the weight reduction and improvement of insulin action reportedly produced by high-protein, low-fat diets in humans [47–49].

Glucoregulation by Other Amino Acids

Although the regulation of hepatic glucose production by leucine in the MBH was the first report of an amino acid regulating carbohydrate metabolism via a central mechanism, recently other two non-BCAAs have been reported to exert a similar central regulatory action. These amino acids are proline and histidine. The action of proline was recently reported by our group [28], and like leucine [27], proline uses a metabolic sensing mechanism. There are however several distinctive differences at the biochemical and cellular level. As to the case of histidine, recently reported to inhibit hepatic glucose production, its mechanism does not seem to involve metabolic conversion in the brain but rather a ligand-receptor interaction [50].

Conclusions

The experimental data discussed here suggest the existence of a novel brain-liver circuit that couples the metabolism of leucine in the MBH with the regulation of glucose metabolism in the liver (summarized in Fig. 7.5). In the postabsorptive state, this circuit is engaged by leucine in response to elevations of its circulating levels and contributes to the maintainance of physiological levels of blood glucose by reducing hepatic glucose production. In susceptible individuals, the faltering of this sensing mechanism would lead to the disregulation of glycemic control. Thus, the novel biochemical mechanisms for central leucine sensing that we have just discussed may be helpful for designing more rational approaches to nutritional as well as pharmacological interventions for the treatment (or prevention) of diabetes and other insulin-resistant metabolic disorders. Additionally, regardless of the specific mechanisms of action, all these recent studies clearly show that BCAAs as well as other amino acids are emerging as new players in the regulation of various physiological and metabolic processes. These novel functions are in addition to their role as metabolic substrates for energy production and as building blocks for proteins. Future studies will most certainly unveil new regulatory functions of amino acids and provide new drug targets for the treatment of a variety of disorders.

References

1. Colombani J, Raisin S, Pantalacci S, Radimerski T, Montagne J, Leopold P. A nutrient sensor mechanism controls Drosophila growth. Cell. 2003;114:739–49.
2. Cutler NS, Pan X, Heitman J, Cardenas ME. The TOR signal transduction cascade controls cellular differentiation in response to nutrients. Mol Biol Cell. 2001;12:4103–13.
3. Nobukuni T, Joaquin M, Roccio M, et al. Amino acids mediate mTOR/raptor signaling through activation of class 3 phosphatidylinositol 3OH-kinase. Proc Natl Acad Sci U S A. 2005;102:14238–43.
4. Rossetti L, Rothman DL, DeFronzo RA, Shulman GI. Effect of dietary protein on in vivo insulin action and liver glycogen repletion. Am J Physiol. 1989;257:E212–9.
5. Patti ME, Brambilla E, Luzi L, Landaker EJ, Kahn CR. Bidirectional modulation of insulin action by amino acids. J Clin Invest. 1998;101:1519–29.
6. Nuttall FQ, Gannon MC, Saeed A, Jordan K, Hoover H. The metabolic response of subjects with type 2 diabetes to a high-protein, weight-maintenance diet. J Clin Endocrinol Metab. 2003;88:3577–83.
7. Gannon MC, Nuttall FQ, Saeed A, Jordan K, Hoover H. An increase in dietary protein improves the blood glucose response in persons with type 2 diabetes. Am J Clin Nutr. 2003;78:734–41.
8. Cota D, Proulx K, Smith KA, et al. Hypothalamic mTOR signaling regulates food intake. Science. 2006;312:927–30.
9. Blouet C, Jo YH, Li X, Schwartz GJ. Mediobasal hypothalamic leucine sensing regulates food intake through activation of a hypothalamus-brainstem circuit. J Neurosci. 2009;29:8302–11.
10. Zhang Y, Guo K, LeBlanc RE, Loh D, Schwartz GJ, Yu YH. Increasing dietary leucine intake reduces diet-induced obesity and improves glucose and cholesterol metabolism in mice via multimechanisms. Diabetes. 2007;56:1647–54.

11. Guo K, Yu YH, Hou J, Zhang Y. Chronic leucine supplementation improves glycemic control in etiologically distinct mouse models of obesity and diabetes mellitus. Nutr Metab (Lond). 2010;7:57.
12. Floyd Jr JC, Fajans SS, Conn JW, Knopf RF, Rull J. Stimulation of insulin secretion by amino acids. J Clin Invest. 1966;45:1487–502.
13. Ohneda A, Parada E, Eisentraut AM, Unger RH. Characterization of response of circulating glucagon to intraduodenal and intravenous administration of amino acids. J Clin Invest. 1968;47:2305–22.
14. Roden M, Perseghin G, Petersen KF, et al. The roles of insulin and glucagon in the regulation of hepatic glycogen synthesis and turnover in humans. J Clin Invest. 1996;97:642–8.
15. Felig P. Amino acid metabolism in man. Annu Rev Biochem. 1975;44:933–55.
16. Flegal KM, Carroll MD, Ogden CL, Johnson CL. Prevalence and trends in obesity among US adults, 1999-2000. JAMA. 2002;288:1723–7.
17. Ginter E, Simko V. Type 2 diabetes mellitus, pandemic in 21st century. Adv Exp Med Biol. 2012;771:42–50.
18. Schwartz MW, Woods SC, Porte Jr D, Seeley RJ, Baskin DG. Central nervous system control of food intake. Nature. 2000;404:661–71.
19. Obici S, Zhang BB, Karkanias G, Rossetti L. Hypothalamic insulin signaling is required for inhibition of glucose production. Nat Med. 2002;8:1376–82.
20. Schwartz MW, Porte Jr D. Diabetes, obesity, and the brain. Science. 2005;307:375–9.
21. Hu Z, Cha SH, Chohnan S, Lane MD. Hypothalamic malonyl-CoA as a mediator of feeding behavior. Proc Natl Acad Sci U S A. 2003;100:12624–9.
22. He W, Lam TK, Obici S, Rossetti L. Molecular disruption of hypothalamic nutrient sensing induces obesity. Nat Neurosci. 2006;9:227–33.
23. Lam TK, Pocai A, Gutierrez-Juarez R, et al. Hypothalamic sensing of circulating fatty acids is required for glucose homeostasis. Nat Med. 2005;11:320–7.
24. Lam TK, Gutierrez-Juarez R, Pocai A, Rossetti L. Regulation of blood glucose by hypothalamic pyruvate metabolism. Science. 2005;309:943–7.
25. Obici S, Feng Z, Morgan K, Stein D, Karkanias G, Rossetti L. Central administration of oleic acid inhibits glucose production and food intake. Diabetes. 2002;51:271–5.
26. Lam TK, Gutierrez-Juarez R, Pocai A, et al. Brain glucose metabolism controls the hepatic secretion of triglyceride-rich lipoproteins. Nat Med. 2007;13:171–80.
27. Su Y, Lam TK, He W, et al. Hypothalamic leucine metabolism regulates liver glucose production. Diabetes. 2012;61:85–93.
28. Arrieta-Cruz I, Su Y, Knight CM, Lam TK, Gutierrez-Juarez R. Evidence for a role of proline and hypothalamic astrocytes in the regulation of glucose metabolism in rats. Diabetes. 2013;62:1152–8.
29. Smith QR. Transport of glutamate and other amino acids at the blood-brain barrier. J Nutr. 2000;130:1016S–22.
30. Choi YH, Fletcher PJ, Anderson GH. Extracellular amino acid profiles in the paraventricular nucleus of the rat hypothalamus are influenced by diet composition. Brain Res. 2001;892:320–8.
31. Lynch CJ, Halle B, Fujii H, et al. Potential role of leucine metabolism in the leucine-signaling pathway involving mTOR. Am J Physiol Endocrinol Metab. 2003;285:E854–63.
32. Suryawan A, Hawes JW, Harris RA, Shimomura Y, Jenkins AE, Hutson SM. A molecular model of human branched-chain amino acid metabolism. Am J Clin Nutr. 1998;68:72–81.
33. Gao Z, Young RA, Li G, et al. Distinguishing features of leucine and alpha-ketoisocaproate sensing in pancreatic beta-cells. Endocrinology. 2003;144:1949–57.
34. Harris RA, Paxton R, DePaoli-Roach AA. Inhibition of branched chain alpha-ketoacid dehydrogenase kinase activity by alpha-chloroisocaproate. J Biol Chem. 1982;257:13915–8.
35. Pocai A, Lam TK, Gutierrez-Juarez R, et al. Hypothalamic K(ATP) channels control hepatic glucose production. Nature. 2005;434:1026–31.
36. Seghers V, Nakazaki M, DeMayo F, Aguilar-Bryan L, Bryan J. Sur1 knockout mice. A model for K(ATP) channel-independent regulation of insulin secretion. J Biol Chem. 2000;275:9270–7.
37. Colombo JP, Cervantes H, Kokorovic M, Pfister U, Perritaz R. Effect of different protein diets on the distribution of amino acids in plasma, liver and brain in the rat. Ann Nutr Metab. 1992;36:23–33.
38. Peters JC, Harper AE. Adaptation of rats to diets containing different levels of protein: effects on food intake, plasma and brain amino acid concentrations and brain neurotransmitter metabolism. J Nutr. 1985;115:382–98.
39. Block KP, Harper AE. High levels of dietary amino and branched-chain alpha-keto acids alter plasma and brain amino acid concentrations in rats. J Nutr. 1991;121:663–71.
40. Tremblay F, Krebs M, Dombrowski L, et al. Overactivation of S6 kinase 1 as a cause of human insulin resistance during increased amino acid availability. Diabetes. 2005;54:2674–84.
41. She P, Reid TM, Bronson SK, et al. Disruption of BCATm in mice leads to increased energy expenditure associated with the activation of a futile protein turnover cycle. Cell Metab. 2007;6:181–94.
42. Sherwin RS. Effect of starvation on the turnover and metabolic response to leucine. J Clin Invest. 1978;61: 1471–81.

43. Nair KS, Matthews DE, Welle SL, Braiman T. Effect of leucine on amino acid and glucose metabolism in humans. Metabolism. 1992;41:643–8.
44. Newgard CB, An J, Bain JR, et al. A branched-chain amino acid-related metabolic signature that differentiates obese and lean humans and contributes to insulin resistance. Cell Metab. 2009;9:311–26.
45. Wang TJ, Larson MG, Vasan RS, et al. Metabolite profiles and the risk of developing diabetes. Nat Med. 2011;17:448–53.
46. Morgan K, Obici S, Rossetti L. Hypothalamic responses to long-chain fatty acids are nutritionally regulated. J Biol Chem. 2004;279:31139–48.
47. McAuley KA, Hopkins CM, Smith KJ, et al. Comparison of high-fat and high-protein diets with a high-carbohydrate diet in insulin-resistant obese women. Diabetologia. 2005;48:8–16.
48. Boden G, Sargrad K, Homko C, Mozzoli M, Stein TP. Effect of a low-carbohydrate diet on appetite, blood glucose levels, and insulin resistance in obese patients with type 2 diabetes. Ann Intern Med. 2005;142:403–11.
49. McAuley KA, Smith KJ, Taylor RW, McLay RT, Williams SM, Mann JI. Long-term effects of popular dietary approaches on weight loss and features of insulin resistance. Int J Obes (Lond). 2006;30:342–9.
50. Kimura K, Nakamura Y, Inaba Y, et al. Histidine augments the suppression of hepatic glucose production by central insulin action. Diabetes. 2013;62:2266.

Chapter 8
Leucine and Resveratrol: Experimental Model of Sirtuin Pathway Activation

Antje Bruckbauer and Michael B. Zemel

Key Points

- Leucine is an activator of Sirt1 signaling.
- Low-dose resveratrol combined with leucine acts synergistically to stimulate fat oxidation, improve insulin sensitivity and increase overall energy metabolism.
- The synergistic effects of the resveratrol/leucine combination are mediated by activation of the AMPK/Sirtuin pathway.
- The resveratrol/leucine combination counteracts the effects of over-nutrition, which inhibits the AMPK/sirtuin pathway.
- The combination of resveratrol/leucine mimics, in part, the effects of exercise by stimulating irisin secretion in muscle cells, thus promoting cross-talk between muscle cells and adipose tissue.
- The synergistic effects of resveratrol/leucine are exerted with low concentrations of the individual compounds that can be easily achieved via oral administration.

Keywords Leucine • Resveratrol • Sirt1 • AMPK • Energy metabolism • Insulin resistance • Synergy • Irisin • Mitochondrial biogenesis

Abbreviations

Sirt1	Silent information regulator protein 1
NAD^+	Nicotinamide adenine dinucleotide
PGC-1α	Peroxisome proliferator-activated receptor-γ co-activator-1 alpha
FOXO	Forkhead box O transcription factor
cAMP	Cyclic adenosine monophosphate

A. Bruckbauer, M.D., Ph.D. (✉)
NuSirt Sciences Inc., Knoxville, TN, USA
e-mail: abruckbauer@nusirt.com

M.B. Zemel, Ph.D.
NuSirt Sciences Inc., Knoxville, TN, USA

University of Tennessee, Knoxville, TN, USA
e-mail: mzemel@nusirt.com

R. Rajendram et al. (eds.), *Branched Chain Amino Acids in Clinical Nutrition: Volume 1*, Nutrition and Health, DOI 10.1007/978-1-4939-1923-9_8,

AMPK	AMP-activated protein kinase
mTOR	Mammalian target of rapamycin
BCATm	Mitochondrial branched chain amino acid transferase
BCKDm	Mitochondrial branched chain α-keto-acid dehydrogenase
BCAA	Branched chain amino acid
HMB	β-Hydroxy-β-methylbutyrate
KIC	α-Keto-isocaproate
C_{max}	Maximum concentration
DIO	Diet-induced obese
PET/CT	Positron Emission Tomography—Computed Tomography
RER	Respiratory exchange ratio
$HOMA_{IR}$	Homeostatic assessment of insulin resistance
FDG	Fluorine-18-deoxy-glucose
IL-6	Interleukin 6
MCP-1	Monocyte chemoattractant protein-1
CRP	C-reactive protein
LKB1	Liver kinase B1
TCA cycle	Tricarboxylic acid cycle
FNDC5	Fibronectin type III domain containing 5
UCP1	Uncoupling protein 1
Cox7a	Cytochrome c oxidase subunit VIIa
CM	Conditioned media

Introduction

The beneficial effects of leucine and resveratrol on health are largely dependent on activation of the sirtuin pathway. The silent information regulator (Sir) proteins, also called sirtuins, are a conserved family of histone and protein deacetylases that use NAD^+ as a co-substrate [1]. This NAD^+-dependence links sirtuins to the metabolic activity of cells, thus regulating numerous physiologic and metabolic pathways. Accordingly, they have been implicated in the prevention of metabolic disorders such as insulin resistance, diabetes, cardiovascular disease, cancer as well as aging [2].

In mammals there are seven sirtuins that differ in their tissue specifity, subcellular localization and enzymatic function [3]. The most extensively studied sirtuin is Sirt1 which deacetylates a variety of proteins, including the peroxisome proliferator-activated receptor-γ co-activator (PGC1α) and the forkhead box O (FOXO) transcription factor, thus playing an important role in regulating mitochondrial biogenesis, lipid and glucose metabolism and overall energy homeostasis [3]. Activation of Sirt1 has also been reported as a key mediator of the beneficial effects of caloric restriction on lifespan extension in yeast, worms and mammals [3]. Thus, Sirt1 activators may have therapeutic potential to protect against metabolic diseases and therefore it is of great interest to find safe and effective strategies to activate the sirtuin pathway.

Resveratrol

Resveratrol is a polyphenol found in the skin of red grapes and other fruits. In addition to its antioxidant capacity, it has been reported as a Sirt1 activator that mimics the effects of caloric restriction on life span, oxidative and inflammatory stress, insulin sensitivity and adiposity [4, 5]. However, the underlying mechanism of action is still debated. While recent evidence suggests that this Sirt1 activation may be an indirect effect mediated by inhibiting cAMP phosphodiesterase, resulting in upregulation of AMPK and increased levels of NAD^+ [6], others have shown that this may be the case only at high concentration (50 μM) while lower concentration result in direct Sirt1 activation [7]. Although activation of Sirt1 by resveratrol has been linked by some to the fluorophore used in the activity assay [8, 9], recent data demonstrate direct activation and show that the fluorophore substitutes for hydrophobic amino acids endogenously present in the substrate [10]. These observed concentration-dependent differences in the mode of resveratrol action (indirect vs direct) may explain its dose- and time-dependent effects and different outcomes found in some studies. For example, anti-inflammatory and anti-oxidant effects were found in the low micromolar range while others such as pro-apoptotic and anti-cancer effects require higher concentrations (>50 μM to mM range) [11, 12]. Similarly, biphasic responses were found with regards to weight gain in mice fed a high-fat diet supplemented with low or high dose of resveratrol [13, 14]. However, even the low micromolar concentrations used in in vitro studies are difficult to obtain by humans, since resveratrol is rapidly metabolized and undergoes extensive enterohepatic sulfation and glucuronidation resulting in up to 20-fold higher plasma levels of conjugated metabolites [15]. Accordingly, an oral intake of 25 mg of trans-resveratrol per day resulted only in a peak plasma concentration of about 7 μg/L (~0.026 μM) of resveratrol in human subjects [16]. Thus, human plasma concentrations are usually 10- to 100-fold lower than those used in cell studies (μM range) [15]. Moreover, there are still many unanswered questions such as about the concentration necessary to achieve physiological effects in humans, the bioavailability and function of resveratrol metabolites, inter-individual variation and variation of tissue specifity [17]. Therefore, results from cell and animals studies cannot be readily translated and some may even not be relevant for humans.

Leucine

In addition to its primary role as a substrates for protein synthesis, leucine has a unique role in participating in signaling pathways, including regulation of muscle protein synthesis via both mTOR-dependent and independent pathways, as well as exerting an anti-proteolytic effect in muscle and other tissues [18, 19]. Leucine has been reported to attenuate adiposity and promote weight loss during energy restriction while preventing muscle loss [20, 21]. Chronic leucine supplementation in combination with a caloric restricted diet for 6 weeks in rats resulted in significant lower amount of body fat and prevention of muscle loss compared to the control restricted diet group [20]. Similarly, twofold leucine supplementation in drinking water in high-fat diet fed mice reduced weight gain up to 32 % and decreased adiposity by 25 % without energy restriction while no effects were seen in regular chow fed mice. These changes were accompanied by decreased hyperglycemia and improved insulin sensitivity as well as decreased hypercholesterolemia [21].

These findings appear incongruent with the observation that plasma branched chain amino acids including leucine are increased in conditions such as obesity, insulin resistance and type 2 diabetes [22–25]. Moreover, activation of the mTOR pathway may promote insulin resistance under high-fat feeding conditions [22, 25, 26]. However, this rise in plasma amino acid concentration appears to be secondary to aberrant branched chain amino acid catabolism and is therefore a result, not a cause, of the insulin resistance [35]. Studies have shown that obesity and insulin resistance are associated with down-regulation of the first two catabolic enzymes (a simplified overview of the leucine catabolism is depicted in Fig. 8.1), the mitochondrial branched chain amino acid transferase (BCATm) and the branched chain α-keto-acid dehydrogenase (BCKD), in liver and adipose tissue [27–29]. While the BCATm expression in liver is negligible, liver BCKD activity is highly regulated and the rate-limiting enzyme for total oxidation. Therefore, a reduction in BCKD activity contributes substantially to increased plasma BCAA concentrations [27]. However, these changes

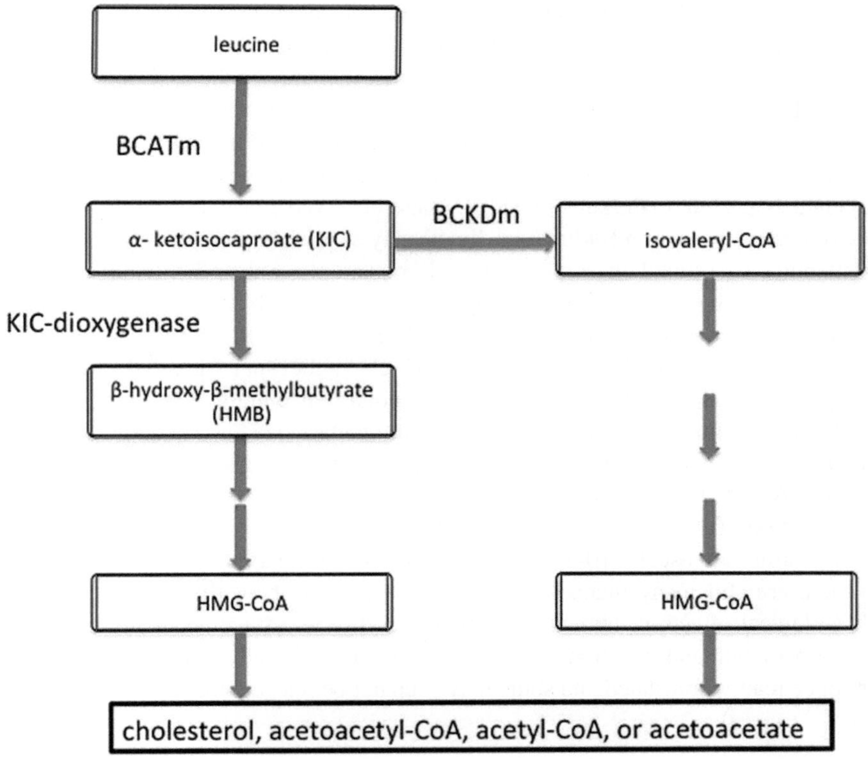

Fig. 8.1 Leucine catabolism. Leucine obtained from the diet is transaminated to α-ketoisocaproate (KIC) by the mitochondrial branched chain aminoacid transferase (BCATm), predominately extrahepatically. KIC may follow two pathways: either it is further metabolized to isovaleryl-CoA via the mitochondrial multi-enzyme complex branched chain keto acid dehydrogenase (BCKDm), or it can be converted to HMB by the cytosolic enzyme KIC dioxygenase in the liver. About 5–10 % of ingested leucine is converted to HMB under physiological conditions. Both, HMB and isovaleryl-CoA can be further metabolized in multiple reactions to β-hydroxy-methylglutaryl-CoA (HMB-CoA), in which either is used for cholesterol synthesis or further metabolized to acetoacetyl-CoA, acetyl-CoA or acetoacetate (adapted from [78])

in BCAA metabolism in obese and insulin-resistant states do not infer a causal relationship, and many studies indicate that high-protein/BCAA diets restore BCKD activity and improve glucose and insulin tolerance [21, 28, 30, 31].

High-protein diets have been demonstrated to preserve lean body mass loss and to promote fat loss under energy restriction [32, 33] and leucine appears to be a key mediator of these effects [31, 34]. Therefore it seems plausible that leucine increases fatty acid oxidation as a result from the higher energetic cost of protein synthesis and turnover. Consistent with this concept, leucine promotes energy partitioning from adipocytes to skeletal myotubes in a co-culture system resulting in net reductions in adipocyte lipid storage and increases in muscle fatty acid oxidation [35, 36]. These effects were associated with an increase in mitochondrial biogenesis and mediated, at least in part, by activation of Sirt1-dependent pathways [36]. Moreover, leucine and its metabolites β-hydroxy-β-methylbutyrate (HMB) and α-keto-isocaproate (KIC) appear to activate Sirt1 directly, as shown in a cell-free system [37]. Although it may be possible that the Sirt1 activation in this activity assay is dependent on the fluorophore used in that assay [38], it was recently demonstrated that hydrophobic amino acids such as leucine may substitute for these fluorophores in an endogenous system [10], as discussed later. Therefore, leucine may act as an allosteric activator for Sirt1 to increase enzyme activity.

Leucine-Resveratrol-Synergy

Based on the ability of both leucine and resveratrol, to stimulate sirtuin-dependent pathways in higher concentrations, we sought to evaluate possible synergistic actions of both on sirtuin-dependent downstream metabolic effects, thereby lowering the necessary concentrations of the individual compounds to a level which can be readily achieved via diet or supplement. The resveratrol concentration in these experiments is about 10- to 100-fold lower than in other comparable studies and is similar to human plasma concentrations (C_{max}) achieved after single ingestion of less than 0.5 g or repeated dose of 150 mg resveratrol [15, 39]. This low concentration was not sufficient to produce any metabolic effects independently while the combination with leucine or with its metabolite HMB exhibited similar effects to those achieved with a 20-fold higher dose.

Using this combination of low-dose resveratrol with leucine or HMB, we found significant synergistic effects on lipid and glucose metabolism, improved insulin sensitivity and a reduction of inflammatory stress.

Effects on Lipid Metabolism

The resveratrol/leucine combination elicited increases in fatty acid oxidation under both low-glucose and high-glucose conditions in cultured muscle cells and adipocytes (Fig. 8.2). Notably, the increase in fat oxidation under high-glucose conditions, which simulates glycemic stress, was about 6 times higher than under low-glucose conditions [40]. Similar results were also found in vivo after feeding diet-induced obese (DIO) mice a high-fat diet supplemented with low-dose resveratrol/leucine or HMB for 6 weeks [40]. Fat oxidation in muscle, measured by palmitate uptake via PET/CT, and heat

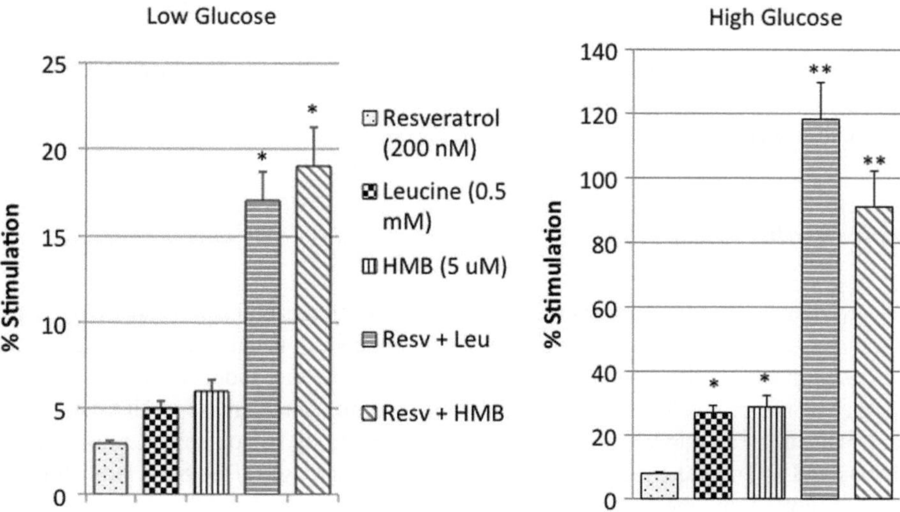

Fig. 8.2 Effects of resveratrol, leucine and HMB on fat oxidation under low- and high-glucose conditions. Differentiated mouse adipocytes were incubated with indicated treatments for 4 h in low-glucose (5 mM) or high-glucose (25 mM) media to mimic physiologic or hyperglycemic conditions. Fatty acid oxidation was measured using [^{3}H]-palmitate. Data are presented as mean ± SE (n=6) and are expressed as % stimulation over control, where low-glucose control = 193 ± 39 cpm/ng DNA and high-glucose control = 302 ± 24 cpm/ng DNA. Stars above the bars indicate significant difference * compared to control (p < 0.05), ** compared to control, Leucine, and HMB (p < 0.005) (adapted from [40])

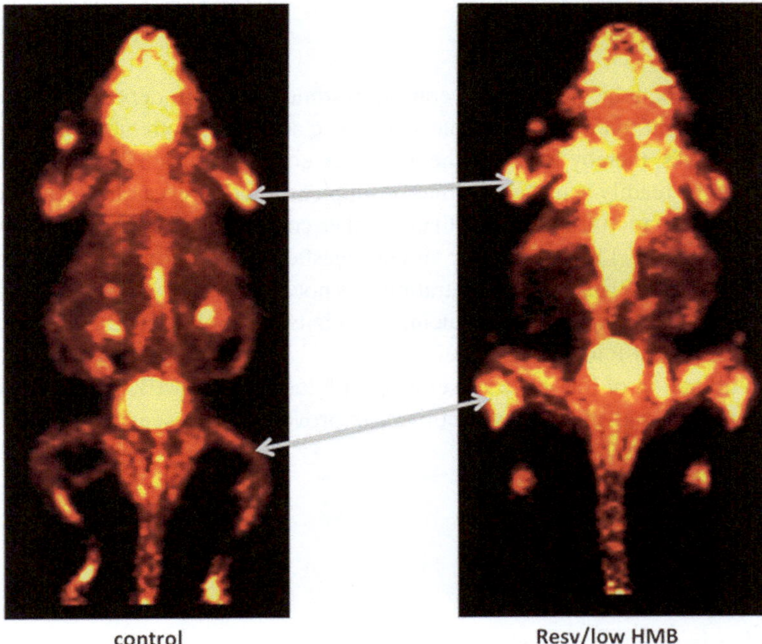

Fig. 8.3 Resveratrol-HMB synergy in low-glucose uptake using FDG-PET. Diet-induced obese mice were fed a high-fat diet (control) or high-fat diet supplemented with a combination of low-dose resveratrol (12.5 mg/kg diet) and HMB (2 g/kg diet) for 6 weeks. At the end of the treatment period, whole body glucose uptake was measured via PET/CT. Glucose was labeled with fluorine-18 (Fluorine-18-deoxy glucose, FDG) and injected into the tail vein of the mouse after an overnight fast. PET/CT images were acquired using an Inveon trimodality (PET/SCPECT/CT platform (Siemens Medical Solution, Knoxville, TN). *Left side*: respresentative image of control diet group. *Right side*: representative image of resveratrol-HMB diet group. *Arrow heads point* to muscle tissue of fore- and hindleg with increased uptake in Resv/HMB group [40]

production was increased while respiratory exchange ratio (RER) was decreased, indicating a whole body shift towards fat oxidation. This was associated with significant reductions in body weight, weight gain and visceral adipose tissue mass.

Effects on Insulin Sensitivity

The resveratrol/leucine combination exerted also synergistic effects on insulin sensitivity. Although the combination did not show any effect on fasting plasma glucose concentrations, fasting plasma insulin levels were significantly decreased in these insulin-resistant, diet-induced obese mice treated for 6 weeks with low resveratrol/leucine. This resulted in a significant decrease in $HOMA_{IR}$ (homeostatic assessment of insulin resistance), a marker of insulin sensitivity. Also skeletal muscle Fluorine-18-deoxy-glucose (^{18}FDG) uptake measured via PET/CT was increased (Fig. 8.3), indicating stimulated muscle glucose uptake and metabolism [40].

Effects on Inflammatory Stress

Low-dose resveratrol combined with leucine or HMB synergistically reduced inflammatory stress in DIO-mice treated for 6 weeks with this combination [40]. The circulating plasma concentrations of IL-6, MCP-1 and CRP were significantly reduced while the individual components did not show any effects.

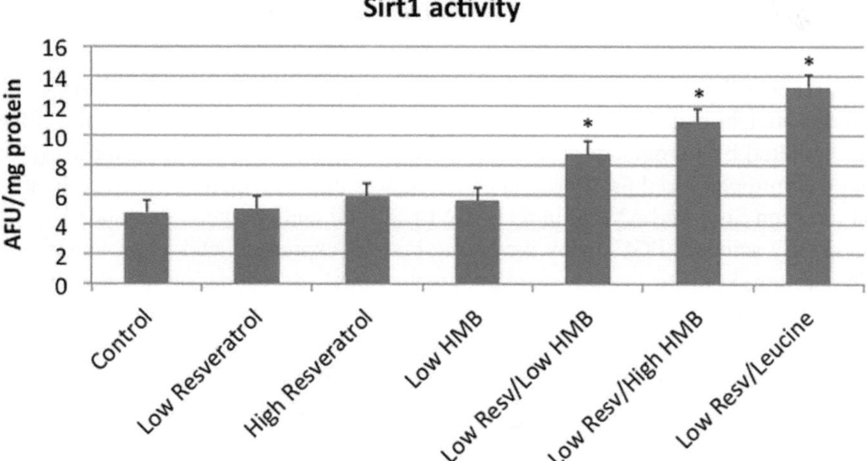

Fig. 8.4 Effects of resveratrol, leucine and HMB on adipose tissue Sirt1 activity in diet-induced obese mice. Diet-induced obese mice were fed a high-fat diet (control) or a high-fat diet supplemented with indicated treatments for 6 weeks. At the end of the treatment period, Sirt1 activity in white adipose tissue lysates was measured. Data are presented as means ± SE (n = 9–10). Stars above the bars indicate significant difference compared to control (p < 0.02) (adapted from [40]). *Treatment*: Low resveratrol: 12.5 mg/kg diet, High resveratrol: 225 mg/kg diet, Low HMB: 2 g/kg diet, High HMB: 10 g/kg diet, Leucine: 24.2 g/kg diet

Although the high-dose resveratrol showed some reduction in these markers, the combination was more effective. Moreover, the anti-inflammatory cytokine adiponectin was increased in response to low-dose resveratrol in combination with either HMB or leucine, while the individual components at these doses exerted no significant effect.

Effects on AMPK/Sirtuin Pathway

The above-mentioned effects were associated with up-regulation of the sirtuin pathway. Both Sirt1 and mitochondrial Sirt3 activity were increased in response to the low-dose resveratrol/leucine combination in mouse adipocytes, while the individual treatments in this concentration did not have any effect. Also Sirt1 activity in white adipose tissue of the DIO-mice was up-regulated in response to 6 week supplementation (Fig. 8.4) [40]. In addition, the resveratrol/leucine or HMB combination produced a significant increase in AMPK activity in mouse adipocytes [41].

Although the exact underlying mechanism of this activation of the Sirtuin pathway is not fully understood, the synergistic effects between resveratrol and leucine/HMB are most likely mediated by allosteric activation of Sirt1 by leucine. As noted earlier in this review, we found leucine and its metabolites β-hydroxy-β-methylbutyrate (HMB) and α-keto-isocaproate (KIC) to activate Sirt1 directly in a cell-free system [37]. Although resveratrol's direct activation of Sirt1 had been linked to the fluorophore used in the activity assay [38], recent evidence demonstrates that hydrophobic amino acids, which are endogenously present, may substitute for the fluorophore [10]. It has also been suggested that resveratrol and other resveratrol-like substances bind on top of the Sirtuin-bound substrate peptide which leads, depending on the fit between substrate and resveratrol-like substance, either to closure of the active catalytic site of the sirtuin and stabilization of the enzyme/substrate resulting in activation or to a non-productive complex resulting in inhibition [42]. Thus, depending on the substrate-peptide the fluorophore is necessary to connect the substrate with resveratrol. Therefore, the highly hydrophobic amino acid leucine may replace the fluorophore and act as an allosteric activator of Sirt1.

Effects of AMPK/Sirtuin Pathway Activation on Energy Metabolism

Sirt1, Sirt3 and AMPK have been demonstrated to modulate glucose and lipid metabolism and to regulate energy metabolism by promoting mitochondrial biogenesis and respiration. Most of these effects are mediated by peroxisome proliferator-activated receptor γ co-activator 1α (PGC-1α), the key regulator of mitochondrial proliferation and metabolism [43] which is regulated by a finely tuned interaction between Sirt1 and AMPK in response to cellular and energetic demands.

AMPK directly activates PGC-1α by increasing both its expression and its protein phosphorylation [44, 45]. In addition, PGC-1α is a downstream target of Sirt1; deacetylation of cytosolic PGC-1α promotes its translocation to the nucleus, where it induces the up-regulation of multiple genes required for mitochondrial biogenesis and respiration [46]. Furthermore, there is a bidirectional interaction of Sirt1 and AMPK; Sirt1 activates AMPK by deacetylation and activation of its upstream kinase LKB1, and AMPK activates Sirt1 by increasing cellular NAD^+ [47, 48]. Moreover, the sirtuin Sirt3 has recently been demonstrated to regulate multiple mitochondrial regulatory proteins as well as proteins involved in TCA cycle, fatty acid oxidation and oxidative phosphorylation [49]. Sirt3, in turn, is activated by PGC-1α as well as indirectly by AMPK due to the increased cellular NAD^+ levels [46]. Thus, this interdependent relationship between AMPK, Sirt1 and Sirt3 leads to overlapping effects on metabolic outcomes. Notably, high-fat diet down-regulates Sirt1 and Sirt3 and up-regulates acetyltransferases leading to an overall hyperacetylation of proteins including PGC-1α [50, 51]. Accordingly, mitochondrial loss and/or dysfunction play a key role in metabolic disorders such as insulin resistance, type II diabetes and cardiovascular disease [52–54]. A summary of these interacting effects is depicted in Fig. 8.5.

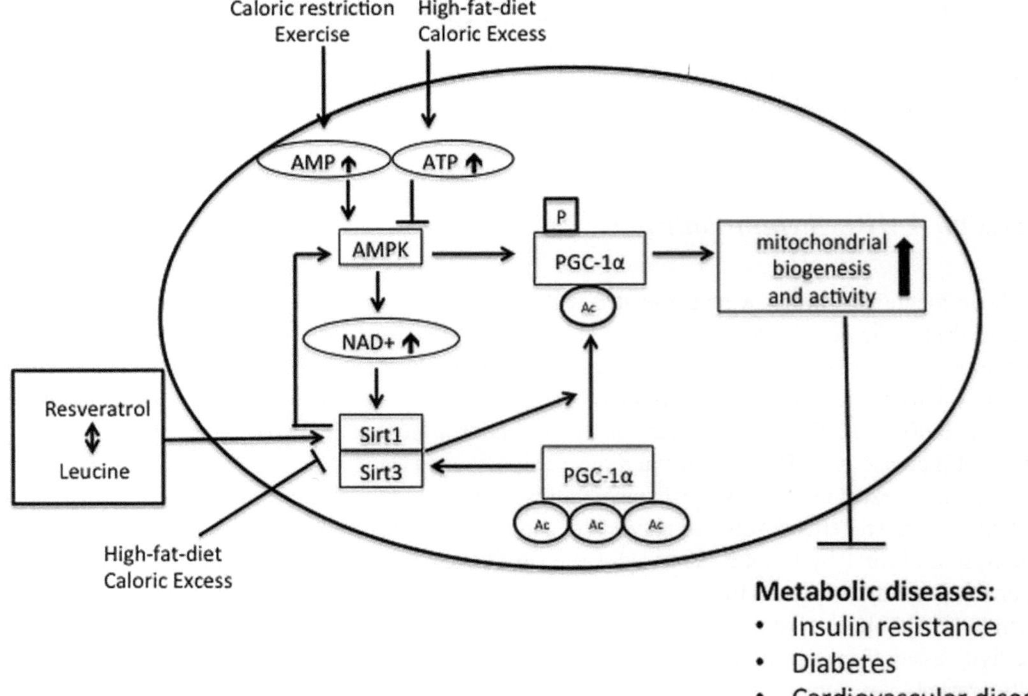

Fig. 8.5 Summary of the synergistic effects of Resveratrol/Leucine combination on AMPK/Sirtuin pathway. AMPK and the sirtuin Sirt1 and Sirt3 act together as nutrient sensors to stimulate PGC-1α-dependent mitochondrial biogenesis and activity, thus exerting beneficial effects in prevention and improvement of metabolic disease. High-fat diet and caloric excess inhibit AMPK and Sirt1/Sirt3 activity leading to hyperacetylation and thereby decreased PGC-1α activity. In contrast, low-dose resveratrol in combination with leucine synergistically activates Sirt1 and Sirt3, counteracting the inhibitory effects of overnutrition

In support of this concept, expression of PGC-1α, as well as other genes involved in mitochondrial biogenesis, were down-regulated in skeletal muscle by high-fat diets in both mice and humans [55]. Further, Sirt1 expression was down-regulated in visceral adipose tissue from morbidly obese patients with severe hepatic steatosis compared to patients with slight or moderate steatosis, indicating that Sirt1 may have a protective role against the development of nonalcoholic fatty liver disease [56]. Sirt1 heterozygous knockout mice (SIRT1$^{+/-}$) exhibit increased body fat content, increased hepatic steatosis and reduced energy expenditure on high-fat diet but not on a low-fat diet [57], and similarly liver-specific Sirt1 knockout mice developed hepatic steatosis and had impaired lipid homoestasis upon high-fat diet feeding [58]. In contrast, modest global overexpression of Sirt1 protected mice against metabolic damage from a high-fat diet by lowering inflammation, improving glucose tolerance and preventing hepatic steatosis [59], and increased Sirt1 expression induced by diet improved insulin and glucose tolerance in mice [60]. Similarly, Sirt3 expression is decreased in muscle from diabetic mice and after chronic high-fat diet, and Sirt3 knockout mice displayed a lower rate of oxygen consumption and impaired insulin sensitivity and mitochondrial function in skeletal muscle [61]. Finally, lower levels of AMPK activity were found in visceral fat of obese insulin-resistant patients and were associated with a higher expression of inflammatory genes compared to insulin sensitive obese controls [62], and global AMPKα1 deficiency in mice leads to higher weight gain, inflammation and insulin resistance under high-fat diet feeding [63]. On the other hand, AMPK stimulation by AICAR treatment restored palmitate-induced insulin resistance in human skeletal myotubes from lean and obese [64] and AMPK stimulation by metformin ameliorated high-fat induced insulin resistance in rats [65].

Effects on Irisin Secretion

The beneficial effects of regular exercise on overall health, and specifically on insulin resistance, are well known. These effects are also associated with changes in AMPK activity, Sirt1 and Sirt3 activity and result in PGC-1α-dependent increases in mitochondrial biogenesis and metabolism to fulfill the higher metabolic demands [66–68]. However, these changes are not only found in skeletal muscle but also in other tissues such as white adipose tissue, resulting in "browning" with increased mitochondrial content and activity [69, 70]. Bostrom et al. [71] recently demonstrated that the myokine irisin mediates some of these systemic effects in response to exercise. Irisin is cleaved and released in the circulation from FNDC5, a type I membrane protein that is expressed in response to exercised-induced PGC-1α stimulation. A 65 % and 100 % increase of irisin in blood was detected after 3 weeks of free wheel running in mice and 10 weeks of endurance training in humans, respectively [71]. Treating primary subcutaneous white adipocytes with FNDC5 protein promoted a sevenfold induction of UCP1 and three other brown fat genes as well as an increase in mitochondrial content and oxygen consumption, consistent with a brown-fat-like phenotype and increased thermogenesis. Similarly, increasing FNDC5 by adenoviral transfection in diet-induced obese mice resulted in higher density of mitochondria in adipocytes, increased mitochondrial gene expression such as UCP1 and Cox7a, increased oxygen consumption as well as improved insulin sensitivity and reduced body weight [71]. Although FNDC5 seems to be mainly expressed in skeletal muscle, it was also recently suggested that white adipose tissue expresses FNDC5 and secretes irisin, and that obese animals over-secret this hormone, suggesting the phenomenon of irisin resistance [72].

Irisin's role in humans is still not fully elucidated; however, current data suggest an important metabolic role and, thus, it may be a potential therapeutic target for metabolic diseases. For example, a positive correlation with muscle mass and body mass index was found in non-diabetic subjects, while irisin was down-regulated after surgically induced weight loss [73, 74]. However, irisin blood levels and FNDC5 expression in muscle and adipose tissue were reduced in obese insulin-resistant patients compared to normal weight controls [74, 75].

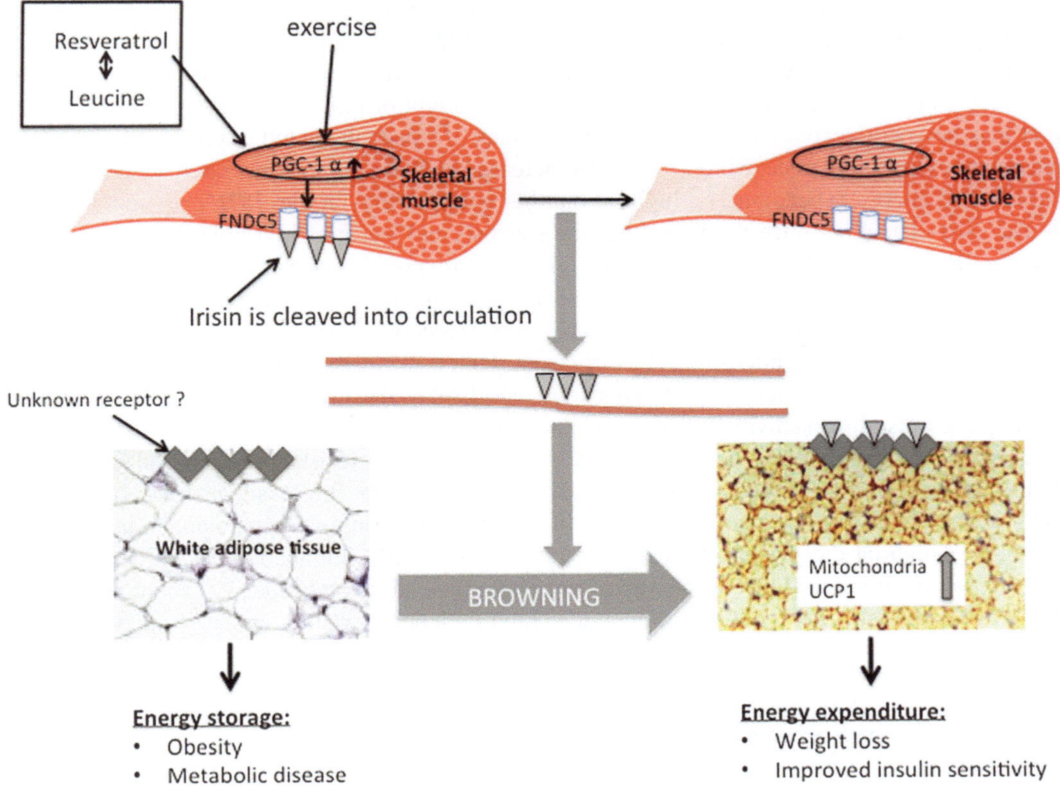

Fig. 8.6 Proposed model of irisin secretion and muscle adipose cross-talk by the Resveratrol/Leucine combination. The resveratrol/leucine combination mimics the effects of exercise by activating PGC-1α-induced upregulation of FNDC5 in skeletal muscle and subsequent cleavage to release irisin into the circulation. Irisin binds to unknown receptors in white adipose tissue, resulting in "browning" of white adipose tissue, characterized by an increase in mitochondrial biogenesis and upregulation of UPC1. This results in an increase in overall energy expenditure and improvements in metabolic disturbances, such as insulin sensitivity

Based on the overlapping effects of the Resveratrol/leucine or HMB combination and exercise on metabolic outcomes in vitro and in vivo, we considered the possibility that this combination could also induce cross-talk between muscle cells and adipose tissue mediated by irisin secretion, as depicted in Fig. 8.6. Indeed, when cultured mouse adipocytes were incubated with conditioned media (CM), which was generated by treating differentiated C2C12 muscle cells for 48 h with resveratrol/leucine or HMB combinations, both AMPK and Sirt1 activity was significantly increased in adipocytes by up to 60 and 100 % compared to the CM-control group, respectively. Also the combination induced a ~2-fold increase in FNDC5 expression in muscle cells and a ~40 % increase in irisin secretion in the incubation media [76]. Similar results were found in vivo; irisin was increased nearly twofold in plasma from diet-induced obese mice fed a diet supplemented with a combination of low-dose resveratrol (12.5 mg/kg diet) and leucine (24.2 g/kg diet) for 6 weeks which was also accompanied by a 344 % increase in UCP1 expression and a 200 % increase in PGC1α expression in white adipose tissue from these mice, suggesting adipose browning. Moreover, a clinical study with a resveratrol/leucine containing supplement showed a ~50 % increase in human plasma irisin levels after 4 weeks of treatment while there was no change in the placebo group [77]. These results are comparable to those reported by Bostrom after exercise in humans and mice.

Conclusions

The AMPK/Sirtuin pathway has been shown to be an important regulator of energy metabolism and therefore is a potential target for prevention and therapy of metabolic diseases such as obesity, insulin resistance and diabetes. The combination of low-dose resveratrol with leucine increases energy metabolism by stimulating lipid oxidation and improving insulin sensitivity in muscle and adipose tissue, as well as reducing obesity-associated inflammatory stress. These effects are mediated by stimulation of Sirt1 and Sirt3 as well as by up-regulation of AMPK. Moreover, this combination mimics, at least in part, the effects of exercise on muscle by stimulation of irisin release and promoting cross-talk between muscle and adipose tissue. These effects are found with low concentrations of the individual compounds, which independently exert no significant effect and are readily achieved via oral administration.

References

1. Michan S, Sinclair D. Sirtuins in mammals: insights into their biological function. Biochem J. 2007;404:1–13.
2. Martinez-Pastor B, Mostoslavsky R. Sirtuins, metabolism, and cancer. Front Pharmacol. 2012;3:22.
3. Houtkooper RH, Pirinen E, Auwerx J. Sirtuins as regulators of metabolism and healthspan. Nat Rev Mol Cell Biol. 2012;13:225–38.
4. Timmer S, Auwerx J, Schrauwen P. The journey of resveratrol from yeast to human. Aging (Albany NY). 2012; 4:146–58.
5. Pearson KJ, Baur JA, Lewis KN, et al. Resveratrol delays age-related deterioration and mimics transcriptional aspects of dietary restriction without extending life span. Cell Metab. 2008;8:157–68.
6. Park SJ, Ahmad F, Philp A, et al. Resveratrol ameliorates aging-related metabolic phenotypes by inhibiting cAMP phosphodiesterases. Cell. 2012;148:421–33.
7. Price NL, Gomes AP, Ling AJ, et al. SIRT1 is required for AMPK activation and the beneficial effects of resveratrol on mitochondrial function. Cell Metab. 2012;15:675–90.
8. Borra MT, Smith BC, Denu JM. Mechanism of human SIRT1 activation by resveratrol. J Biol Chem. 2005; 280:17187–95.
9. Pacholec M, Bleasdale JE, Chrunyk B, et al. SRT1720, SRT2183, SRT1460, and resveratrol are not direct activators of SIRT1. J Biol Chem. 2010;285:8340–51.
10. Hubbard BP, Gomes AP, Dai H, et al. Evidence for a common mechanism of SIRT1 regulation by allosteric activators. Science. 2013;339:1216–9.
11. Mukherjee S, Dudley JI, Das DK. Dose-dependency of resveratrol in providing health benefits. Dose Response. 2010;8:478–500.
12. Calabrese EJ, Mattson MP, Calabrese V. Resveratrol commonly displays hormesis: occurrence and biomedical significance. Hum Exp Toxicol. 2010;29:980–1015.
13. Lagouge M, Argmann C, Gerhart-Hines Z, et al. Resveratrol improves mitochondrial function and protects against metabolic disease by activating SIRT1 and PGC-1alpha. Cell. 2006;127:1109–22.
14. Cho SJ, Jung UJ, Choi MS. Differential effects of low-dose resveratrol on adiposity and hepatic steatosis in diet-induced obese mice. Br J Nutr. 2012;1–10.
15. Boocock DJ, Faust GE, Patel KR, et al. Phase I dose escalation pharmacokinetic study in healthy volunteers of resveratrol, a potential cancer chemopreventive agent. Cancer Epidemiol Biomarkers Prev. 2007;16:1246–52.
16. Goldberg DM, Yan J, Soleas GJ. Absorption of three wine-related polyphenols in three different matrices by healthy subjects. Clin Biochem. 2003;36:79–87.
17. Smoliga JM, Baur JA, Hausenblas HA. Resveratrol and health–a comprehensive review of human clinical trials. Mol Nutr Food Res. 2011;55:1129–41.
18. Stipanuk MH. Leucine and protein synthesis: mTOR and beyond. Nutr Rev. 2007;65:122–9.
19. Zanchi NE, Nicastro H, Lancha Jr AH. Potential antiproteolytic effects of L-leucine: observations of in vitro and in vivo studies. Nutr Metab (Lond). 2008;5:20.
20. Donato Jr J, Pedrosa RG, Cruzat VF, et al. Effects of leucine supplementation on the body composition and protein status of rats submitted to food restriction. Nutrition. 2006;22:520–7.
21. Zhang Y, Guo K, LeBlanc RE, et al. Increasing dietary leucine intake reduces diet-induced obesity and improves glucose and cholesterol metabolism in mice via multimechanisms. Diabetes. 2007;56:1647–54.

22. Wang TJ, Larson MG, Vasan RS, et al. Metabolite profiles and the risk of developing diabetes. Nat Med. 2011; 17:448–53.
23. McCormack SE, Shaham O, McCarthy MA, et al. Circulating branched-chain amino acid concentrations are associated with obesity and future insulin resistance in children and adolescents. Pediatr Obes. 2013;8:52–61.
24. Mochida T, Tanaka T, Shiraki Y, et al. Time-dependent changes in the plasma amino acid concentration in diabetes mellitus. Mol Genet Metab. 2011;103:406–9.
25. Newgard CB, An J, Bain JR, et al. A branched-chain amino acid-related metabolic signature that differentiates obese and lean humans and contributes to insulin resistance. Cell Metab. 2009;9:311–26.
26. Melnik BC. Leucine signaling in the pathogenesis of type 2 diabetes and obesity. World J Diabetes. 2012; 3:38–53.
27. She P, Van Horn C, Reid T, et al. Obesity-related elevations in plasma leucine are associated with alterations in enzymes involved in branched-chain amino acid metabolism. Am J Physiol Endocrinol Metab. 2007;293:E1552–63.
28. Kuzuya T, Katano Y, Nakano I, et al. Regulation of branched-chain amino acid catabolism in rat models for spontaneous type 2 diabetes mellitus. Biochem Biophys Res Commun. 2008;373:94–8.
29. Bajotto G, Murakami T, Nagasaki M, et al. Decreased enzyme activity and contents of hepatic branched-chain alpha-keto acid dehydrogenase complex subunits in a rat model for type 2 diabetes mellitus. Metabolism. 2009;58:1489–95.
30. Wang B, Kammer LM, Ding Z, et al. Amino acid mixture acutely improves the glucose tolerance of healthy overweight adults. Nutr Res. 2012;32:30–8.
31. Eller LK, Saha DC, Shearer J, et al. Dietary leucine improves whole-body insulin sensitivity independent of body fat in diet-induced obese Sprague-Dawley rats. J Nutr Biochem. 2013;24(7):1285–94.
32. Tang M, Armstrong CL, Leidy HJ, et al. Normal vs. high-protein weight loss diets in men: Effects on body composition and indices of metabolic syndrome. Obesity (Silver Spring). 2013;21:E204–10.
33. Wycherley TP, Moran LJ, Clifton PM, et al. Effects of energy-restricted high-protein, low-fat compared with standard-protein, low-fat diets: a meta-analysis of randomized controlled trials. Am J Clin Nutr. 2012;96:1281–98.
34. Freudenberg A, Petzke KJ, Klaus S. Comparison of high-protein diets and leucine supplementation in the prevention of metabolic syndrome and related disorders in mice. J Nutr Biochem. 2012;23:1524–30.
35. Sun X, Zemel M. Leucine and calcium regulate fat metabolism and energy partitioning in murine adipocytes and muscle cells. Lipids. 2007;42:297–305.
36. Sun X, Zemel MB. Leucine modulation of mitochondrial mass and oxygen consumption in skeletal muscle cells and adipocytes. Nutr Metab (Lond). 2009;6:26.
37. Bruckbauer A, Zemel MB. Effects of dairy consumption on SIRT1 and mitochondrial biogenesis in adipocytes and muscle cells. Nutr Metab (Lond). 2011;8:91.
38. Kaeberlein M, McDonagh T, Heltweg B, et al. Substrate-specific activation of sirtuins by resveratrol. J Biol Chem. 2005;280:17038–45.
39. Almeida L, Vaz-da-Silva M, Falcao A, et al. Pharmacokinetic and safety profile of trans-resveratrol in a rising multiple-dose study in healthy volunteers. Mol Nutr Food Res. 2009;53 Suppl 1:S7–15.
40. Bruckbauer A, Zemel MB, Thorpe T, et al. Synergistic effects of leucine and resveratrol on insulin sensitivity and fat metabolism in adipocytes and mice. Nutr Metab (Lond). 2012;9:77.
41. Bruckbauer A, Baggett B, Zemel MB. Synergistic effects of polyphenols and β-hydroxy- β-butyrate (HMB) on energy metabolism. FASEB J. 2013;27:637.23.
42. Gertz M, Nguyen GT, Fischer F, et al. A molecular mechanism for direct sirtuin activation by resveratrol. PLoS One. 2012;7:e49761.
43. Lin J, Wu H, Tarr PT, et al. Transcriptional co-activator PGC-1 alpha drives the formation of slow-twitch muscle fibres. Nature. 2002;418:797–801.
44. Jager S, Handschin C, St-Pierre J, et al. AMP-activated protein kinase (AMPK) action in skeletal muscle via direct phosphorylation of PGC-1alpha. Proc Natl Acad Sci U S A. 2007;104:12017–22.
45. Irrcher I, Ljubicic V, Kirwan AF, et al. AMP-activated protein kinase-regulated activation of the PGC-1alpha promoter in skeletal muscle cells. PLoS One. 2008;3:e3614.
46. Brenmoehl J, Hoeflich A. Dual control of mitochondrial biogenesis by sirtuin 1 and sirtuin 3. Mitochondrion. 2013;13(6):755–61.
47. Lan F, Cacicedo JM, Ruderman N, et al. SIRT1 modulation of the acetylation status, cytosolic localization, and activity of LKB1. Possible role in AMP-activated protein kinase activation. J Biol Chem. 2008;283:27628–35.
48. Canto C, Gerhart-Hines Z, Feige JN, et al. AMPK regulates energy expenditure by modulating NAD+metabolism and SIRT1 activity. Nature. 2009;458:1056–60.
49. Rardin MJ, Newman JC, Held JM, et al. Label-free quantitative proteomics of the lysine acetylome in mitochondria identifies substrates of SIRT3 in metabolic pathways. Proc Natl Acad Sci U S A. 2013;110:6601–6.
50. Green MF, Hirschey MD. SIRT3 weighs heavily in the metabolic balance: a new role for SIRT3 in metabolic syndrome. J Gerontol A Biol Sci Med Sci. 2013;68:105–7.

51. Hirschey MD, Shimazu T, Jing E, et al. SIRT3 deficiency and mitochondrial protein hyperacetylation accelerate the development of the metabolic syndrome. Mol Cell. 2011;44:177–90.
52. Kim JA, Wei Y, Sowers JR. Role of mitochondrial dysfunction in insulin resistance. Circ Res. 2008;102:401–14.
53. Schrauwen-Hinderling VB, Roden M, Kooi ME, et al. Muscular mitochondrial dysfunction and type 2 diabetes mellitus. Curr Opin Clin Nutr Metab Care. 2007;10:698–703.
54. Ren J, Pulakat L, Whaley-Connell A, et al. Mitochondrial biogenesis in the metabolic syndrome and cardiovascular disease. J Mol Med. 2010;88:993–1001.
55. Sparks LM, Xie H, Koza RA, et al. A high-fat diet coordinately downregulates genes required for mitochondrial oxidative phosphorylation in skeletal muscle. Diabetes. 2005;54:1926–33.
56. Costa Cdos S, Hammes TO, Rohden F, et al. SIRT1 transcription is decreased in visceral adipose tissue of morbidly obese patients with severe hepatic steatosis. Obes Surg. 2010;20:633–9.
57. Xu F, Gao Z, Zhang J, et al. Lack of SIRT1 (Mammalian Sirtuin 1) activity leads to liver steatosis in the SIRT1+/- mice: a role of lipid mobilization and inflammation. Endocrinology. 2010;151:2504–14.
58. Purushotham A, Schug TT, Xu Q, et al. Hepatocyte-specific deletion of SIRT1 alters fatty acid metabolism and results in hepatic steatosis and inflammation. Cell Metab. 2009;9:327–38.
59. Pfluger PT, Herranz D, Velasco-Miguel S, et al. Sirt1 protects against high-fat diet-induced metabolic damage. Proc Natl Acad Sci U S A. 2008;105:9793–8.
60. Chen S, Li J, Zhang Z, et al. Effects of resveratrol on the amelioration of insulin resistance in KKAy mice. Can J Physiol Pharmacol. 2012;90:237–42.
61. Jing E, Emanuelli B, Hirschey MD, et al. Sirtuin-3 (Sirt3) regulates skeletal muscle metabolism and insulin signaling via altered mitochondrial oxidation and reactive oxygen species production. Proc Natl Acad Sci U S A. 2011; 108:14608–13.
62. Gauthier MS, O'Brien EL, Bigornia S, et al. Decreased AMP-activated protein kinase activity is associated with increased inflammation in visceral adipose tissue and with whole-body insulin resistance in morbidly obese humans. Biochem Biophys Res Commun. 2011;404:382–7.
63. Zhang W, Zhang X, Wang H, et al. AMP-activated protein kinase alpha1 protects against diet-induced insulin resistance and obesity. Diabetes. 2012;61:3114–25.
64. Bikman BT, Zheng D, Reed MA, et al. Lipid-induced insulin resistance is prevented in lean and obese myotubes by AICAR treatment. Am J Physiol Regul Integr Comp Physiol. 2010;298:R1692–9.
65. Liu Y, Wan Q, Guan Q, et al. High-fat diet feeding impairs both the expression and activity of AMPKa in rats' skeletal muscle. Biochem Biophys Res Commun. 2006;339:701–7.
66. Little JP, Safdar A, Bishop D, et al. An acute bout of high-intensity interval training increases the nuclear abundance of PGC-1alpha and activates mitochondrial biogenesis in human skeletal muscle. Am J Physiol Regul Integr Comp Physiol. 2011;300:R1303–10.
67. Hood MS, Little JP, Tarnopolsky MA, et al. Low-volume interval training improves muscle oxidative capacity in sedentary adults. Med Sci Sports Exerc. 2011;43:1849–56.
68. Pucci B, Villanova L, Sansone L, et al. Sirtuins: the molecular basis of beneficial effects of physical activity. Intern Emerg Med. 2013;8 Suppl 1:S23–5.
69. Little JP, Safdar A, Benton CR, et al. Skeletal muscle and beyond: the role of exercise as a mediator of systemic mitochondrial biogenesis. Appl Physiol Nutr Metab. 2011;36:598–607.
70. Stallknecht B, Vinten J, Ploug T, et al. Increased activities of mitochondrial enzymes in white adipose tissue in trained rats. Am J Physiol. 1991;261:E410–4.
71. Bostrom P, Wu J, Jedrychowski MP, et al. A PGC1-alpha-dependent myokine that drives brown-fat-like development of white fat and thermogenesis. Nature. 2012;481:463–8.
72. Roca-Rivada A, Castelao C, Senin LL, et al. FNDC5/irisin is not only a myokine but also an adipokine. PLoS One. 2013;8:e60563.
73. Huh JY, Panagiotou G, Mougios V, et al. FNDC5 and irisin in humans: I. Predictors of circulating concentrations in serum and plasma and II mRNA expression and circulating concentrations in response to weight loss and exercise. Metabolism. 2012;61:1725–38.
74. Liu JJ, Wong MD, Toy WC, et al. Lower circulating irisin is associated with type 2 diabetes mellitus. J Diabetes Complications. 2013;27(4):365–9.
75. Moreno-Navarrete JM, Ortega F, Serrano M, et al. Irisin is expressed and produced by human muscle and adipose tissue in association with obesity and insulin resistance. J Clin Endocrinol Metab. 2013;98:E769–78.
76. Bruckbauer A, Baggett B, Zemel MB. Synergistic effects of leucine and its metabolites with polyphenols on irisin in myotubes and diet-induced obese mice [abstract]. FASEB J. 2013;27:637.11.
77. Lipps J, Hagan S, Clark M, et al. Resveratrol synergy in pre-diabetes [Abstract 78-LB]. Diabetes. 2013;62:LB22.
78. Zanchi NE, Gerlinger-Romero F, Guimaraes-Ferreira L, et al. HMB supplementation: clinical and athletic performance-related effects and mechanisms of action. Amino Acids. 2011;40:1015–25.

Chapter 9
Branched Chain Amino Acids and Blood Ammonia

Gitte Dam, Peter Ott, Niels Kristian Aagaard, Lise Lotte Gluud, and Hendrik Vilstrup

Key Points

- The pathogenesis of hepatic encephalopathy is related to increased blood ammonia concentration.
- During hyperammonemia muscle tissue becomes an alternative organ for ammonia detoxification.
- BCAA is believed to supply carbon skeletons for the TCA cycle and enhance muscle ammonia detoxification.
- BCAA supplementation increases skeletal muscle ammonia removal by accelerated glutamine formation.
- Acute ingestion of BCAA increases the arterial ammonia concentration.

Keywords Ammonia • Branched chain amino acids • Muscle metabolism • Tricarboxylic acid cycle • Cirrhosis • Hepatic encephalopathy

Abbreviations

HE	Hepatic encephalopathy
GS	Glutamine synthetase
TCA cycle	Tricarboxylic acid cycle
BCAA	Branched chain amino acids
BCKA	Branched chain keto acids
BCKDH	Branched chain α-keto acid dehydrogenase

G. Dam, M.D., Ph.D. (✉) • P. Ott, M.D., D.M.Sc. • N.K. Aagaard, M.D., Ph.D. • H. Vilstrup, M.D., D.M.Sc.
Department of Hepatology and Gastroenterology, Aarhus University Hospital, Aarhus, Denmark
e-mail: gitte_dam@hotmail.com; gitdam@rm.dk; peterott@rm.dk;
nielaaga@rm.dk; hendrik.vilstrup@aarhus.rm.dk

L.L. Gluud, M.D., D.M.Sc.
Department of Medicine, Copenhagen University Hospital Gentofte, Gentofte, Denmark
e-mail: liselottegluud@yahoo.dk

R. Rajendram et al. (eds.), *Branched Chain Amino Acids in Clinical Nutrition: Volume 1*, Nutrition and Health, DOI 10.1007/978-1-4939-1923-9_9,

Introduction

Virtually all organs are involved in the metabolism of ammonia and arterial ammonia levels are determined by an interaction and balance between ammonia-producing and ammonia-removing organs [1]. In health, ammonia transport and metabolism are regulated to maintain low plasma concentrations (normal venous range 10–50 μmol/L). Ammonia is toxic to the organism and the body therefore depends on metabolism and removal of ammonia. The urea cycle plays a crucial role, as urea is the main end-product of nitrogen metabolism and is definitively removed from the body by renal excretion. Alternatively, ammonia can temporarily be incorporated into different amino acids. Glutamine plays a central role in this nitrogen traffic by serving as a non-toxic transporter of useful nitrogen [2]. Glutamine supplies nitrogen from glutamine-producing organs (muscle and kidney) to organs capable of consuming glutamine (liver, kidney and gut). Different enzymes in these organs play key roles in the inter-organ ammonia metabolism. Glutamine synthetase (GS) catalyzes the condensation of ammonia and glutamate in the cytosol to form glutamine [3], whereas glutaminase has the opposite action in the mitochondria. Glutamine and glutamate in the organs therefore serve as nitrogen donors and acceptors, respectively.

Under normal conditions, detoxification of ammonia predominantly takes place in the liver, whereas the major ammonia producing organs are the gut and the kidney. The healthy kidney is capable of producing ammonia which can be either released into the systemic circulation or excreted into the urine. This process depends on the acid–base state of the body rather than on the need for ammonia elimination. The brain contributes insignificantly in the healthy state but can turn into a net ammonia removing organ during hyperammonemia. Data on lung, heart and skin ammonia uptake is scarce and inconclusive. Lung tissue contains GS, ammonia has been detected in exhaled gas from the lungs and lung ammonia uptake and glutamine release has been demonstrated in pigs with acute liver failure [4].

In healthy subjects, muscle tissue does not contribute significantly to ammonia removal [5]. However, ammonia homeostasis is profoundly altered in hyperammonemic situations where muscle tissue becomes the main alternative organ, besides liver, for at least temporary detoxification of ammonia [6–9].

Branched chain amino acids (BCAA; Isoleucine, leucine and valine) have attracted particular interest as they are believed to support this muscle ammonia detoxification and they have, therefore, been utilized as a therapeutic nutritional supplement in patients with hyperammonemia. Elevation of blood ammonia concentration occurs in a variety of situations. Physiologically, ammonia is increased during strenuous exercise. Pathophysiologically ammonia is increased in hepatocellular dysfunction, development of porto-systemic shunt circulation and in urea cycle disorders. Liver disease represents the field in which the BCAA have been most extensively studied. Their potential ammonia lowering effect has been intensely investigated in hepatic encephalopathy (HE). The pathogenesis of HE is believed to be related to increased blood ammonia concentration.

Organ Contribution to Ammonia Metabolism

Liver

The role of the liver in maintaining normal ammonia metabolism is crucial and urea synthesis is a unique liver function that is essential to the health and survival of the individual. During normal physiological conditions ammonia detoxification in the liver takes place at two anatomically aligned sites. The urea-cycle operates in the peri-portally located hepatocytes where ammonia and bicarbonate are condensed in the mitochondria to yield carbamoylphosphate, initiating and feeding the urea-cycle [10]. Urea-synthesis is quantitatively by far the major mechanism of ammonia removal in mammals and is the only ammonia removal mechanism that is regulated by the metabolic excess of nitrogen. Any ammonia escaping detoxification in the urea-cycle will be trapped downstream in the last row of the perivenous hepatocytes that are the only ones that express GS and convert ammonia and glutamate into glutamine [11]. In healthy subjects, the liver has a huge capacity for increasing urea synthesis after a large protein meal, where ammonia and amino acids are released from the gut into the portal blood. The anatomical interposition of the liver between the intestines and the systemic blood thus makes the liver able to scavenge all ammonia produced in the gut. In patients with liver cirrhosis, the hepatic clearance of ammonia is limited by diminished urea synthesis as part of the decreased functional liver mass [12, 13], loss of normal perisinusoidal expression of GS [14] and by porto-systemic shunting [15]. Consequently, ammonia reaches the systemic circulation, the arterial ammonia concentration increases and ammonia will exert unusual and potentially dangerous effects on the organs of the body.

Gut

The gut is considered to be the main source of systemic ammonia production. About 50 % of intestinal ammonia arises directly from metabolism of dietary protein after feeding. In healthy subjects, ingestion of amino acids does not induce hyperammonemia as the ammonia is cleared by the liver. However in patients with cirrhosis the liver fails to clear the ammonia load after a meal and the arterial ammonia concentration increases [16]. Ingestion of different proteins has a diverse potency to induce hyperammonemia. The potency relies on the amino acid composition of food proteins, which depends on the provenience of the proteins. Meat protein is more ammoniagenic than diary and vegetable protein [16]. Still, the increased splanchnic contribution to hyperammonemia in cirrhosis seems to be primarily a result of intra–and extrahepatic portacaval shunting rather than altered intestinal ammonia production [17].

Enterocytes express high glutaminase and minimal GS activities and utilize glutamine as their main energy source. Glutamine promotes growth of intestinal mucosa; large amounts of glutamine is taken up from either the intestinal lumen or from the blood stream and is converted by glutaminase to ammonia and glutamate which is released into the mesenteric vein [18]. In accordance, the amount of ammonia released from the intestines mirrors the glutamine extraction by the small and large intestines in humans. Likewise, the intestinal ammonia production in stable cirrhosis patients is related to the glutamine uptake [17].

Finally, the colonic luminal bacterial flora metabolize urea and amino acids and thereby releases ammonia [19]. About one-third of all urea produced is hydrolyzed by gut bacterial ureases, giving rise to a large hepatic re-uptake of ammonia and re-synthesis of urea. The amount of urea produced is considerably larger than the amount excreted into the urine. Patient with cirrhosis frequently have bacterial overgrowth and are reported to hydrolyze a larger fraction of their urea synthesis, resulting in a larger ammonia load to the liver and the systemic circulation [20].

Brain

In brain tissue, there is intense activity of both GS and glutaminase, with GS present only in astrocytes and glutaminase also in neurons [21]. Therefore the brain predominantly detoxifies ammonia by production of glutamine. Data are variable, but most studies indicate that ammonia uptake is negligible in healthy subjects at rest. It has recently been shown that brain ammonia uptake depends linearly on the arterial concentration of ammonia [22] resulting in a net uptake of ammonia into the brain during liver failure [22, 23]. Accordingly, liver transplantation immediately stops the cerebral ammonia uptake in fulminant hepatic failure [24]. The accumulation of glutamine in astrocytes during hyperammonemia and the limitation of glutamine transport out of the cells seem to lead to osmotic stress and cellular overhydration. This event has been associated with the development of HE [25].

Kidney

The kidneys contain both glutaminase and GS [26] and the main overall substrate for kidney ammonia production is glutamine [27]. In the fasting state, alanine may, however, be the main ammonia precursor, also in patients with cirrhosis [28, 29].

In healthy individuals, the ammonia produced by the kidney can be released into the systemic circulation or excreted into the urine. During normal steady state, the ratio is about one-third into the urine and two-thirds into the blood stream [27]. The urinary ammonia excretion is a means of excreting protons and generating bicarbonate in acidosis [27]. This means that the renal ammonia excretion is primarily regulated by the acid–base status, metabolic acidosis increasing the ammonia excretion and alkalosis reducing it. The lifesaving adaptations allow the kidney to remove large excess of H^+. However, the previous paradigms of passive, lipid-phase NH_3 diffusion and NH_4^+ trapping in the acidic urine have now been replaced by more complicated models in which transporter-mediated movement of NH_3 and NH_4^+ are fundamental components of renal ammonia physiology [30].

Artificial hyperammonemia can push the balance in favor of urinary ammonia excretion [31], and under such circumstances the kidneys may in fact even turn into an organ of net ammonia removal from the body [28]. This renal potential may be overridden by severe hyperammonemia. In experimental acute ischemic liver failure, no more ammonia appeared in the urine than after establishment of a portacaval shunt which lead to only moderate hyperammonemia [28]. It is not known whether this is due to incipient multiorgan failure or differences in regulation.

Muscle

Muscle tissue is the largest organ in the body and in spite of its low GS activity, muscle plays a substantial role in whole-body ammonia metabolism [32]. Originally Lockwood, by a ^{13}N-ammonia PET method, found 50 % of the arterial ammonia to be metabolized by skeletal muscle [33]. In contrast, more recent studies found no muscle uptake in healthy subjects. In one such study, there was no significant ammonia uptake across the leg in healthy subjects, as measured simultaneously by the ^{13}N-ammonia PET method and directly by arterio-venous concentration differences and flow measurements [5]. There was, however, a significant muscle ammonia uptake in patients with liver disease (Fig. 9.1) [5].

Hence, when the liver fails, muscle tissue may become an important alternative pathway for ammonia detoxification.

Metabolism of BCAA

Under normal conditions, a protein meal is followed by a shift from hepato-splanchnic amino acid uptake to amino acid release. The BCAA are unique as they escape splanchnic uptake by the gut and the liver and are therefore among the predominant amino acids released to the periphery after feeding

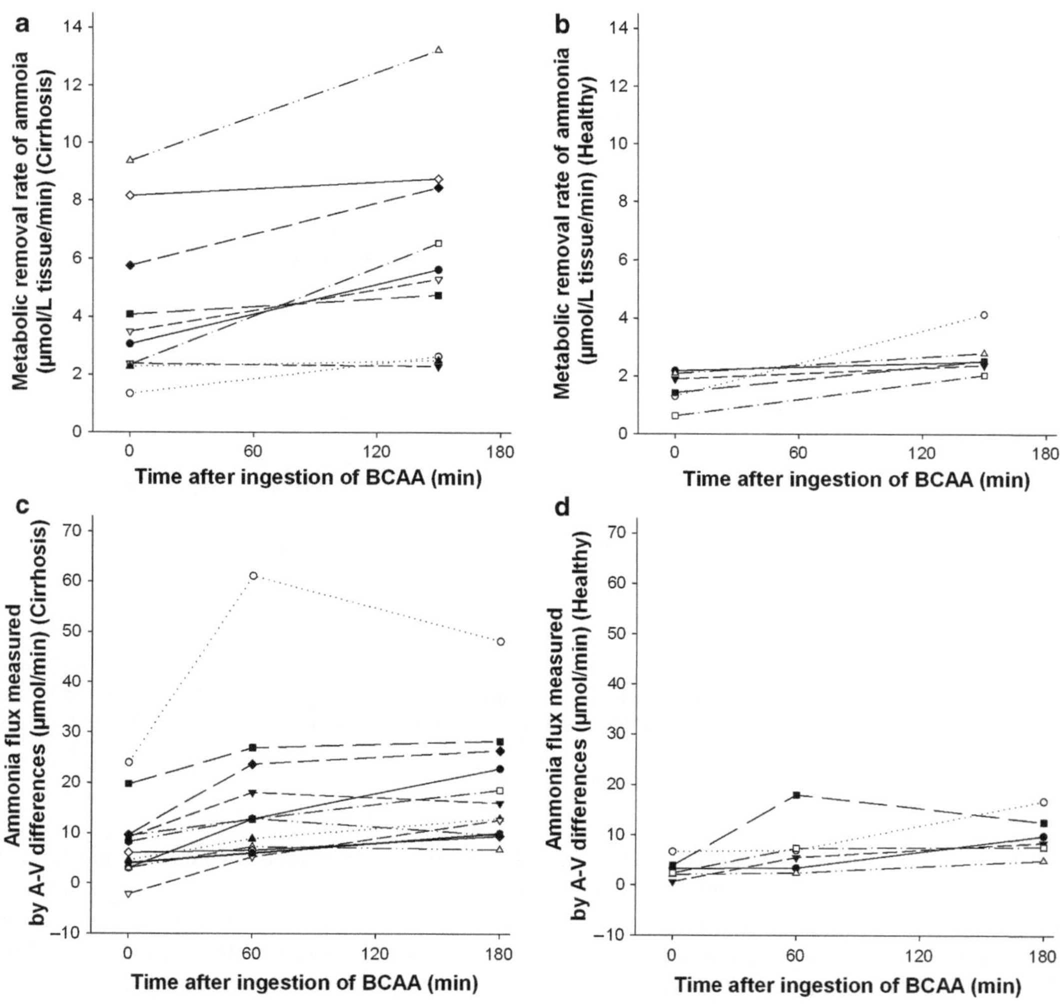

Fig. 9.1 Muscle ammonia uptake at baseline and after ingestion of BCAA. Metabolic removal rate of ammonia measured by ^{13}N-ammonia PET; (**a**) patients with cirrhosis ($n=10$) and (**b**) healthy subjects ($n=6$). Ammonia flux (uptake) measured by A–V differences; (**c**) patients with cirrhosis ($n=12$) and (**d**) healthy subjects ($n=6$). The figure was reprinted with permission [5]

[34, 35]. BCAA are primarily utilized by skeletal muscle under insulin control to cover about 10 % of the energy requirements [36].

The catabolic pathway for BCAA is located in the mitochondria. The first two steps are common to all three BCAAs. They comprise first the transamination of BCAA to branched chain ketoacids (BCKA), which is catalyzed by branched chain aminotransferase [37]. The second reaction is the irreversible oxidative decarboxylation of BCKA to form Acyl-CoA derivatives, catalyzed by the branched chain α-keto acid dehydrogenase (BCKDH) complex. These reactants are converted into either Acetyl-CoA or succinyl-CoA that can enter the tricarboxylic acid (TCA) cycle [37].

Oxidation of BCAA, besides serving as energy source, also yields precursors for other amino acids and play important roles in promoting protein synthesis by activating (phosphorylating) the regulatory mechanisms of mRNA translation [38, 39]. When BCAA are energized, the BCKA provide carbon skeletons for the TCA cycle. BCAA hereby give rise to the two TCA cycle intermediates, Acetyl-CoA and Succinyl-CoA. The detoxification of high ammonia concentration via glutamine synthesis depletes the glutamate pools, and the ensuing increased synthesis of glutamate secondarily deprives

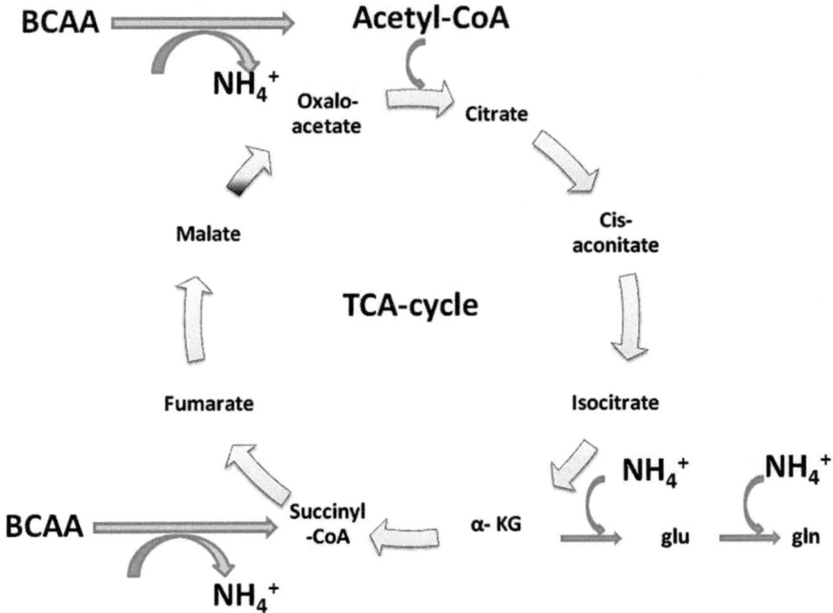

Fig. 9.2 Schematic representation of the TCA-cycle. BCAA supply the TCA-cycle with carbon skeletons and thereby enhance the production of alfa-ketoglutarate (α-KG). α-KG can be reductively aminated to glutamate (Glu) and glutamate may be amidated to glutamine (Gln). The figure was reprinted with permission from "Springer" (Metab Brain Dis 2013. PMID: 23315357)

the TCA cycle of carbon skeletons (Fig. 9.2). In that situation, metabolism of BCAA theoretically can replenish carbon skeletons for cycle reactants to keep the TCA cycle running and produce energy. If so, the metabolism of BCAA would reconstitute alfa-ketoglutarate (anaplerosis) and maintain the glutamate pool, allowing the subsequent amidation with free ammonia to form glutamine. Thus, BCAA might, at least transiently, reduce the ammonia level, because formation of glutamine from alfa-keto-glutarate removes two nitrogen atoms while deamination of BCAA produces only one. Consequently, there will be a net fixation of one ammonia molecule for each BCAA metabolized.

Experimental data support this line of thought. BCAA metabolism in muscle is associated with metabolism of ammonia [40]. Brain metabolism of BCAA with relation to detoxification of ammonia is described in detail in the chapter written by Arne Schousboe. Studies investigating amino acid fluxes across the kidney are scarce but kidney BCAA metabolism will be discussed briefly followed by a review of the studies dealing with muscle ammonia and BCAA metabolism and its relation to HE.

Renal BCAA Metabolism

In kidney there is a low BCKDH complex activity which shows that little BCKA is used to form Acyl-CoA derivatives in the renal parenchyma. The activity of BCAA-transferase is high and BCAAs are hence mainly involved in transamination processes. This results in production of ammonia. BCAA are also to some degree involved in ammoniagenesis during metabolic acidosis. It has been shown that acidosis stimulates BCAA oxidation in renal tubule cells by increasing both the amount and activation state of BCKDH-complex [41].

In dogs, the kidney significantly takes up BCAA after an amino acid meal [42]. In normal man and patients with chronic liver disease in the post-absorptive phase, the kidneys releases leucine, while there is no net exchange of valine or isoleucine [43]. The effects of an intragastric dose of amino acids mimicking the composition of hemoglobin, was given with or without intragastric administration of isoleucine [44]. There was a substantial increase in renal ammoniagenesis which was not ameliorated by co-administration of isoleucine. This is the only available study on the effects of one or all of the BCAAs on renal ammonia metabolism. The study suggests that the kidney, under certain conditions, may be an important ammonia producing organ. It does, however, not indicate that the ammonia production is reduced by isoleucine.

From these scarce data it can only be concluded that the kidneys take up BCAA and produce ammonia in the absorptive phase and release leucine during the fasted state. However the available studies do not allow speculations on a possible effect of BCAA on renal ammonia metabolism.

Muscle BCAA and Ammonia Metabolism with a Special Focus on Hepatic Encephalopathy

HE is a reversible neuropsychiatric disorder associated with acute or chronic liver failure [45]. Patients with HE virtually always have higher blood ammonia levels than normal and usually higher than cirrhosis patients without HE [22]. Ammonia is believed to play a key role as 'the first hit' in the pathogenesis of HE and treatment and prevention strategies are aimed at lowering blood ammonia.

The treatment with BCAA is one such intervention. Pavlov and his colleagues in Saint Petersburg were the first to discover that intake of different proteins had effects on the confusional state of porto-caval shunted dogs. In 1893 they saw that dogs that were fed with meat developed ataxia and coma. They called it the "meat intoxication syndrome". The derangement was reversed in dogs shifted to a milk and bread diet [46]. Dairy proteins have a higher percentage of BCAA which later give rise to the idea that BCAA may improve HE.

More than 60 years later it was shown that patients with liver cirrhosis have an altered amino acid plasma pool, including low BCAA levels [47]. The concentrations of most other amino acids are elevated, particularly of glutamine and also the aromatic amino acids (AAA; phenylalanine, tyrosin and tryptophan). These observations were confirmed in several publications and were summarized as the ratio between BCAA and AAA, the so-called Fischers's ratio. A low ratio was shown to correlate with a high grade of HE [48, 49]. The mechanism of the decreased BCAA levels in cirrhosis has been debated for decades. Hyperinsulinemia and accelerated starvation are both suggested to have a modifying influence [50, 51] and most authors agree that skeletal muscle metabolism is important for the phenomenon. The co-occurrence of high ammonia suggests that ammonia in some way may induce muscle BCAA catabolism and lead to the low plasma concentration [40].

Accordingly, Iob in 1966 demonstrated an increased whole-body clearance of BCAA in patients with cirrhosis [52] and the clearance was later shown to relate to the ammonia concentration [51]. Marchesini was, however, unable to explain the low BCAA by an increased clearance, but found a decreased endogenous release of BCAA into the plasma pool [53]. It was then found in rats that ammonia infusion increases glutamine and decreases glutamate and BCAA concentrations in both skeletal muscle and plasma [54]. Similarly, the muscles of patients with advanced cirrhosis take up more BCAA with higher ammonia levels, shown by higher muscular a-v BCAA gradients [40]. In 2011 Holecek studied the importance of ammonia on the reduction of BCAA in cirrhosis, with the aim of showing whether there is an effect of ammonia on the BCAA metabolism in muscle [55]. By using an in-vitro technique they controlled for indirect effects such as alterations in glucose, insulin and pH. They showed that in both red and white muscle high levels of ammonia stimulates the oxidation of leucine and the release of glutamine whereas the release of BCAA is inhibited. This supports that

Table 9.1 Flux of BCAA across leg muscle before and after intake of BCAA (μmol/min)

	Patients with cirrhosis ($n = 12$)			Healthy subjects ($n = 6$)		
	Baseline	1 h[a]	3 h[a]	Baseline	1 h[a]	3 h[a]
Leucine	−1.5 ± 6.5	261 ± 66.4*	−52.0 ± 44.1	−0.3 ± 2.3	272 ± 90.8*	92.3 ± 32.2*
Valine	1.8 ± 2.4	230 ± 63.9*	−20.8 ± 37.6	−2.1 ± 2.5	178 ± 57.0*	38.3 ± 10.4*
Isoleucine	1.6 ± 4.3	173 ± 35.2*	−12.0 ± 63.2	−0.1 ± 1.5	170 ± 55.3*	65.1 ± 26.1*
Total BCAA	1.93 ± 14.9	664.0 ± 142*	−84.7 ± 110	−2.5 ± 6	620 ± 202*	196 ± 67*

*$p < 0.05$, compared to baseline
Values are given as mean ± SEM
Positive flux indicates uptake; negative release
No significant difference In BCAA flux was observed at baseline between patients and healthy subjects
The table was reprinted with permission [5]
[a]1 and 3 h after intake of BCAA

ammonia may lower the plasma levels of BCAA by diverting their carbon skeletons towards glutamine synthesis in muscle. In contrast, we found no overall difference in muscle BCAA uptake between patients with cirrhosis and high ammonia levels and healthy persons (Table 9.1). However, the findings may be blurred by the fact that cirrhosis patients frequently suffer from muscle wasting and our measurements were not corrected for muscle mass, so the muscle BCAA uptake were still probably higher per kg muscle.

Taken together, the available data makes it meaningful to assume that the low plasma BCAA concentration in liver disease is caused by a higher skeletal muscle metabolism of BCAA which is in turn stimulated by hyperammonemia.

Does Muscle BCAA Metabolism Stimulate Muscle Ammonia Metabolism?

BCAA supplementation improves the manifestations of HE and seem to prevent recurrent HE [56]. Furthermore, BCAA reduce ammonia and improve nitrogen balance in long-term studies on cirrhosis patients [57] and seem also to lower the risk of death and progression to liver failure [58, 59]. The mechanisms of these beneficial effects are unknown in detail and probably complex. Recent thinking focuses on the effects of BCAA on skeletal muscle metabolism and ammonia removal. The hypothesis is that BCAA supplementation increases skeletal muscle ammonia removal by accelerated glutamine formation (Fig. 9.2). This would be central, because skeletal muscle removal of ammonia becomes important in patients with failing liver. In acute and chronic liver failure skeletal muscle takes part in regulation of ammonia by taking up ammonia and releasing glutamine [6–9]. This effect may actually be larger than that of the cirrhotic liver [17]. Hence, skeletal muscle plays an important role in ammonia detoxification in patients with liver dysfunction.

Despite the theoretically and also experimentally well founded presumptions, we still lack data showing that ingestion or infusion of BCAA treatment actually has an effect on ammonia metabolism in skeletal muscle in patients with cirrhosis. Most evidence is indirect and direct data based on experimentally controlled effects of BCAA supplementation on ammonia and amino acid organ fluxes are limited. The acute effects of BCAA on ammonia and amino acid metabolism in human muscle were investigated by a short-term intervention in patients with cirrhosis and healthy controls [5]. Data showed that ingestion of BCAA increased the muscular uptake of BCAA (Table 9.1). Furthermore, BCAA increased the ammonia uptake across the thigh muscle measured by both the ^{13}N-ammonia PET method and by means of a-v differences in both healthy persons and patients with cirrhosis (Fig. 9.2) [5]. As expected this was followed by a massive muscle release of glutamine [5]. This finding directly supports the hypothesis that BCAA may be used to replenish α-ketoglutarate and thereby

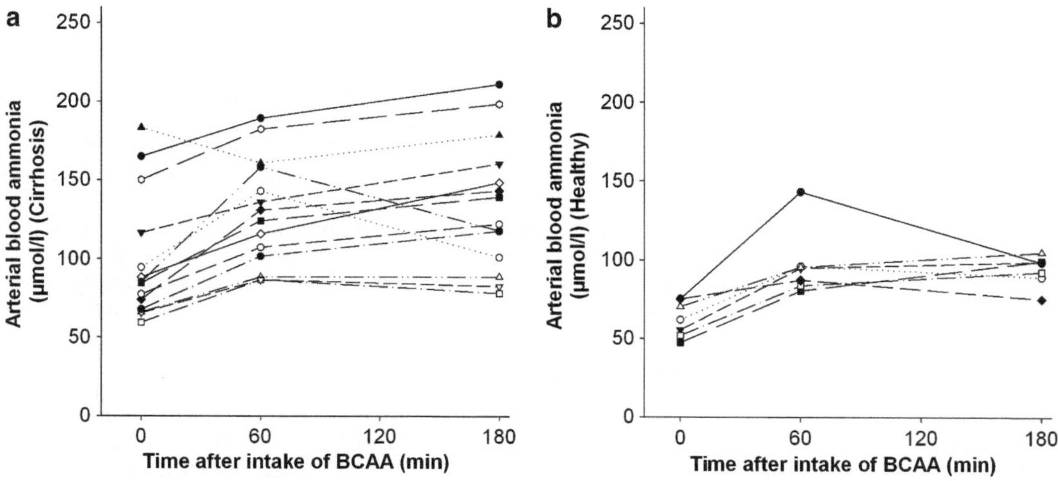

Fig. 9.3 Arterial blood concentrations of ammonia at baseline and after ingestion of BCAA. Arterial ammonia in (**a**) patients with cirrhosis ($n=14$) and (**b**) healthy subjects ($n=7$). The figure was reprinted with permission [5]

facilitate glutamine production and increase muscle removal of ammonia. However, the BCAA load also caused an acute and substantial increase in blood ammonia concentration by more than 30 % (Fig. 9.3) [5], a finding which actually agrees with Marchesini's BCAA clearance study from 1987 where the venous ammonia concentration doubled after infusion of BCAA [53]. We ascribe this increase to deamidation in other organs such as intestines and kidneys of the glutamine released from the muscles [9, 18]. This underlines that the ammonia detoxification in muscle by production of glutamine still leaves the body with the need for definitive elimination of ammonia.

Our findings are in apparent conflict with reports from long-term therapeutic trials where BCAA supplementation is reported to reduce blood ammonia [57]. The beneficial long-term effects of BCAA could be related to effects other than the ammonia lowering. BCAA enhance muscle mass build-up [60] and exert strong anabolic effects via stimulation of protein synthesis. The mechanism involves activation of anabolic signaling pathways, which increase the translation of mRNA [39]. Furthermore, it has been demonstrated in partial hepatectomy studies in rats that BCAA accelerate liver regeneration [61]. This effect involves increased liver protein synthesis via stimulation of hepatocyte growth factor.

Conclusions

BCAA are metabolized extra-hepatically, in renal and primarily muscle tissue. Studies on the effects of BCAA metabolism in the kidneys is limited whereas there is evidence to support that muscle BCAA metabolism increases in hyperammonemic states. BCAA metabolism in muscle tissue, driven by ammonia, contributes towards the lower BCAA plasma levels in patients with liver disease. However, the effect of BCAA on muscle ammonia detoxification is not definitively settled. Ingestion of BCAA enhances muscle removal of ammonia and production of glutamine but also causes an acute increase in blood ammonia. This emphasizes that supplementation with BCAA may just stimulate the futile cycle of circulation of nitrogen between muscle, gut, kidney and liver. Further studies are needed to examine both the short-term effects of BCAA on inter-organ amino acid and ammonia metabolism and the effect of long-term treatment with BCAA on muscle mass and ammonia levels.

References

1. Meijer AJ, Lamers WH, Chamuleau RA. Nitrogen metabolism and ornithine cycle function. Physiol Rev. 1990; 70:701–48.
2. Lacey JM, Wilmore DW. Is glutamine a conditionally essential amino acid? Nutr Rev. 1990;48:297–309.
3. Liaw SH, Kuo I, Eisenberg D. Discovery of the ammonium substrate site on glutamine synthetase, a third cation binding site. Protein Sci. 1995;4:2358–65.
4. Ytrebo LM, Sen S, Rose C, Ten Have GA, Davies NA, Hodges S, et al. Interorgan ammonia, glutamate, and glutamine trafficking in pigs with acute liver failure. Am J Physiol Gastrointest Liver Physiol. 2006;291:373–81.
5. Dam G, Keiding S, Munk OL, Ott P, Buhl M, Vilstrup H, et al. Branched-chain amino acids increase arterial ammonia in spite of enhanced intrinsic muscle ammonia metabolism in patients with cirrhosis and healthy subjects. Am J Physiol Gastrointest Liver Physiol. 2011;301:269–77.
6. Bessman SP, Bradley JE. Uptake of ammonia by muscle; its implications in ammoniagenic coma. N Engl J Med. 1955;253:1143–7.
7. Chatauret N, Desjardins P, Zwingmann C, Rose C, Rao KV, Butterworth RF. Direct molecular and spectroscopic evidence for increased ammonia removal capacity of skeletal muscle in acute liver failure. J Hepatol. 2006;44:1083–8.
8. Ganda OP, Ruderman NB. Muscle nitrogen metabolism in chronic hepatic insufficiency. Metabolism. 1976;25:427–35.
9. Clemmesen JO, Kondrup J, Ott P. Splanchnic and leg exchange of amino acids and ammonia in acute liver failure. Gastroenterology. 2000;118:1131–9.
10. Mitchell S, Ellingson C, Coyne T, Hall L, Neill M, Christian N, et al. Genetic variation in the urea cycle: a model resource for investigating key candidate genes for common diseases. Hum Mutat. 2009;30:56–60.
11. Kaiser S, Gerok W, Haussinger D. Ammonia and glutamine metabolism in human liver slices: new aspects on the pathogenesis of hyperammonaemia in chronic liver disease. Eur J Clin Invest. 1988;18:535–42.
12. Hamberg O, Nielsen K, Vilstrup H. Effects of an increase in protein intake on hepatic efficacy for urea synthesis in healthy subjects and in patients with cirrhosis. J Hepatol. 1992;14:237–43.
13. Vilstrup H. Synthesis of urea after stimulation with amino acids: relation to liver function. Gut. 1980;21:990–5.
14. Fleming KE, Wanless IR. Glutamine synthetase expression in activated hepatocyte progenitor cells and loss of hepatocellular expression in congestion and cirrhosis. Liver Int. 2013;33(4):525–34.
15. Syrota A, Paraf A, Gaudebout C, Desgrez A. Significance of intra- and extrahepatic portasystemic shunting in survival of cirrhotic patients. Dig Dis Sci. 1981;26:878–85.
16. Rudman D, Galambos JT, Smith 3rd RB, Salam AA, Warren WD. Comparison of the effect of various amino acids upon the blood ammonia concentration of patients with liver disease. Am J Clin Nutr. 1973;26:916–25.
17. Olde Damink SW, Jalan R, Redhead DN, Hayes PC, Deutz NE, Soeters PB. Interorgan ammonia and amino acid metabolism in metabolically stable patients with cirrhosis and a TIPSS. Hepatology. 2002;36:1163–71.
18. Windmueller HG, Spaeth AE. Uptake and metabolism of plasma glutamine by the small intestine. J Biol Chem. 1974;249:5070–9.
19. Weber Jr FL, Veach GL. The importance of the small intestine in gut ammonium production in the fasting dog. Gastroenterology. 1979;77:235–40.
20. Hansen BA, Vilstrup H. Increased intestinal hydrolysis of urea in patients with alcoholic cirrhosis. Scand J Gastroenterol. 1985;20:346–50.
21. Cooper AJ, Plum F. Biochemistry and physiology of brain ammonia. Physiol Rev. 1987;67:440–519.
22. Dam G, Keiding S, Munk OL, Ott P, Vilstrup H, Bak LK, et al. Hepatic encephalopathy is associated with decreased cerebral oxygen metabolism and blood flow, not increased ammonia uptake. Hepatology. 2012;1:258–65.
23. Clemmesen JO, Larsen FS, Kondrup J, Hansen BA, Ott P. Cerebral herniation in patients with acute liver failure is correlated with arterial ammonia concentration. Hepatology. 1999;29:648–53.
24. Glassford NJ, Farley KJ, Warrillow S, Bellomo R. Liver transplantation rapidly stops cerebral ammonia uptake in fulminant hepatic failure. Crit Care Resusc. 2011;13:113–8.
25. Norenberg MD, Rao KV, Jayakumar AR. Mechanisms of ammonia-induced astrocyte swelling. Metab Brain Dis. 2005;20:303–18.
26. Welbourne TC, Childress D, Givens G. Renal regulation of interorgan glutamine flow in metabolic acidosis. Am J Physiol. 1986;251:859–66.
27. Halperin ML, Kamel KS, Ethier JH, Stinebaugh BJ, Jungas RL. Biochemistry and physiology of ammonium excretion. 2nd ed. New York: Raven Press Ltd; 1992.
28. Welters CF, Deutz NE, Dejong CH, Soeters PB. Enhanced renal vein ammonia efflux after a protein meal in the pig. J Hepatol. 1999;31:489–96.
29. Olde Damink SW, Dejong CH, Deutz NE, Redhead DN, Hayes PC, Soeters PB, Jalan R. Kidney plays a major role in ammonia homeostasis after portasystemic shunting in patients with cirrhosis. Am J Physiol Gastrointest Liver Physiol. 2006;291:189–94.

30. Weiner ID, Verlander JW. Role of NH3 and NH4+ transporters in renal acid-base transport. Am J Physiol Renal Physiol. 2011;300:11–23.
31. Owen EE, Johnson JH, Tyor MP. The effect of induced hyperammonemia on renal ammonia metabolism. J Clin Invest. 1961;40:215–21.
32. Lund P. A radiochemical assay for glutamine synthetase, and activity of the enzyme in rat tissues. Biochem J. 1970;118:35–9.
33. Lockwood AH, McDonald JM, Reiman RE, Gelbard AS, Laughlin JS, Duffy TE, Plum F. The dynamics of ammonia metabolism in man. Effects of liver disease and hyperammonemia. J Clin Invest. 1979;63:449–60.
34. Wahren J, Felig P. Influence of protein ingestion on the amino acid metabolism in diabetes mellitus. Journ Annu Diabetol Hotel Dieu. 1976;7–20.
35. Deferrari G, Garibotto G, Robaudo C, Sala M, Tizianello A. Splanchnic exchange of amino acids after amino acid ingestion in patients with chronic renal insufficiency. Am J Clin Nutr. 1988;48:72–83.
36. Shinnick FL, Harper AE. Branched-chain amino acid oxidation by isolated rat tissue preparations. Biochim Biophys Acta. 1976;437:477–86.
37. Sears DD, Hsiao G, Hsiao A, Yu JG, Courtney CH, Ofrecio JM, et al. Mechanisms of human insulin resistance and thiazolidinedione-mediated insulin sensitization. Proc Natl Acad Sci U S A. 2009;106:18745–50.
38. Ferrando AA, Williams BD, Stuart CA, Lane HW, Wolfe RR. Oral branched-chain amino acids decrease whole-body proteolysis. JPEN J Parenter Enteral Nutr. 1995;19:47–54.
39. Atherton PJ, Smith K, Etheridge T, Rankin D, Rennie MJ. Distinct anabolic signalling responses to amino acids in C2C12 skeletal muscle cells. Amino Acids. 2010;38:1533–9.
40. Hayashi M, Ohnishi H, Kawade Y, Muto Y, Takahashi Y. Augmented utilization of branched-chain amino acids by skeletal muscle in decompensated liver cirrhosis in special relation to ammonia detoxication. Gastroenterol Jpn. 1981;16:64–70.
41. Wang X, Price SR. Differential regulation of branched-chain alpha-ketoacid dehydrogenase kinase expression by glucocorticoids and acidification in LLC-PK1-GR101 cells. Am J Physiol Renal Physiol. 2004;286:504–8.
42. Kuhlmann MK, Kopple JD. Amino acid metabolism in the kidney. Semin Nephrol. 1990;10:445–57.
43. Tizianello A, Deferrari G, Garibotto G, Robaudo C, Lutman M, Passerone G, Bruzzone M. Branched-chain amino acid metabolism in chronic renal failure. Kidney Int Suppl. 1983;16:17–22.
44. Olde Damink SW, Jalan R, Deutz NE, Dejong CH, Redhead DN, Hynd P, et al. Isoleucine infusion during "simulated" upper gastrointestinal bleeding improves liver and muscle protein synthesis in cirrhotic patients. Hepatology. 2007;45:560–8.
45. Ferenci P, Lockwood A, Mullen K, Tarter R, Weissenborn K, Blei AT. Hepatic encephalopathy–definition, nomenclature, diagnosis, and quantification: final report of the working party at the 11th World Congresses of Gastroenterology, Vienna, 1998. Hepatology. 2002;35:716–21.
46. Hahn M, Massen O, Nencki M, Pavlov J. Die Eck'sche fistel zwischen der unteren hohlvene und der pfortadre und ihre folgen für den organismus. Archiv für Experimentelle Pathologie und Pharmakologie. 1893;32:161–210.
47. Müting D, Wortmann V. Amino acid metabolism in liver diseases. Deutsch Med Wochenschr. 1956;81:1853–6.
48. Fischer JE, Rosen HM, Ebeid AM, James JH, Keane JM, Soeters PB. The effect of normalization of plasma amino acids on hepatic encephalopathy in man. Surgery. 1976;80:77–91.
49. Morgan MY, Milsom JP, Sherlock S. Plasma ratio of valine, leucine and isoleucine to phenylalanine and tyrosine in liver disease. Gut. 1978;19:1068–73.
50. Marchesini G, Forlani G, Zoli M, Angiolini A, Scolari MP, Bianchi FB, Pisi E. Insulin and glucagon levels in liver cirrhosis. Relationship with plasma amino acid imbalance of chronic hepatic encephalopathy. Dig Dis Sci. 1979;24:594–601.
51. Yamato M, Muto Y, Yoshida T, Kato M. Clearance rate of plasma Branched-chain amino acids correlates significantly with blood ammonia level in patients with liver cirrhosis. Int Hepatol Commun. 1995; 3:91–96.
52. Iob V, Coon WW, Sloan M. Altered clearance of free amino acids from plasma of patients with cirrhosis of the liver. J Surg Res. 1966;6:233–9.
53. Marchesini G, Bianchi GP, Vilstrup H, Checchia GA, Patrono D, Zoli M. Plasma clearances of branched-chain amino acids in control subjects and in patients with cirrhosis. J Hepatol. 1987;4:108–17.
54. Leweling H, Breitkreutz R, Behne F, Staedt U, Striebel JP, Holm E. Hyperammonemia-induced depletion of glutamate and branched-chain amino acids in muscle and plasma. J Hepatol. 1996;25:756–62.
55. Holecek M, Kandar R, Sispera L, Kovarik M. Acute hyperammonemia activates branched-chain amino acid catabolism and decreases their extracellular concentrations: different sensitivity of red and white muscle. Amino Acids. 2011;40:575–84.
56. Gluud LL, Dam G, Borre M, Les I, Cordoba J, Marchesini G, Aagaard NK, Vilstrup H. Lactulose, rifaximin or branched chain amino acids for hepatic encephalopathy: what is the evidence? Metab Brain Dis. 2013;28(2):221–5. PMID: 23275147.

57. Marchesini G, Dioguardi FS, Bianchi GP, Zoli M, Bellati G, Roffi L, et al. Long-term oral branched-chain amino acid treatment in chronic hepatic encephalopathy. A randomized double-blind casein-controlled trial. The Italian Multicenter Study Group. J Hepatol. 1990;11:92–101.
58. Marchesini G, Bianchi G, Merli M, Amodio P, Panella C, Loguercio C, et al. Nutritional supplementation with branched-chain amino acids in advanced cirrhosis: a double-blind, randomized trial. Gastroenterology. 2003; 124:1792–801.
59. Muto Y, Sato S, Watanabe A, Moriwaki H, Suzuki K, Kato A, et al. Effects of oral branched-chain amino acid granules on event-free survival in patients with liver cirrhosis. Clin Gastroenterol Hepatol. 2005;3:705–13.
60. Tischler ME, Desautels M, Goldberg AL. Does leucine, leucyl-tRNA, or some metabolite of leucine regulate protein synthesis and degradation in skeletal and cardiac muscle? J Biol Chem. 1982;257:1613–21.
61. Tomiya T, Inoue Y, Yanase M, Arai M, Ikeda H, Tejima K, et al. Leucine stimulates the secretion of hepatocyte growth factor by hepatic stellate cells. Biochem Biophys Res Commun. 2002;297:1108–11.

Chapter 10
Use of $^{2}H_{3}$-Leucine to Monitor Apoproteins

Asha V. Badaloo, Marvin Reid, and Farook Jahoor

Key Points

- $^{2}H_{3}$-leucine is a stable isotopomer of leucine.
- The rate of incorporation of $^{2}H_{3}$-leucine into apoproteins of lipoproteins is used to calculate their in vivo rates of synthesis and catabolism.
- In older studies, lipoprotein synthesis rates were inferred from rates of catabolism using radioactive labelled lipoproteins.
- A major advantage of the use of $^{2}H_{3}$-leucine in stable isotope tracer methodology is that stable isotopes are not radioactive and are safe for use in children and during pregnancy.
- The main early use of $^{2}H_{3}$-leucine to monitor apoproteins was related to exploring the role of plasma lipoproteins in the development of atherosclerosis and cardiovascular disease.
- The methodology using $^{2}H_{3}$-leucine as tracer has been extended to exploring mechanisms underlying shifts in concentrations of different lipoproteins in various clinical states such as human immunodeficiency virus infection (HIV) and malnutrition related fatty liver.

Keywords $^{2}H_{3}$-leucine • Stable isotope • Apoprotein • Lipoprotein • Cardiovascular risk • HIV • Fatty liver • Malnutrition

Abbreviations

CM	Chylomicrons
VLDL	Very low density lipoproteins
LDL	Low density lipoproteins
HDL	High density lipoproteins
IDL	Intermediate density lipoproteins
FSR	Fractional synthesis rate

A.V. Badaloo, B.Sc., M.Sc., Ph.D. (✉) • M. Reid, M.B.B.S., Ph.D.
Tropical Metabolism Research Unit, University of the West Indies, Mona, Kingston, Jamaica
e-mail: asha.badaloo@uwimona.edu.jm; avbadaloo@gmail.com; marvin.reid@uwimona.edu.jm

F. Jahoor, B.Sc., M.Sc., Ph.D.
Baylor College of Medicine, Houston, TX, USA
e-mail: fjahoor@bcm.edu

R. Rajendram et al. (eds.), *Branched Chain Amino Acids in Clinical Nutrition: Volume 1*, Nutrition and Health, DOI 10.1007/978-1-4939-1923-9_10,

FCR	Fractional catabolic rate
CAD	Coronary artery disease
NIDDM	Non-insulin-dependent diabetes mellitus
α-KICA	Alpha-keto isocaproic acid
HIV	Human immunodeficiency virus infection
ART	Antiretroviral therapy
TAG	Triacylglycerol
NLD	Normolipidaemic without lipodystrophy
LD	Dyslipidaemic with lipodystrophy
FDB	Familial defective apolipoprotein B-100
SAM	Severe acute malnutrition

Introduction

Apoproteins are proteins that combine with non-protein groups referred to as prosthetic groups to form conjugated proteins. The apoproteins that have been mostly monitored using $^{2}H_{3}$-leucine are those incorporated into lipoproteins. Specifically, these apoproteins are referred to as apolipoproteins distinguishing them from other apoproteins such as apoenzymes. Lipoproteins are molecular complexes of lipids and specific apoproteins. They are important in the inter-organal transport of lipids for metabolism. $^{2}H_{3}$-leucine is a stable isotopomer of the branched chain amino acid leucine. As a precursor amino acid for the synthesis of apoproteins, it is frequently used as a tracer to measure the kinetics of the apoprotein moieties of blood lipoproteins in vivo, providing valuable insights about the mechanisms underlying changes in the relative amounts of lipoproteins in various clinical and pathological states. The focus of this chapter will be on the use of $^{2}H_{3}$-leucine as a tracer to measure the kinetics of apolipoproteins. A general outline of the composition of lipoproteins and their role in lipid transport is presented in the introduction, followed by a brief historical perspective of methods used to assess their metabolism. This leads to two other sections, starting with details of the $^{2}H_{3}$-leucine tracer method mainly used to measure apolipoproteins kinetics and then a review of relevant research in which the $^{2}H_{3}$-leucine tracer technique was used in different physiological and pathological states.

Composition and General Functions of Lipoproteins

Lipoproteins transport most of the body's lipids (cholesterol, phospholipids and triacylglycerols) through the lymphatic and circulatory systems and also participate in lipid metabolism. They are classified based on their density. The main lipoprotein particles are chylomicrons (CM), very low density lipoproteins (VLDL), low density lipoproteins (LDL) and high density lipoproteins (HDL). They are in a constant state of integrated synthesis and degradation, and removal from plasma entailing the formation of other intermediate particles, namely chylomicron remnants and intermediate density lipoproteins (IDL).

Lipoproteins consist of a neutral lipid core surrounded by more polar lipids and one or more types of apoproteins. The five main classes of apolipoproteins are apoA, apoB, apoC, apoE and apo(a) with most having subclasses such as apoB-48 and apoB-100. Table 10.1 gives a summary of plasma lipoproteins, their origin, and main lipid pattern and apoprotein composition. There are other classes of apoproteins with no clearly established functions. Apolipoproteins are essential for maintaining the structure and solubility of the lipoproteins. In addition, they have specific structural domains and cell targeting signals that are recognized by cell receptors, and they determine the activities of a range of other proteins such as lipases and lipid transfer factors.

Table 10.1 Plasma lipoproteins, their origin, main lipid pattern and apoprotein composition

Lipoproteins	Origin	Main lipid pattern	Apoprotein
CM	Intestine	TAG>C	Apo B-48, apo C-11, apo E
CM remnants	From CM in plasma	C & CE>TAG	ApoB-48, apoE
VLDL	Liver	TAG>C & CE	ApoB-100, apoC-11, apoE
LDL	From VLDL in plasma	C & CE>TAG	ApoB-100
HDL	liver	CE	ApoA, apoC-11, apoE

TAG triacylglycerol, *C* cholesterol, *CE* cholesteryl ester. Adapted from Champe and Harvey [49]

CM are formed in the intestinal mucosal cells. They carry dietary lipids and other lipids made in the mucosal cells to the peripheral tissues. VLDL particles are synthesized in the liver and are secreted into the blood circulation where their main function is to transport triacylglycerols (TAG) from the liver to peripheral tissues. HDL, also synthesized and secreted from the liver removes free cholesterol from extrahepatic tissues and immediately esterify it to form cholesteryl esters. TAG, which is the major component of VLDL, is exchanged with cholesteryl esters of HDL resulting in conversion of VLDL to the intermediate density lipoprotein (IDL) and then to LDL. Also, HDL carries cholesteryl esters mostly to the liver or steriodogenic organs. In the liver, cholesteryl esters are degraded and the released cholesterol can be converted to bile acids, secreted into bile or repackaged into lipoproteins. Triacylglycerols in HDL are degraded by hepatic lipases reforming small HDL particles which restart the uptake of cholesterol from cells.

LDL is removed from the circulation by the LDL-receptor endocytosis pathway that has a regulatory effect on intracellular cholesterol levels. However, LDL chemically modified by oxidation or acetylation can be taken up by circulating macrophages which possess receptors having broad binding specificity. In this process there is no control of intracellular cholesterol levels and excessive uptake of modified LDL causes the formation of "foam" cells which contribute to atherosclerotic plaque formation.

Historical Perspective of Monitoring Apolipoprotein Kinetics

Evidence that plasma lipoproteins play a central role in the development of atherosclerosis and cardiovascular disease has been the catalyst for an interest in measuring their rates of production and clearance. In particular, there is much evidence that elevated concentration of LDL and decreased concentration of HDL are associated with an increased risk of coronary artery disease [1, 2] which is a major global cause of disability and death.

Early attempts to assess lipoprotein synthesis beginning in the 1970s were inferred from rates of catabolism using radioactive labeled lipoproteins [3, 4]. This assumes a steady state where production and clearance are equal, which may not be the case. The method involves isolating the apolipoprotein of interest from a participant to be radioiodinated and re-injected into the participant. A major disadvantage of this approach is that lipoproteins and apoproteins can be modified during isolation and radioiodination. Also, exposure to radiation limits use of the method in young children and pregnant women and in conducting multiple measurements in the same individual such as before and after an intervention. The opportunity to use other methods to overcome some of the problems associated with using radioactive tracers became possible with the availability of highly sensitive gas chromatography mass spectrometers and synthetic stable isotopic forms of amino acids.

The general principle of this new method is that the fractional synthesis rate (FSR) of a product (apoprotein) can be determined following infusion of a stable isotope labeled amino acid tracer (the precursor) that is incorporated into the product. As a precursor of the apolipoproteins, stable isotopically labeled leucine, particularly $^2H_3$- leucine, was shown to be a suitable tracer [3, 5, 6] and it has been frequently used to measure synthesis rates and to derive catabolic rates of several apolipoproteins. This kinetic information has been used to assess how variations in apolipoprotein turnover rates influence the function of the parent lipoprotein(s). The isotopic technique gives a direct measure of production rather than inferring from rate of catabolism. A major advantage of this method is due to the fact that stable isotopes are not radioactive and therefore are safe for use in children and during pregnancy.

Isotopes of an element contain different number of neutrons and hence varying mass. Isotopomers of a molecule refer to different number or arrangement of isotopically labeled positions within the molecule. Within $^2H_3$-leucine, three hydrogen atoms are replaced with the stable isotope deuterium which has a mass of 2 compared to mass of 1 of nearly all naturally occurring hydrogen atoms. The role as a tracer depends on the difference in mass of the tracer (heavier) compared to its naturally occurring form (tracee) and can be detected by a mass spectrometer in biological samples. Isotopic enrichment is a general term used for the tracer to tracee ratio or amount of tracer relative to tracee.

Measurement of the Synthesis Rates of Plasma Apolipoproteins Using $^2H_3$-Leucine

In most studies, FSR of apoproteins derived in the liver is calculated from the rate of $^2H_3$-leucine incorporation into the proteins over time using a prime-constant infusion of the tracer and applying steady state kinetics. To a lesser extent, a bolus dose of $^2H_3$-leucine and non-steady kinetic models have been used to assess turnover rates of apolipoproteins (e.g. [7, 8]). Details of general principles underlying the use of isotopic tracers in metabolic research are given by Wolfe and Chinkes [9]. The prime-constant infusion method is addressed in this section.

Theoretical Considerations of Using the Prime-Constant Infusion Method

The method is based on applying the steady state kinetic equation to the conversion of the isotopic tracer (e.g. $^2H_3$-leucine) as a precursor into a product (e.g. apolipoprotein). It is assumed that the product comprises a single homogenous pool that is turning over with a single order rate constant (10). Thus if the pool is in a metabolic constant state the fractional turnover rate is equal to FSR. Based on this, the underlying theory for FSR calculation is based on the linear increase in the isotopic enrichment of the apoprotein (product) as it asymptotically approach plateau enrichment, while the isotopic enrichment of the precursor ($^2H_3$-leucine) is at a plateau (Fig. 10.1). According to the kinetic equation fitting this single exponential curve [5, 10] the fractional turnover rate or FSR is equal to the initial

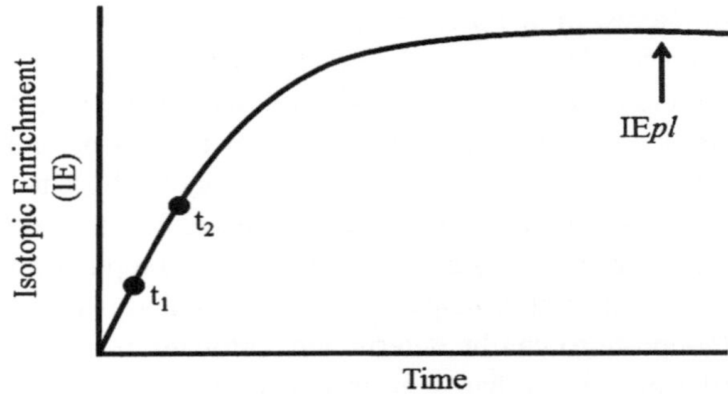

Fig. 10.1 Illustration of increase in isotopic enrichment of apoprotein-bound leucine from time t_1 to t_2 during the rise to a plateau. IEpl is the plateau isotopic enrichment of the precursor pool

rate of change of product enrichment (slope) divided by the plateau precursor enrichment. A few investigators, for example Ouguerram and colleagues [11, 12] have used computer software for simulation, analysis and modeling to perform analysis of kinetic data obtained using primed constant infusion of $^2H_3$-leucine.

As indicated, the constant infusion is to produce a constant isotopic enrichment in the precursor amino acid pool; but it may take a long time to reach an isotopic equilibrium if the pool size of the precursor is large relative to its rate of turnover. This may lead to significant recycling of the tracer back into the precursor pool [13] that will alter the kinetics of the protein. Therefore, a prime-constant infusion of tracer is important to achieve isotopic equilibrium in the precursor pool in as short a time as possible. The appropriate priming dose is determined empirically based on the premise that at isotopic equilibrium, the rate of appearance of unlabelled precursor is equal to the isotopic tracer infusion rate [9].

Choice of Tracer

It is ideal to use a tracer amino acid that does not have a large gradient between extra- and intracellular enrichment. Usually, an essential amino acid such as leucine is preferred because de novo synthesis of non-essential amino acids will greatly dilute their intracellular enrichments. In fact, Lichtenstein et al. [5] reported that they originally selected to administer a primed-constant infusion of deuterated leucine ($^2H_3$-leucine) to measure synthesis of apoB-100 in VLDL and LDL, and apoA-1in HDL for several reasons. These are: (1) leucine is a dietary essential amino acid. Hence the enrichment of leucine pools in the liver are not affected by endogenous synthesis; (2) as a branched chain amino acid, leucine is metabolized to a large extent in muscle, reducing the possibility of label recycling in the liver; (3) leucine is converted to its keto acid (alpha-ketoisocaproate) and is rapidly released into plasma, resulting in less effect of the tracer on other amino acid pools in the liver; and (4) deuterated leucine is commercially available and is relatively affordable in the quantities required for this technique. In the study by Lichtenstein et al. [5], deuterated leucine, valine and lysine gave similar results in normal healthy individuals. Another consideration is that the abundance of leucine in the apolipoprotein is appropriate to determine enrichment in a small amount of sample.

Sampling the Precursor Pool

The most direct measurement of the isotopic enrichment of the precursor pool for protein synthesis is to measure the enrichment of the amino acyl-charged tRNA. This is difficult to measure accurately, and cannot be applied when the tissue is not available in human studies. Furthermore, more tissue is needed than is commonly available. Several investigations have indicated indirect surrogates of the actual precursor pool enrichment [9, 10]. These markers can be plasma metabolites of the tracer that are derived from the tissue of interest. One approach has been to use the enrichment of plasma alpha-keto isocaproic acid (α-KICA) to estimate the intrahepatic leucine precursor pool enrichment. However, α-KICA enrichment may not be an accurate marker of the hepatic leucine pool under all conditions because studies show that there may be significant differences between the steady state isotopic enrichment of plasma α-KICA and VLDL-apoB-100-bound leucine [14, 15]. For example, Reeds et al. [15] infused $^2H_3$-leucine intravenously in adult human subjects in the fed and fasted states and found that enrichment of plasma α-KICA was ~30 % less than that of VLDL-apoB-100-bound leucine in the fed state.

Later, the use of plasma VLDL-apoB-100 enrichment to represent intra-hepatic enrichment of the tracer was widely adopted for plasma proteins that are made in the liver. This approach was based on the theory that rapidly turning over hepatic proteins such as VLDL-apoB-100 will reach an isotopic equilibrium after 6 to 8 h of tracer infusion at which time its enrichment will be equal to that of the amino acid precursor pool from which all plasma proteins are derived. Subsequent studies revealed that VLDL-apoB-100-bound tracer amino acid reaches a plateau in 4–6 h and requires only ~0.5 ml of plasma at timed intervals for GCMS analysis [16]. Furthermore, reports from primed-constant infusion tracer studies in fasted pigs show that VLDL-apoB-100-bound tracer enrichment is the same as the hepatic t-RNA enrichments [14, 17]. Hence, the plateau enrichment of isolated plasma VLDL-apoB-100 is a reasonable surrogate for enrichment in the hepatic precursor pool from which the protein is synthesized. A concern is that gut derived VLDL-apoB-48 may co-precipitate with VLDL-apolipoprotein B-100 during extraction, especially in the fed state. Therefore, pure apo B-100 must be further extracted from the lipoprotein [16].

Typical Procedures Using Prime-Constant $^2H_3$-Leucine Infusion

Dose and duration of infusion time can vary depending on the nature of the study. Therefore, pilot measurements should be conducted to identify plateau enrichment in the precursor pool. If possible, the measurements are made after a 10 h overnight fast. The prime-constant $^2H_3$-leucine tracer methodology for measuring rates of synthesis of plasma apolipoproteins and of other hepatic derived plasma proteins is similar. A practical example for measuring the synthesis rate of prealbumin in adults using the steady state isotopic enrichment of VLDL-apoB-100 to estimate the isotopic enrichment of the hepatic precursor pool has been described in detail by Jahoor [16]. A similar method used to measure VLDL-apo B-100 rate of synthesis in children admitted to hospital for treatment of severe acute malnutrition (SAM) is described herein. The same procedure was used in the severely malnourished state [18] and in the well-nourished state on recovery from malnutrition [19]. According to the ward protocol for dietary treatment, the children are fed boluses of milk feeds every 3 h or 4 h each day over 24 h, but not during measurements. The isotope infusion was performed over a 6-h period starting 2 h after the last bolus meal. Feeds missed during the study were given over the remainder of the 24 h.

Tracer Infusion and Sample Collection

As described before [16, 18, 19], a catheter is inserted in each arm of the participant after applying topical anaesthetic (EMLA cream; Astra Pharmaceuticals Ltd, Langley, United Kingdom). One catheter is used for infusion of the $^2H_3$-leucine and the other for blood sampling. All blood samples are collected in pre-chilled tubes containing Na_2EDTA and a cocktail of sodium azide, merthiolate, and soybean trypsin inhibitor. After taking a 2-ml blood sample for baseline isotopic measurements, a 10 μmol kg^{-1} priming dose of $^2H_3$-leucine is administered and followed immediately by a continuous infusion at 10 μmol kg^{-1} h^{-1} for 6 h. Additional 1-mL blood samples are taken hourly during the infusion. The blood samples are centrifuged immediately at 1000 g for 15 min at 4 °C and the plasma is removed and stored at −70 °C for later analyses.

Sample Analysis

VLDL is isolated from plasma by ultracentrifugation followed by isopropanol precipitation [14, 20], and its apoB-l00 is precipitated, isolated, and acid hydrolyzed to amino acids [16]. The amino acids released from the apoprotein are purified by cation exchange chromatography and the tracer/tracee ratio of the protein-derived leucine is determined by negative chemical ionization gas chromatography–mass spectrometry. Leucine is converted to the n-propyl ester, heptaflourobutyramide derivative, and its isotope ratio is determined by monitoring ions at *m/z* 349 to 352 as previously described [14, 21]. Each sample is analyzed on the gas chromatography-mass spectrometer in triplicate and the average value used in the calculation.

Calculation

The FSR of VLDL-apoB-100 is calculated from the rate of incorporation of $^2H_3$-leucine into the apoprotein during the rise to a plateau using the following equation:

$$\mathrm{FSR}(\%/\mathrm{h}) = \left[(\mathrm{IE}_{t2} - \mathrm{IE}_{t1}) \times 100 / \mathrm{IE}_{pl} \times t_2 - t_1 \right]$$

where $\mathrm{IE}_{t2} - \mathrm{IE}_{t1}$ is the increase in isotopic enrichment of VLDL-apoB-100-bound leucine over the period $t_2 - t_1$ h of the infusion, and $\mathrm{IE}_{\mathrm{pl}}$ is the plateau isotopic enrichment of VLDL-apoB-100-bound leucine.

FSR represents the fraction of the apoprotein pool synthesized per unit of time. The absolute synthetic rate is calculated as the product of FSR and the protein pool size. The apoprotein pool mass is the product of plasma volume and concentration of the apoprotein.

Monitoring of Apolipoproteins in Different Physiological and Pathological States Using $^2H_3$-Leucine

This section reviews some relevant studies showing how direct measurements of apolipoprotein kinetics using $^2H_3$-leucine have been useful in filling gaps in understanding the mechanisms underlying defects in lipoprotein metabolism. The studies were carried out in the areas of cardiovascular and related disease, human immunodeficiency virus infection (HIV) and malnutrition related fatty liver. In particular, the kinetic measurements have been useful in providing insight in mechanisms underlying shifts in concentrations of different lipoproteins. The ability to measure rates of synthesis and catabolism of any metabolite provides an evaluation of balance between production and catabolism. Such information can be useful for the approach to treatment. Typically, the methods entail a primed-constant infusion of $^2H_3$-leucine as described before.

Apolipoproteins in Cardiovascular and Related States

Reduced HDL and elevated LDL concentrations are associated with hypercholesterolaemia, atherosclerosis and risk of coronary artery disease (CAD) as a result of lipid deposition and plaque formation. CAD is a major health problem worldwide. Also, atherosclerosis is an important complication of non-insulin-dependent diabetes mellitus (NIDDM).

HDL-Apolipoproteins apo-A1 and apo-A11

The major protein of HDL is apoA-I and is found in two distinct HDL subpopulations, HDL-apolipoprotein-AI, which contains apoA-I only, and HDL-apolipoprotein-AI:AII which contains apoA-I and apoA-II. Low levels of apoA-I have been shown in normotriglyceridaemic patients with CAD [22] whereas others found low levels of both apoA-I and apoA-II in patients with CAD and reduced HDL cholesterol [23, 24]. At puberty, HDL cholesterol levels diverge between sexes resulting in higher levels of HDL cholesterol and apoA-I in women compared to men. However, whereas concentrations of HDL cholesterol and apoA-I are similar in premenopausal and postmenopausal women [25, 26], postmenopausal women have higher levels of LDL and triglyceride-rich lipoprotein than premenopausal women. This change is related to risk of women developing CAD as they age. These observations led Tilly-Kiesi et al. [27] to investigate the role of apoA-1 kinetics in the mechanism accounting for the sex difference in plasma HDL cholesterol concentration in postmenopausal females and in older males matched for plasma triglyceride and total cholesterol levels. ApoA-I synthesis rates in both HDL-apolipoprotein-AI and HDL-apolipoprotein-AI:AII were higher in females than in males. Therefore, Tilly-Kiesi et al. [27] suggested that lower HDL cholesterol concentration in males compared to females is attributable to the production rate and not the catabolic rate of apoA-1. On the other hand, Frenais et al. [28] found that a decrease in plasma apoA-1 level in NIDDM was due to an increase of HDL-apoA-I fractional catabolic rate (FCR).

Furthermore, based on the suggestion that lipoprotein lipase deficiency is a cause of low plasma HDL-cholesterol, Pérez-Méndez et al. [29] investigated the role of apoAI and apoA-II in two heterozygous carriers (male and female) of lipoprotein lipase deficiency. Both carriers had low plasma HDL-cholesterol levels and only the male patient was hypertriglyceridaemic. There was no difference in the apoA-1 FCR between the carriers and the 5 controls, whereas its production rate was low in the male and normal in the female patient. ApoA-II production rate in both patients was similar to controls but ApoA-II FCR in patient 1 was about 20 % lower than the mean of the control group but was normal in patient 2. FCR of VLDL-apoB was 4-times slower in patient 1 compared to patient 2. The researchers concluded that the results did not support the proposed enhanced FCR of apoA-I in lipoprotein lipase deficient patients and suggest the need to reconsider the effects of lipoprotein lipase activity on HDL metabolism.

Also, the discovery of heterogeneity within lipoproteins has expanded the exploration to identify specific aberration of dyslipidaemia. It had been shown that carriers of an apolipoprotein A-I variant (apoA-I (Zaragoza) L144R) have HDL-cholesterol levels below the 5th percentile for age and sex, low apoA-I concentrations and no evidence of coronary artery disease. They have higher percentage of HDL triglyceride and lower percentage of HDL esterified cholesterol compared to control subjects. On this basis, Recalde et at. [30] measured kinetics of apoA-I and apoA-II in two patients carrying the apoA-I (Zaragoza). The results revealed that FCR of both apoA-I and apoA-II were markedly elevated in both carriers of apoA-I (Zaragoza) when compared with controls without affecting production of these apolipoproteins.

VLDL-Apolipoprotein

VLDL transports predominantly TAG from the liver to the blood stream. ApoB-100 is the major apoprotein in VLDL. This lipoprotein has 2 subclasses: $VLDL_1$ and $VLDL_2$ particles. Compared to the small dense $VLDL_2$ particles, $VLDL_1$ are large and buoyant because they contain more TAG. $VLDL_1$ is increased in NIDDM and is proposed to be the major determinant of plasma TAG in this condition. Using bolus doses of $^2H_3$-leucine and $^2H_3$-glycerol, Adiels et al. [7] investigated the mechanism for the increase in $VLDL_1$ in NIDDM. They found that overproduction of VLDL in NIDDM was related to enhanced secretion of $VLDL_1$-apoB-100 and TAG similar in size and composition to non-diabetic controls. Fifty-five percent of the variation in $VLDL_1$ –TAG production rate was explained by plasma glucose, insulin and fatty acids. Production of $VLDL_2$ was not affected by NIDDM. Hence, the investigators proposed that hyperglycaemia is the driving force for overproduction of $VLDL_1$. Indeed, with respect to treatment, Ouguerram et al. [12] showed that the beneficial effect of long-chain n-3 PUFA supplementation in diabetic subjects was through a reduction in VLDL-apoB-100 concentration in association with a decrease in $VLDL_1$ production rate. Also, there was an increase in the VLDL conversion rate to IDL with no change in FCR.

To examine the role of insulin in regulating lipid metabolism in the liver, Malmstrom et al. [8] explored differential effect of insulin or acipimox on VLDL-apolipoprotein production in normal adult men aged about 40 years. They use $^2H_3$-leucine to trace apoB kinetics in $VLDL_1$ and $VLDL_2$ subclasses, and a non-steady-state multicompartmental model for data analysis. The study was designed to explore the effect of hyperinsulinaemia versus lowering of free fatty acids by acipimox on the production of VLDL subclasses. With acipomox, production of $VLDL_2$ increased but $VLDL_1$ decreased, hence overall VLDL production did not change. The total production rate of VLDL-apoB was reduced with insulin infusion in association with decreasing production of large TAG-rich $VLDL_1$ particles. Therefore the authors postulated that "insulin has a direct suppressive effect on the production of VLDL-apoB in the liver, independent of the availability of free fatty acids" [8].

Familial Defective Apolipoprotein B-100

During the transition from VLDL to LDL, the intermediate density lipoprotein (IDL) is formed. ApoB-100 is a major apoprotein constituent of VLDL, IDL and LDL. Familial defective apolipoprotein B-100 (FDB) is an inherited disorder caused by an amino acid substitution of apoB-100 leading to its defective binding to LDL receptor and accumulation of LDL cholesterol in the plasma. Pietzch et al. [31] investigated kinetics of apoB-100 containing lipoproteins in FDB heterozygotes using labeling of apoB-100 with [ring-$^{13}C_6$] phenylalanine and $^2H_3$-leucine with compartmental analysis using the SAAM31 program. They reported that their results suggest that "relatively small increase of LDL concentrations in FDB is due to an increased clearance of LDL precursor particles via the LDL-receptor and apoA-receptors as well as a decreased conversion of IDL to LDL" [31]. These findings were similar to that of Schaefer et al. [32] using a different approach of primed constant infusion of $^2H_3$-leucine to assess apoB-100 and ApoE kinetics in a patient homozygous for FDB. They found decreased production of LDL apoB-100, low concentration and increased catabolism of VLDL-ApoE.

HIV

Antiretroviral therapy (ART) in HIV patients leads to a reduction in morbidity and mortality, but is associated with dyslipidaemia, lipodystrophy, insulin resistance and risk of cardiovascular disease with increased exposure [33]. The dyslipidaemia is characterized by elevated serum TAG and reduced HDL-cholesterol with a predominance of small dense LDL. A relationship between hypertriglyceridaemia and increased hepatic secretion of VLDL-TAG into plasma and/or reduced TAG clearance from plasma is expected. One of the early studies to investigate the metabolism of VLDL to LDL in dyslipidaemic HIV patients receiving antiretroviral combination therapy was by Schmitz et al. [34] using $^2H_3$-leucine as tracer. Controls were healthy normolipidaemic subjects. They measured the kinetics of apoB containing lipoproteins ($VLDL_1$, $VLDL_2$, IDL and LDL). The HIV patients had higher concentrations of TAG and cholesterol and exhibited insulin resistance compared to the controls. In the HIV patients, total apoB synthesis increased and was shifted toward triglyceride rich $VLDL_1$; and transfer of apoB from the buoyant $VLDL_1$ to denser $VLDL_2$ was reduced. These results suggested that elevated TAG in HIV-infected patients with ART are primarily due to reduced rates of VLDL transfer into denser lipoproteins implying a lower rate of lipoprotein lipase-mediated delipidation. Later Ouguerram et al. [35] also investigated apoB containing lipoproteins but in a more comprehensive study and in two groups of HIV patients receiving ART. One group was normolipidaemic without lipodystrophy (NLD) and the other was dyslipidaemic with lipodystrophy (LD). The data by Ouguerram et al. [35] (Table 10.2) support the findings of Schmitz et al. [34]. They found an increase in VLDL, a decrease in IDL and no difference in LDL production rate. The LD subjects also showed marked decrease in transformation of VLDL to IDL, IDL to LDL and a decrease in FCR of VLDL,

Table 10.2 Apolipoprotein B100 containing lipoprotein metabolism in HIV male patients on combined antiretroviral therapy

	Production rate (mg kg^{-1} h^{-1})			Fractional catabolic rate (h^{-1})		
Group	VLDL	IDL	LDL	VLDL	IDL	LDL
LD	1.24±0.33*	0.20±0.10*	0.38±0.19	0.199±0.132*	0.110±0.08*	0.010±0.005*
NLD	0.80±0.21	0.48±0.24	0.45±0.25	0.555±0.398	0.523±0.275	0.025±0.014

Data (mean ± SD) adapted from Ouguerram et al. [35]. Data was analyzed using computer software for simulation, analysis and modeling (SAAMII). LD; dyslipidaemic with lipodystrophy, $n=7$, age; 38±10 year. NLD; normolipidaemic without lipodystrophy, $n=7$, age; 43±4 year

*$P < 0.05$

and LDL. They also had higher plasma concentration of TAG and total cholesterol and were insulin resistant. Therefore, Ouguerram et al. [35] suggested that the metabolic disturbances of apoB100 observed in ART related lipodystrophy is consecutive to induced insulin resistance because the pattern of results is similar to that observed using the same protocol in insulin resistant subjects.

Childhood Malnutrition

The accumulation of excess TAG in the liver is a common feature of childhood severe acute malnutrition (SAM) [36, 37] and it is associated with poor prognosis especially when the lipid content of the liver is greater than 40 % of liver weight [38]. Fatty liver is more prevalent in oedematous malnutrition, which is the form that is generally more difficult to treat. Therefore, there has been interest in understanding the pathogenesis of the hepatic steatosis so that treatment modalities can be designed to accelerate the removal of excess lipid from the liver. Hence, the role of VLDL has been investigated because it transports TAG from the liver to the circulatory system. Two lines of indirect evidence led to the general belief that impaired synthesis of the apoB-100 in VLDL results in reduced formation and secretion of VLDL and hence, in accumulation of liver TAG. First, plasma TAG and cholesterol concentrations in some children with SAM and fatty livers were shown to be low at admission and increased during treatment when it was believed that VLDL synthesis improved [39–41]. However, low TAG has not been a consistent finding in the severely malnourished state [42, 43]. The other premise for slow VLDL synthesis was based on a slower rate of incorporation of radiolabeled glycine into the apoproteins of LDL and VLDL in rats fed a low-protein 6 % casein diet [44] and of the greater amount of radiolabeled oleic acid incorporated into TAG of plasma lipoprotein after rats fed a protein-free diet for 15 to 17 days developed fatty liver and were injected with LDL apolipoprotein [44, 45]. This led to further assumption that protein deficiency in SAM results in reduced VLDL-apoB-100 synthesis. After a long period accepting as fact that impaired synthesis of apoB-100 was the underlying cause of the fatty liver in SAM, we measured VLDL-apoB-100 rate of synthesis directly using $^{2}H_{3}$-leucine as tracer and found no impairment in its synthesis rate [18]. Instead, VLDL-apoB-100 synthesis rate in children with SAM, aged 7 – 18 months, increased as liver fat increased suggesting a compensatory response (see Fig. 10.2). Lower ratio of liver attenuation to spleen attenuation by computerized tomography indicates more fatty liver [46]. Also, plasma VLDL-apoB-100 (Fig. 10.3) and cholesterol concentrations were positively related to the degree of liver fat supporting enhanced secretion of liver lipid in the children with greater degrees of hepatic steatosis.

Furthermore, as shown in Table 10.3, measurement of VLDL-apoB-100 synthesis rate showed no difference between children (aged 4–20 months) with oedematous and non-oedematous SAM or from the initial malnourished stage to when they had recovered from malnutrition [19]. Overall the use of tracer methodology with $^{2}H_{3}$-leucine shows that VLDL- apoB-100 synthesis is not reduced when children develop either oedematous or non-oedematous SAM. Other factors such as high rate of lipolysis and reduced whole body lipid oxidation seem to play a role in the development of hepatic steatosis in SAM [47, 48].

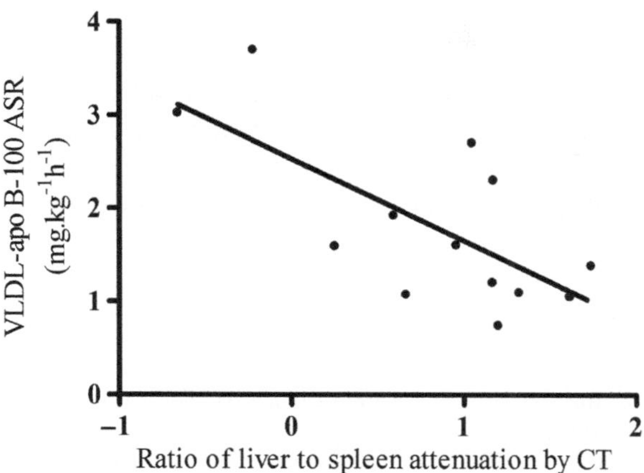

Fig. 10.2 Relationship between VLDL-apo B-100 absolute synthesis rate (ASR) and ratio of liver to spleen attenuation by computerized tomography (CT) in children with severe acute malnutrition. mean slope: -0.88 ± 0.28 ($r^2 = 0.48$, $p < 0.02$)

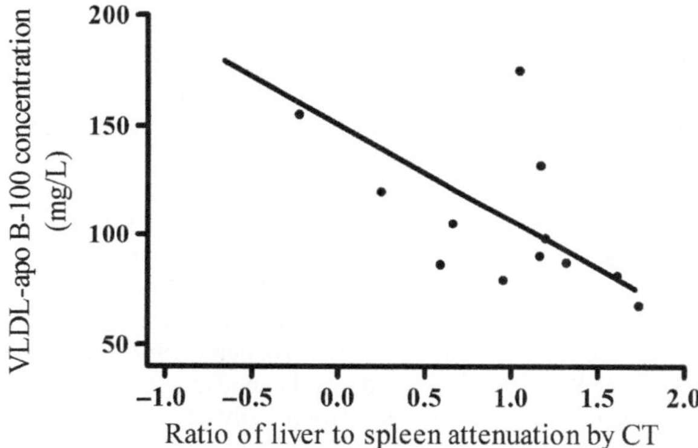

Fig. 10.3 Relationship between VLDL-apo B-100 concentration and the ratio of liver to spleen attenuation by computerized tomography (CT) in children with severe acute malnutrition. Mean slope: -44 ± 12 ($r^2 = 0.53$, $p < 0.02$)

Table 10.3 VLDL apo B-100 absolute synthesis rates during rehabilitation of children diagnosed with edematous and non-edematous malnutrition

	Absolute synthesis rates of VLDL-apo B-100 mg kg^{-1} h^{-1}; mean (95 % CI)	
	Oedematous $n = 15$	Non-oedematous $n = 7$
Severely malnourished	1.71 (1.36, 2.14)	1.33 (1.02, 1.74)
Mid catch-up growth	1.77 (1.51, 2.08)	1.47 (1.08, 2.01)
Recovery	1.35 (1.17, 1.57)	1.37 (1.02, 1.84)

Mid catch-up growth: on replacing 50 % of weight deficit. Recovery: attaining at least 90 % of reference weight for length. Adapted from Badaloo et al. [19]

Conclusions

The use of $^2H_3$-leucine in stable isotopic tracer methodology to monitor apolipoproteins kinetics is well established. In contrast to radioisotopes, a significant advantage of stable isotopes is that they are safe for use in humans of all age and during pregnancy. Before the technique to measure in vivo rates of synthesis and catabolism was developed, much was assumed based on concentrations of lipoproteins and on other indirect evidence. Thus direct measurements of apolipoproteins kinetics using $^2H_3$-leucine have contributed not only new information about mechanisms underlying shifts in their concentrations in different clinical states such as in the development of atherosclerosis and cardiovascular disease; but have assisted in confirming if assumptions based on indirect evidence were valid. In particular, use of $^2H_3$-leucine in making direct measurement of VLDL-apoB-100 synthesis revealed that hepatic steatosis in malnourished children is not due to a reduction in the synthesis of VLDL-apoB-100 when they become malnourished, as was the belief for a long time. Overall the use of stable isotope tracer methods with $^2H_3$-leucine has the potential to explore mechanisms of alteration in lipoprotein metabolism in various clinical states that can be important in developing treatment.

References

1. Bush TL, Fried LP, Barrett-Connor E. Cholesterol, lipoproteins, and coronary heart disease in women. Clin Chem. 1988;34:B60–70.
2. Castelli WP, Garrison RJ, Wilson PW, et al. Incidence of coronary heart disease and lipoprotein cholesterol levels. The Framingham Study. JAMA. 1986;256:2835–8.
3. Cohn JS, Wagner DA, Cohn SD, et al. Measurement of very low density and low density lipoprotein apolipoprotein (Apo) B-100 and high density lipoprotein Apo A-I production in human subjects using deuterated leucine. Effect of fasting and feeding. J Clin Invest. 1990;85:804–11.
4. Ramakrishnan R. Studying apolipoprotein turnover with stable isotope tracers: correct analysis is by modeling enrichments. J Lipid Res. 2006;47:2738–53.
5. Lichtenstein AH, Cohn JS, Hachey DL, et al. Comparison of deuterated leucine, valine, and lysine in the measurement of human apolipoprotein A-I and B-100 kinetics. J Lipid Res. 1990;31:1693–701.
6. Patterson BW, Hachey DL, Cook GL, et al. Incorporation of a stable isotopically labeled amino acid into multiple human apolipoproteins. J Lipid Res. 1991;32:1063–72.
7. Adiels M, Boren J, Caslake MJ, et al. Overproduction of VLDL1 driven by hyperglycemia is a dominant feature of diabetic dyslipidemia. Arterioscler Thromb Vasc Biol. 2005;25:1697–703.
8. Malmstrom R, Packard CJ, Caslake M, et al. Effects of insulin and acipimox on VLDL1 and VLDL2 apolipoprotein B production in normal subjects. Diabetes. 1998;47:779–87.
9. Wolfe RR, Chinkes DL, editors. Isotope tracers in metabolic research. New Jersey, NJ: Wiley; 2005.
10. Patterson BW. Use of stable isotopically labeled tracers for studies of metabolic kinetics: an overview. Metabolism. 1997;46:322–9.
11. Ouguerram K, Magot T, Zair Y, et al. Effect of atorvastatin on apolipoprotein B100 containing lipoprotein metabolism in type-2 diabetes. J Pharmacol Exp Ther. 2003;306:332–7.
12. Ouguerram K, Maugeais C, Gardette J, et al. Effect of n-3 fatty acids on metabolism of apoB100-containing lipoprotein in type 2 diabetic subjects. Br J Nutr. 2006;96:100–6.
13. Halliday D, McKeran RO. Measurement of muscle protein synthetic rate from serial muscle biopsies and total body protein turnover in man by continuous intravenous infusion of L-(alpha-15 N)lysine. Clin Sci Mol Med. 1975; 49:581–90.
14. Jahoor F, Burrin DG, Reeds PJ, et al. Measurement of plasma protein synthesis rate in infant pig: an investigation of alternative tracer approaches. Am J Physiol. 1994;267:R221–7.
15. Reeds PJ, Hachey DL, Patterson BW, et al. VLDL apolipoprotein B-100, a potential indicator of the isotopic labeling of the hepatic protein synthetic precursor pool in humans: studies with multiple stable isotopically labeled amino acids. J Nutr. 1992;122:457–66.
16. Jahoor F. The measurement of protein kinetics with stable isotopes tracers. In: Abrahms SA, Wong WW, editors. Stable isotopes in human nutrition laboratory methods and research applications. Cambridge, MA: CABI Publishing; 2003. p. 11–3.
17. Baumann PQ, Stirewalt WS, O'Rourke BD, et al. Precursor pools of protein synthesis: a stable isotope study in a swine model. Am J Physiol. 1994;267:E203–9.
18. Badaloo A, Reid M, Soares D, et al. Relation between liver fat content and the rate of VLDL apolipoprotein B-100 synthesis in children with protein-energy malnutrition. Am J Clin Nutr. 2005;81:1126–32.
19. Badaloo AV, Forrester T, Reid M, et al. Nutritional repletion of children with severe acute malnutrition does not affect VLDL apolipoprotein B-100 synthesis rate. J Nutr. 2012;142:931–5.
20. Egusa G, Brady DW, Grundy SM, et al. Isopropanol precipitation method for the determination of apolipoprotein B specific activity and plasma concentrations during metabolic studies of very low density lipoprotein and low density lipoprotein apolipoprotein B. J Lipid Res. 1983;24:1261–7.
21. Jahoor F, Wykes L, Del Rosario M, et al. Chronic protein undernutrition and an acute inflammatory stimulus elicit different protein kinetic responses in plasma but not in muscle of piglets. J Nutr. 1999;129:693–9.
22. Puchois P, Kandoussi A, Fievet P, et al. Apolipoprotein A-I containing lipoproteins in coronary artery disease. Atherosclerosis. 1987;68:35–40.

23. Coste-Burel M, Mainard F, Chivot L, et al. Study of lipoprotein particles LpAI and LpAI:AII in patients before coronary bypass surgery. Clin Chem. 1990;36:1889–91.
24. Genest Jr JJ, Bard JM, Fruchart JC, et al. Plasma apolipoprotein A-I, A-II, B, E and C-III containing particles in men with premature coronary artery disease. Atherosclerosis. 1991;90:149–57.
25. Li Z, McNamara JR, Fruchart JC, et al. Effects of gender and menopausal status on plasma lipoprotein subspecies and particle sizes. J Lipid Res. 1996;37:1886–96.
26. Ohta T, Hattori S, Murakami M, et al. Age- and sex-related differences in lipoproteins containing apoprotein A-I. Arteriosclerosis. 1989;9:90–5.
27. Tilly-Kiesi M, Lichtenstein AH, Joven J, et al. Impact of gender on the metabolism of apolipoprotein A-I in HDL subclasses LpAI and LpAI:AII in older subjects. Arterioscler Thromb Vasc Biol. 1997;17:3513–8.
28. Frenais R, Ouguerram K, Maugeais C, et al. High density lipoprotein apolipoprotein AI kinetics in NIDDM: a stable isotope study. Diabetologia. 1997;40:578–83.
29. Perez-Mendez O, Duhal N, Lacroix B, et al. Different VLDL apo B, and HDL apo AI and apo AII metabolism in two heterozygous carriers of unrelated mutations in the lipoprotein lipase gene. Clin Chim Acta. 2006;368:149–54.
30. Recalde D, Velez-Carrasco W, Civeira F, et al. Enhanced fractional catabolic rate of apo A-I and apo A-II in heterozygous subjects for apo A-I(Zaragoza) (L144R). Atherosclerosis. 2001;154:613–23.
31. Pietzsch J, Wiedemann B, Julius U, et al. Increased clearance of low density lipoprotein precursors in patients with heterozygous familial defective apolipoprotein B-100: a stable isotope approach. J Lipid Res. 1996;37:2074–87.
32. Schaefer JR, Scharnagl H, Baumstark MW, et al. Homozygous familial defective apolipoprotein B-100. Enhanced removal of apolipoprotein E-containing VLDLs and decreased production of LDLs. Arterioscler Thromb Vasc Biol. 1997;17:348–53.
33. Friis-Moller N, Reiss P, Sabin CA, et al. Class of antiretroviral drugs and the risk of myocardial infarction. N Engl J Med. 2007;356:1723–35.
34. Schmitz M, Michl GM, Walli R, et al. Alterations of apolipoprotein B metabolism in HIV-infected patients with antiretroviral combination therapy. J Acquir Immune Defic Syndr. 2001;26:225–35.
35. Ouguerram K, Zair Y, Billon S, et al. Disturbance of apolipoprotein B100 containing lipoprotein metabolism in severe hyperlipidemic and lipodystrophic HIV patients on combined antiretroviral therapy: evidences of insulin resistance effect. Med Chem. 2008;4:544–50.
36. McLean AE. Hepatic failure in malnutrition. Lancet. 1962;2:1292–4.
37. Waterlow JC. Protein energy malnutrition. Herts: Smith-Gordon and Compny Ltd.; 2006.
38. Waterlow JC. Amount and rate of disappearance of liver fat in malnourished infants in Jamaica. Am J Clin Nutr. 1975;28:1330–6.
39. Coward WA, Whitehead RG. Changes in serum -lipoprotein concentration during the development of kwashiorkor and in recovery. Br J Nutr. 1972;27:383–94.
40. Flores H, Pak N, Maccioni A, et al. Lipid transport in kwashiorkor. Br J Nutr. 1970;24:1005–11.
41. Truswell AS, Hansen JD, Watson CE, et al. Relation of serum lipids and lipoproteins to fatty liver in kwashiorkor. Am J Clin Nutr. 1969;22:568–76.
42. Agbedana EO, Johnson AO, Taylor GO. Selective deficiency of hepatic triglyceride lipase and hypertriglyceridaemia in kwashiorkor. Br J Nutr. 1979;42:351–6.
43. Dhansay MA, Benade AJ, Donald PR. Plasma lecithin-cholesterol acyltransferase activity and plasma lipoprotein composition and concentrations in kwashiorkor. Am J Clin Nutr. 1991;53:512–9.
44. Seakins A, Waterlow JC. Effect of a low-protein diet on the incorporation of amino acids into rat serum lipoproteins. Biochem J. 1972;129:793–5.
45. Flores H, Sierralta W, Monckeberg F. Triglyceride transport in protein-depleted rats. J Nutr. 1970;100:375–9.
46. Ricci C, Longo R, Gioulis E, et al. Noninvasive in vivo quantitative assessment of fat content in human liver. J Hepatol. 1997;27:108–13.
47. Badaloo AV, Forrester T, Reid M, et al. Lipid kinetic differences between children with kwashiorkor and those with marasmus. Am J Clin Nutr. 2006;83:1283–8.
48. Doherty JF, Golden MH, Brooks SE. Peroxisomes and the fatty liver of malnutrition: an hypothesis. Am J Clin Nutr. 1991;54:674–7.
49. Champe PC, Harvey RA. Biochemistry. Philadelphia: Lippincott Williams & Wilkins; 1994.

Part II
Inherited Defects in Branched Chain Amino Acid Metabolism

Chapter 11
Branched Chain Amino Acid Oxidation Disorders

Ronald J.A. Wanders, Marinus Duran, and Ference Loupatty

Key Points

- Branched chain amino acids are degraded in mitochondria in virtually every human cell except the erythrocyte.
- Genetic deficiencies in virtually every enzyme involved in the breakdown of leucine, isoleucine, and valine are known.
- The clinical signs and symptoms of patients with a genetic defect in the oxidation of leucine, isoleucine, and valine vary widely from virtually asymptomatic to lethal at young age.
- The laboratory diagnosis of patients affected by an inborn error of branched chain amino acid degradation involves metabolite analysis in body fluids using different platforms followed by enzymatic and molecular studies.
- The pathophysiology of the different inborn errors of branched chain amino acid oxidation remains largely unresolved.
- Screening for several of the branched chain amino acid degradation defects is included in newborn screening programs around the world.

Keywords Leucine • Isoleucine • Valine • Amino acids • Inborn errors of metabolism • Mitochondria • Amino oxidation disorders

R.J.A. Wanders, Ph.D. (✉) • M. Duran, Ph.D.
Laboratory of Genetic Metabolic Diseases, Academic Medical Centre, University of Amsterdam, Meibergdreef 9, Amsterdam, 1105 AZ, The Netherlands
e-mail: r.j.wanders@amc.uva.nl; m.e.festen@amc.uva.nl; m.duran@amc.uva.nl

F. Loupatty, Ph.D.
Laboratory of Genetic Metabolic Diseases, Academic Medical Centre, University of Amsterdam, Meibergdreef 9, Amsterdam, 1105 AZ, The Netherlands

Department of Clinical Chemistry, Reinier de Graaf Groep, Postbus 5011, Delft, 2600 GA, The Netherlands
e-mail: loupafj@hotmail.com

R. Rajendram et al. (eds.), *Branched Chain Amino Acids in Clinical Nutrition: Volume 1*, Nutrition and Health, DOI 10.1007/978-1-4939-1923-9_11,

Introduction

Leucine, isoleucine, and valine are neutral, aliphatic amino acids (AAs) which each contain a methyl branch in their side chain. As the human body is unable to synthesize these three branched chain amino acids (BCAAs) de novo, leucine, isoleucine, and valine are essential nutrients. Although they are present in all protein containing food products, the most prominent sources are dairy products, red meat, whey, and egg protein. Humans require 25–65 mg per kg body weight per day as provided by most diets unless a very unbalanced protein restricted diet is taken. The BCAAs are indispensable AAs as building blocks of newly synthesized proteins but also fulfil other functions as signalling molecules for instance. BCAAs obtained in excess are immediately degraded in cells because AAs cannot be stored like carbohydrates (as glycogen) and fatty acids (as triglycerides).

In humans, the BCAAs are degraded in the mitochondrion via the concerted action of a series of enzymes with acetyl-CoA, propionyl-CoA, and acetoacetic acid as end products (see Fig. 11.1). Catabolism starts with the transamination either in the cytosol or in the mitochondrion to produce the different 2-oxo acids after which the 2-oxo acids undergo oxidative decarboxylation to yield the different CoA esters including isobutyryl-CoA, 2-methylbutyryl-CoA, and isobutyryl-CoA from valine, isoleucine, and leucine, respectively. Oxidative decarboxylation of all three 2-oxo acids is catalyzed by the enzyme branched chain 2-oxo acid dehydrogenase (BCKADH) which is deficient in maple syrup urine disease (MSUD) as described in detail elsewhere in this book. The subsequent catabolism is different for each of the CoA esters as shown in Fig. 11.1. Defects in 11 of the 13 enzymes involved in the degradation of the different CoA esters have been described in literature. In this chapter we describe these enzyme deficiencies in detail.

Inborn Errors of Valine Degradation

Isobutyryl-CoA as derived from valine after prior transamination and oxidative decarboxylation is converted into the common intermediate propionyl-CoA via a sequence of five different enzyme reactions. Deficiencies in four out of the five enzymes have been described.

- *Isobutyryl-CoA dehydrogenase (IBD) deficiency (MIM 604773)* was first reported in 1998 by Roe and coworkers [1] as “an unrecognized defect in human valine metabolism” in a 2-year-old female who was well until 12 months of age but was then found to be anaemic with cardiomyopathy in addition. Acylcarnitine analysis revealed elevated C4 carnitine and a low free carnitine level whereas urine organic acids were normal. Later studies revealed that the C4 carnitine was in fact isobutyrylcarnitine. Since the description of the first case by Roe and coworkers in 1998 many additional cases have been reported [2–5]. Screening for IBD deficiency has been introduced in the USA already since many years by means of acylcarnitine analysis in dry blood spots. Ogelsbee et al. have developed a newborn screening follow-up algorithm for the diagnosis of IBD deficiency [5]. This algorithm allows discrimination between IBD deficiency and short-chain acyl-CoA dehydrogenase (SCAD) deficiency which are both characterized by elevated C4 carnitine. Resolution between IBD deficiency and SCAD deficiency is usually done by urine analysis since IBD deficient patients excrete isobutyrylglycine whereas in SCAD deficient patients ethylmalonic acid is excreted. Many patients have been picked up since then with isolated IBD deficiency. Importantly, most patients have remained asymptomatic which explains why in most countries IBD deficiency has not been included in neonatal screening programs.
- The principal metabolites which accumulate in IBD deficient patients are isobutyrylglycine, isobutyrylcarnitine, and isobutyric acid although these metabolites are definitely not found in all patients (Fig. 11.2). The activity of IBD is very low in cells usually used for diagnostic purposes, including

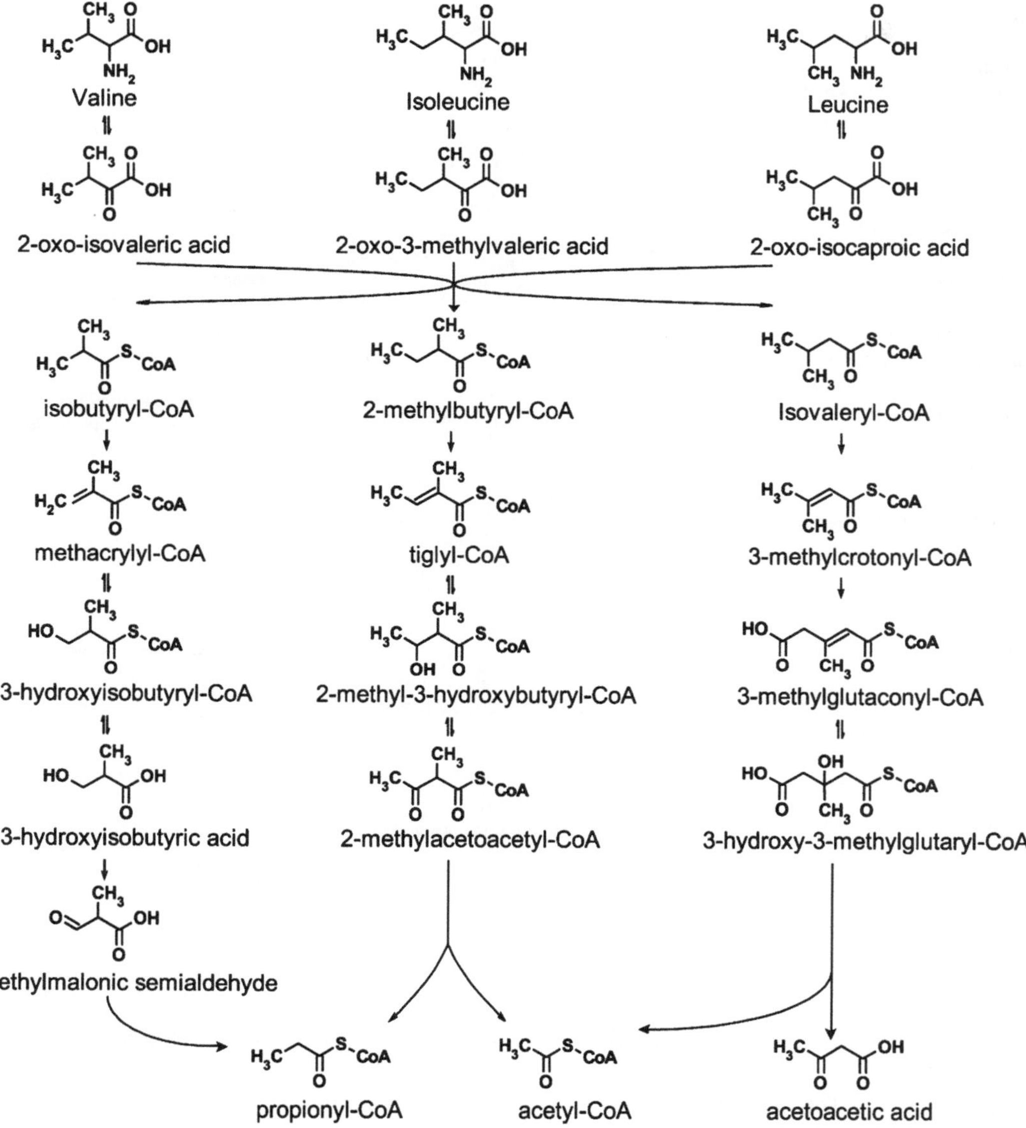

Fig. 11.1 Enzymology of the oxidative pathways involved in the breakdown of the branched chain amino acids valine, leucine, and isoleucine. The structures and names of each of the intermediates are shown

lymphocytes and cultured skin fibroblasts. For this reason molecular analysis of the *ACAD8* gene which codes for isobutyryl-CoA dehydrogenase is the first line of investigation following metabolic studies. In case new mutations are identified, expression studies in Chang cells as done by Pedersen and coworkers should be done to prove pathogenicity [4].

- *3-Hydroxyisobutyryl-CoA hydrolase (HIBCH) deficiency (MIM 250620).* So far, only few patients with 3-hydroxyisobutyryl-CoA hydrolase deficiency have been described in literature. The first patient, as described by Brown et al. had dysmorphic features at birth, poor feeding, failure to thrive, lack of development of motor skills, hypotonia, and absence of neurologic development [6]. The infant died at 3 months. Other abnormalities included vertebral abnormalities, tetralogy of Fallot, and agenesis of the cingulate gyrus and corpus callosum. The deficiency of 3-hydroxyisobutyryl-CoA hydrolase was suspected based on the finding of elevated urinary levels of *S*-2-carboxypropyl-cysteine

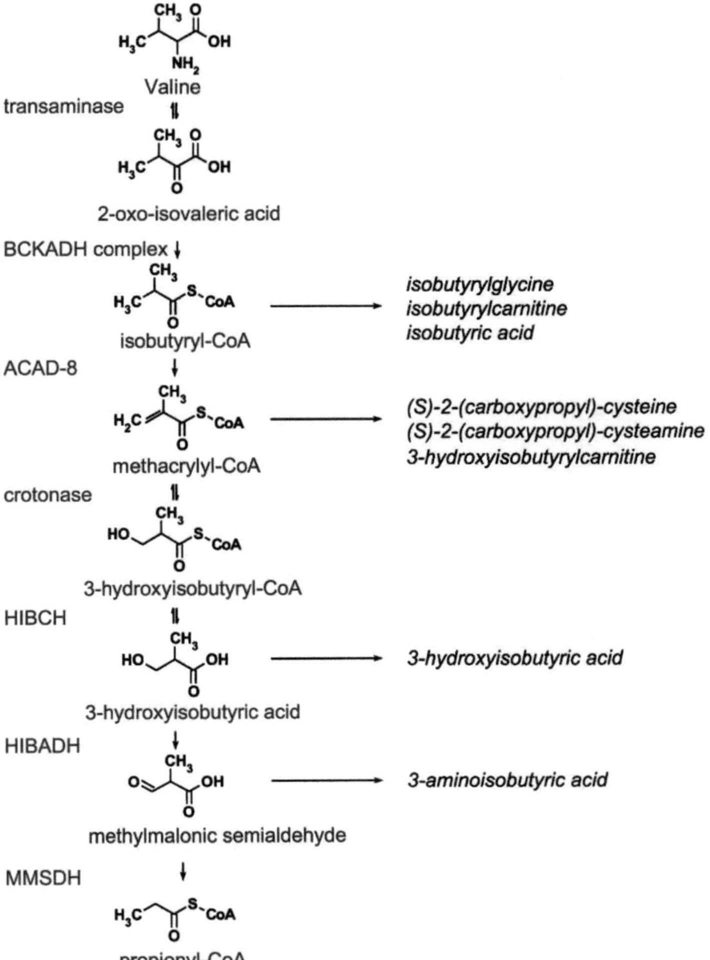

Fig. 11.2 Catabolic pathway of valine. The structures and the names of the intermediates are shown in the *center* with the name of each individual enzyme on the *left* and the main metabolites which accumulate as a direct consequence of the enzyme block on the *right*. Abbreviations used (see text): *ACAD8* isobutyryl-CoA dehydrogenase, *HIBCH* 3-hydroxyisobutyryl-CoA hydrolase, *HIBADH* 3-hydroxyisobutyrate dehydrogenase, *MMSDH* methylmalonate semialdehyde dehydrogenase

and *S*-2-carboxypropyl-cysteamine. The activity of 3-hydroxyisobutyryl-CoA hydrolase was subsequently measured and found to be markedly deficient. We described a second patient with HIBCH deficiency [7]. This patient also showed hypotonia, poor feeding, motor delay, and neurological regression in infancy. Additional features in the newly identified patient included episodes of ketoacidosis, and Leigh-like changes in the basal ganglia on MRI. Except from the two patients described by Brown et al. [6] and Loupatty et al. [7] we have been involved in the identification of additional patients. In these patients there was also severe neurological deterioration.

- The characteristic metabolites which accumulate in case of HIBCH deficiency include methacrylylglycine, 3-hydroxyisobutyric acid, *S*-2-carboxypropyl-cysteine and *S*-2-carboxypropyl-cysteamine. The latter two metabolites are formed from methacrylyl-CoA reacting with either cysteine or cysteamine (Fig. 11.2). The enzyme HIBCH is well expressed in multiple cell types including fibroblasts, which allows the enzyme activity to be determined in fibroblasts from candidate patients. We have set up a simple spectrophotometric assay as described in detail in Loupatty et al. [7]. The enzyme assay we have developed measures the HIBCH-driven hydrolysis of 3-hydroxyisobutyryl-CoA by following the reduction of DTNB by coenzyme A (CoASH) in time at 412 nm. Molecular analysis in the two patients identified have shown clear-cut mutations (see [7] for more details). In the meantime we have been involved in the identification of new patients with HIBCH deficiency who remain to be reported. It is remarkable that the clinical signs and symptoms of HIBCH deficiency are very severe with cranial facial dysmorphia, Leigh-like symptoms, and severe neurological abnormalities, whereas both the defect upstream of HIBCH (IBD-deficiency) as well as downstream (HIBADH deficiency) do not exhibit these features which probably has to do with the extreme toxicity of methacrylic acid which may react with multiple cellular targets.

- *3-Hydroxyisobutyric acid dehydrogenase (HIBADH) deficiency (MIM 236795).* In literature many patients with 3-hydroxyisobutyric aciduria have been described [8]. Patients may show widely different phenotypes ranging from mild vomiting attacks with normal brain and cognitive development to delayed motor development, profound mental retardation and early death. Some patients exhibit dysmorphic features including a small triangular face, low-set ears, long philtrum and microcephaly. Patients with 3-hydroxy-isobutyric aciduria usually excrete elevated amounts of 3-hydroxyisobutyric acid in urine ranging from 60 to 390 mmol/mol creatinine (normal <40). In half of the reported cases, elevated lactic acid levels have been observed. In many of these patients we have performed enzymatic analysis of 3-hydroxyisobutyric dehydrogenase using a specifically developed assay as described in Loupatty et al. [8]. In none of these patients a deficiency of HIBADH was found. Recently, however, we have identified the first two cases of 3-hydroxyisobutyric acid dehydrogenase deficiency in an Italian family with no apparent consanguinity. The index patient was the second child of healthy unrelated Italian parents and was born at 35 weeks of gestation by normal delivery with an APGAR score of 9/10, weight: 2,450 g; length: 46 cm, and head circumference: 33 cm. There was poor sucking, vomiting, and failure to thrive from the beginning. At weaning, he started to refuse meat and showed chronic watery diarrhea. At the end of the first year he showed growth delay and normal motor development, whereas speech development was delayed. Since the first year he presented recurrent respiratory infections. At the age of 2 years he was hospitalized because of anorexia and alopecia. He presented mild dysmorphic features, no clinodactyly of the fifth finger, and no syndactyly. At hospitalization his weight was: 10.5 kg (3rd–5th percentile), height 84.5 cm (25th percentile), and head circumference 50 cm (50th–70th percentile). He presented hypoglycemia after prolonged fasting without metabolic acidosis and ketonuria. He showed mild leucopenia with neutropenia and normal transaminases. There was neither lactic acidemia nor hyperammonemia. Plasma biotinidase, plasma amino acids, and acylcarnitine profiling revealed no abnormalities. Urinary organic acid analysis showed repeated elevation of 3-hydroxyisobutyric acid (2,583–3,550 mmol/mol creatinine) and 2-ethyl-3-hydroxypropionic acid. Echocardiography and abdominal ultrasound were normal. A defect in the metabolism of 3-hydroxyisobutyric acid was suspected and for this reason a therapy with a low-protein diet (1.2 g/kg/day) and carnitine (100 mg/kg/day) was initiated at 3 years of age. Under this regimen 3-hydroxyisobutyric acid and 2-ethyl-3-hydroxypropionic acid decreased but never normalized (852–901 mmol/mol creatinine and 46–50 mmol/mol creatinine respectively). Family compliance was very poor and they refused brain MRS. Speech and attention ameliorated: WPPSI scale: verbal IQ 94, Performance IQ 85, Total IQ 88. At 5 years and 2 months of age weight is 16.8 kg (25th percentile), length 108 cm (25th–50th percentile), and head circumference 53 cm (70th percentile). The alopecia disappeared although areas of patchy hair loss have remained and hair is wiry. There is also a second affected child in the family who was only identified after identification of the first patient which occurred at 14 years and 5 months of age.

  Based on the findings in the two patients with bonafide HIBADH deficiency the characteristic metabolites which accumulate in case of HIBADH deficiency are 3-hydroxyisobutyric acid and 2-ethyl-3-hydroxypropionic acid. The enzyme HIBADH is well expressed in fibroblasts and molecular analysis of the HIBADH-gene has identified clear-cut mutations in the gene involved (Wanders et al., submitted for publication).

- *Methylmalonic semialdehyde dehydrogenase (MMSDH) deficiency (MIM 603178).* The first patient with methylmalonic semialdehyde dehydrogenase (MMSDH) deficiency has been described by Pollitt et al. [9] and Gray et al. [10]. The male patient was hospitalized at the age of 3 weeks with unexplained episodes of diarrhea and vomiting. Subsequent urinary organic acid analysis revealed extremely elevated levels of 3-hydroxyisobutyric acid, which led to the discovery that 3-hydroxypropionic acid, 2-(hydroxymethyl)butyric acid (2-ethylhydracrylic acid), beta-alanine, and both isomers of 3-aminoisobutyric acids were also being excreted in excess. These findings prompted Pollitt et al. [9] to conclude that "this new metabolic disorder is probably due to deficient activities of malonic, methylmalonic, and ethylmalonic semialdehyde dehydrogenases which could either be catalyzed by one enzyme, or by different enzymes with a common subunit". No enzyme studies were done in the index patient. In 1989 Harris and coworkers reported the purification of MMSDH from rat liver and provided convincing evidence suggesting that MMSDH was the single enzyme reacting with malonic, methylmalonic, and ethylmalonic semialdehyde [11]. This report was soon followed by another paper from the same group [12] describing the cDNA cloning of MMSDH. It took another few years before Chambliss et al. [13] reported cloning of the human MMSDH cDNA and corresponding gene. This allowed molecular analysis of MMSDH in the original patient described by Pollitt et al. and Gray et al. [9, 10] which revealed homozygosity for a missense mutation (c.1336G>A; G446R) and deletion of exon 10. Sass et al. [14] reported another two patients with methylmalonic semialdehyde dehydrogenase deficiency. The first patient, a boy, first came to medical attention at 12 months of age because of developmental delay. Upon admission he was noticed to have a relative microcephaly and mildly dysmorphic facial features (bulbous nose with hypoplastic nasae alae and a long philtrum) along with bilateral obstructed nasolacrimal ducts. Plasma amino acid profiling revealed no abnormalities whereas urinary organic acid profiling showed huge excretion of 3-hydroxyisobutyric acid amounting to 7,000 mmol/mol creatinine (normal <100 mmol/mol creatinine) and mildly increased concentrations of 3-hydroxypropionic and 2-ethyl-3-hydroxypropionic acids. He received a low-protein valine-restricted diet, supplemented with carnitine and did make developmental progress. At 26 months of age, however, he developed a febrile illness and the patient ultimately died. At autopsy there was cerebral edema with vacuolization of the white matter, bilateral cerebellar tonsillar swelling, edema, and herniation along with an acute parietal subarachnoid hemorrhage. Patient two was a girl with consanguineous parents reported by Shields et al. in 2001 [15]. In brief the girl was noticed at birth to have bilateral microphthalmia with cataracts and hypotonia. There was facial dysmorphia. Subsequent urinary organic acid analysis revealed excess 3-hydroxyisobutyric acid (1,615 mmol/mol creatinine) as the key metabolite. The child was placed on a diet moderately reduced in protein, but the progress was poor and she showed severe developmental delay. In the patients so far described with methylmalonic semialdehyde dehydrogenase (MMSDH) deficiency there is accumulation of 3-hydroxyisobutyric acid often, but not always, combined with increased urinary excretion of beta-alanine, 3-hydroxypropionic acid, and both isomers of 3-aminoisobutyric acids.

We have recently developed a solid enzymatic assay for MMSDH in fibroblasts which has allowed identification of additional patients with MMSDH deficiency (manuscript in preparation). The enzyme MMSDH is expressed in multiple cell types including fibroblasts and can readily be measured provided the substrate which is unstable can be generated in sufficient amounts. Furthermore, molecular analysis in patients in whom a true enzyme deficiency was determined has shown clear-cut mutations in the *ALDH6A1* gene as first described by Chambliss et al. ([13]; see also Sass et al. [14]).

Inborn Errors of Leucine Degradation

Leucine is first transaminated to 2-oxo-isocaproic acid followed by oxidative decarboxylation by BCKADH to produce isovaleryl-CoA. In contrast to 2-methylbutyryl-CoA, isovaleryl-CoA cannot undergo beta-oxidation due to the methyl group at the 3-position which makes beta-oxidation impossible. Accordingly, isovaleryl-CoA follows a different pathway of oxidation which starts with a dehydrogenation step to generate 3-methylcrotonyl-CoA but is then followed by a carboxylation step as catalyzed by the enzyme 3-methylcrotonyl-CoA carboxylase (3MCC) with 3-methylglutaconyl-CoA as product. Subsequently, there is hydration of the double-bond as catalyzed by the enzyme 3-methylglutaconyl-CoA hydratase. The product of this reaction is 3-hydroxy-3-methylglutaryl-CoA (HMG-CoA) which is split into acetyl-CoA and acetoacetic acid by the enzyme HMG-CoA lyase. Four inborn errors of the distal pathway of leucine degradation have been identified including: (1) isovaleric acidemia; (2) 3-methylcrotonyl-CoA carboxylase deficiency; (3) 3-methylglutaconyl-CoA hydratase deficiency; and (4) 3-hydroxy-3-methylglutaryl-CoA lyase deficiency as described in more detail below.

- *Isovaleric aciduria (MIM 243500).* Isovaleric aciduria is caused by a deficiency of the enzyme isovaleryl-CoA dehydrogenase (IVD). This flavoenzyme catalyses the conversion of isovaleryl-CoA to 3-methylcrotonyl-CoA. Reoxidation of enzyme-bound FADH2 to FAD is mediated by the electron-transfer-protein system consisting of ETF and ETF dehydrogenase. Since its first report by Tanaka and coworkers [16] approximately 100 cases have been reported [17–19], but many more are known to date. Initially, two distinct forms of isovaleric aciduria were discerned including an acute, early-onset form and a much milder presentation. The acute neonatal form is associated with massive metabolic acidosis, encephalopathy, and early death, whereas the chronic form is characterized by episodes of severe ketoacidotic attacks with asymptomatic intervals. More recently, a third phenotype has been identified in which patients manifest mild biochemical abnormalities and are asymptomatic [20]. The phenotypic abnormalities of isovaleric aciduria are supposed to be due to the accumulation of isovaleric acid which is toxic to the central nervous system. Interestingly, during episodes of acute metabolic decompensation the concentration of isovaleric acid can reach levels as high as several hundred times normal values, but due to its rapid conjugation to other compounds isovaleric acid itself is not the hallmark of this disorder. In fact, isovalerylcarnitine in conjunction with extremely low carnitine, and isovalerylglycine are the distinctive metabolites elevated in this disorder in plasma and urine, respectively [21–23] (Fig. 11.3). The majority of patients with isovaleric acidemia today are diagnosed through newborn screening by use of acylcarnitine profiling in dried blood spots which reveals increased amounts of C_5-acylcarnitine, i.e., isovaleryl carnitine. The enzyme isovaleryl-CoA dehydrogenase is a tetramer of four identical 43 kDa subunits and belongs to the family of acyl-CoA dehydrogenases which all donate their electrons to electron-transfer flavoprotein (ETF) which is subsequently reoxidized by ETF dehydrogenase with the electrons ending up in the respiratory chain at the level of ubiquinone. Several methods have been described in literature to measure the activity of glutaryl-CoA dehydrogenase including the ETF-based assay and an assay in which radiolabeled glutaryl-CoA is used and activity is measured by determination of the radioactive CO_2 produced. We have developed an easy nonradioactive HPLC-based method which involves incubation of lymphocyte or fibroblast homogenates in the presence of glutaryl-CoA with ferrocenium hexafluorophosphate as electron acceptor followed by a resolution of the different acyl-CoA esters by HPLC. This method allows the activity of isovaleryl-CoA dehydrogenase to be determined accurately (Wanders et al., in preparation). The gene encoding isovaleryl-CoA dehydrogenase is mapped to chromosome 15q14-q15 and consists of 12 exons that span 15 kb of genomic DNA. A range of different disease-causing mutations have been documented in the numerous patients with isovaleric aciduria. No genotype–phenotype correlation has been found. Reduction of leucine intake and supplementation with either glycine or carnitine are useful strategies in the management of this disorder [24, 25].
- *3-Methylcrotonyl-CoA carboxylase (MCC) deficiency (MIM 210200 and MIM 210210).* Isolated 3-methylcrotonyl-CoA carboxylase (MCC) deficiency is an autosomal recessive disorder of leucine catabolism caused by the deficient activity of the mitochondrial enzyme MCC. The enzyme belongs to the group of biotin-dependent carboxylases which also includes pyruvate carboxylase (PC), propionyl-CoA carboxylase (PCC), and acetyl-CoA carboxylase (ACC). MCC consists of an alpha and a beta subunit assembled into an alpha6, beta6 configuration. The clinical picture of

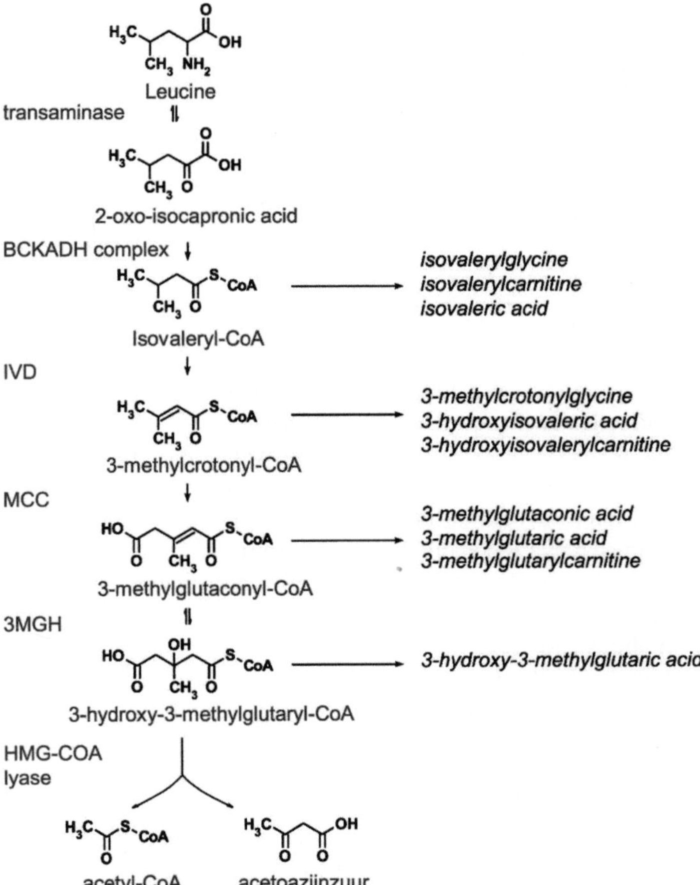

Fig. 11.3 Catabolic pathway of leucine. The structures and names of the intermediates are shown in the *center* with the name of each individual enzyme on the *left* and the main metabolites which accumulate as a direct consequence of the enzyme block on the *right*. Abbreviations used (see text): *BCKADH* branched chain keto acid dehydrogenase complex, *IVD* isovaleryl-CoA dehydrogenase, *MCC* 3-methylcrotonyl-CoA carboxylase, *3MGH* 3-methylglutaconyl-CoA hydratase

MCC deficiency is extremely heterogeneous, even within the same family as exemplified by the reports of Eminoglu et al. [26] and Visser et al. [27]. The phenotype ranges between the severe early-onset form to an adult form with no signs and symptoms. Indeed, some patients rapidly develop an acute metabolic crisis usually triggered by intercurrent infections or a protein-rich diet. Symptoms include vomiting, opisthotonus, involuntary movements, seizures, coma, and apnea, typically associated with metabolic acidosis, hypoglycemia, and in some cases hyperammonemia. Other patients present with neurological abnormalities such as seizures, muscular hypotonia, and/or developmental delay (see [28] for recent review). At the other end of the disease spectrum are MCC deficient individuals showing no signs and symptoms even at older age. This includes MCC deficient parents and siblings identified by family screening once MCC deficiency has been established in an index case. Furthermore, the majority of children diagnosed by newborn screening (NBS) have remained asymptomatic so far [29–32]. Interestingly, a thorough analysis by Stadler et al. [29] revealed that only 10 % of MCC deficient individuals do develop symptoms. Recently, a comprehensive study has been published by Baumgartner and coworkers [28] in a cohort of 88 MCC deficient individuals of which 53 were identified by newborn screening, 26 were diagnosed on the basis of clinical signs and symptoms of a positive family history, and 9 mothers were identified following the positive newborn screening result of their baby. Only 5 of the 53 patients identified by newborn screening presented with acute metabolic decompensations which underscores the conclusions by Stadler and coworkers [29].

- The characteristic metabolites in MCC deficiency are: elevated 3-hydroxy-isovaleric acid, 3-hydroxy-isovaleryl carnitine, 3-methylcrotonyl carnitine, and 3-methylcrotonyl glycine (Fig. 11.3). The enzyme 3MCC is widely expressed which implies that MCC deficiency can readily be confirmed enzymatically by analyses in blood cells and/or fibroblasts.
- As discussed above MCC is a hetero-multimeric enzyme which the alpha and beta subunit encoded by different genes named MCCC1 (formerly called *MCCA*) and MCCC2 (formerly called *MCBB*). MCC deficiency may be caused by mutations in either MCCC1 or MCCC2 with the majority being missense mutations along with small insertions/deletions, nonsense, frameshift, and splice site mutations (see [28] for recent review).
- *3-Methylglutaconic aciduria type 1 (MIM 250950).* The 3-methylglutaconic acidurias represent a heterogeneous group of disorders with elevated 3-methylglutaconic acid as unifying factor. Wortmann et al. [33] has proposed to classify this group into two distinct subgroups. Subgroup 1 contains the primary 3-methylglutaconic acidurias with only one member identified so far which is 3-methylglutaconyl-CoA hydratase deficiency formerly known as 3-methylglutaconic aciduria type 1. Subgroup 2 includes the secondary 3-methylglutaconic acidurias and consists of a number of different defects including Barth syndrome (TAZ deficiency) [34, 35], MEGDEL syndrome (SERAC1 deficiency) [36], Costeff syndrome (OPA3 deficiency) [37], DCMA syndrome (DNAJC19 deficiency) [38], and TMEM70 deficiency [39]. True 3-methylglutaconyl-CoA hydratase deficiency was first described by Duran et al. [40] in two brothers with speech retardation as only abnormality. As true for most of the other disorders of branched chain amino acid degradation, the clinical phenotype associated with 3-methylglutaconyl-CoA hydratase deficiency is very heterogeneous as exemplified recently by Wortmann et al. [41]. Indeed, whereas in some patients the signs and symptoms may be confined to just a mild delay in speech development as in the patients described by Duran et al. [40], several other patients have been reported with severe (leuko)encephalopathy in the neonatal period but also later in adulthood [41].
- The characteristic metabolic abnormalities associated with 3-methylglutaconyl-CoA hydratase deficiency include elevated 3-methylglutaconic acid, 3-methylglutaric acid, and 3-methylglutaryl-carnitine (Fig. 11.3), although these abnormalities are not observed in all patients. The hydratase enzyme can readily be measured in blood cells and/or fibroblasts. The original assay described for this enzyme was laborious and complicated one problem being the lack of a commercial source of the substrate 3-methylglutaconyl-CoA. We have recently found that the enzyme can be readily measured in the reverse direction using commercially available 3-hydroxy-3-methylglutaryl-CoA (HMG-CoA) [42]. The molecular basis of 3-methylglutaconyl-CoA hydratase deficiency has been identified simultaneously by IJlst et al. [43] and by Ly et al. [44].
- *3-Hydroxy-3-methylglutaryl-CoA lyase(HMG-CoA lyase) deficiency (MIM 246450).* The first description of HMG-CoA lyase deficiency was by Faull and coworkers in 1976 [45]. The index patient presented at 7 months of age with vomiting and diarrhea which was rapidly, i.e., within 24 h, followed by additional signs and symptoms including lethargy, dehydration, cyanosis, and apnea. This acute crisis required resuscitation. Upon proper treatment, the patient recovered and has developed very well since then. Following this first patient, many additional patients with HMG-CoA lyase deficiency have been described since then. The clinical presentation of HMG-CoA lyase deficiency is very variable even within families. The disorder may be lethal if not properly treated. The key features associated with HMG-CoA lyase deficiency include severe infantile hypoglycemia, metabolic acidosis, which may be life-threating, with hepatomegaly, lethargy, or coma, and the characteristic absence of ketosis as additional features. Neurological complications may be severe due to the hypoketotic hypoglycemia and acidosis which may result in severe and permanent handicaps. Patients may also manifest Reye-like episodes, characterized by increased levels of ammonia and abnormal liver function tests, fever, hepatomegaly, altered levels of consciousness, and seizures.

The characteristic metabolites of HMG-CoA lyase deficiency include 3-hydroxy-3-methylglutaric acid, 3-methylglutaconic acid, 3-methylglutaric acid, 3-hydroxyisovaleric acid, and 3-hydroxy-3-methylglutarylcarnitine and 3-methylglutarylcarnitine (see Fig. 11.3). The enzyme HMG-CoA lyase is present in multiple cell types including lymphocytes, platelets, and cultured skin fibroblasts. Earlier we have developed a very simple, straightforward spectrophotometric assay which allows HMG-CoA lyase to be determined in different blood cell types (see [46] for review).

Inborn Errors of Isoleucine Degradation

Isoleucine is first transaminated to 2-oxo-3-methylvaleric acid, followed by oxidative decarboxylation by BCKADH to produce 2-methylbutyryl-CoA. Since 2-methylbutyryl-CoA has the methyl group at the 2-position this CoA ester can be beta-oxidized directly. Accordingly, 2-methylbutyryl-CoA is dehydrogenated by the acyl-CoA dehydrogenase called short branched chain acyl-CoA dehydrogenase (SBCAD) to produce tiglyl-CoA. Hydration of tiglyl-CoA by the enzyme crotonase yields 2-methyl-3-hydroxybutyryl-CoA. Subsequently, the enzyme 2-methyl-3-hydroxybutyryl-CoA dehydrogenase generates 2-methylacetoacetyl-CoA lyase which can then undergo thiolytic cleavage by the enzyme beta-ketothiolase to produce propionyl-CoA and acetyl-CoA. Defects in each of the four enzymes involved in the degradation of 2-methylbutyryl-CoA have been described.

- *2-Methylbutyryl-CoA dehydrogenase (SBCAD) deficiency (MIM 610006)*. SBCAD deficiency was first described by Gibson et al. [47] and Andresen et al. [48]. The clinical spectrum of SBCAD deficiency is very wide with some patients showing neurological symptoms comprising retarded motor development, generalized muscular atrophy, and strabismus, whereas in other patients, especially those identified by expanded newborn screening programs, no clinical manifestations have become apparent. Metabolic abnormalities associated with SBCAD deficiency include 2-methylbutyrylglycine, 2-methylglutarylcarnitine, and 2-methylbutyric acid (Fig. 11.4). Most patients, especially those identified by newborn screening, are identified by means of acylcarnitine analysis. Since 2-methylbutyrylcarnitine is a C5 carnitine, discrimination has to be made between SBCAD deficiency and isovaleryl-CoA dehydrogenase deficiency which can be done by means of organic acid analysis.
- The enzyme SBCAD belongs to the family of acyl-CoA dehydrogenases and is expressed in multiple cell types. We have developed a method allowing measurement of SBCAD in fibroblast homogenates which includes incubation of (*S*)-2-methylbutyryl-CoA with ferrocenium hexafluorophosphate as electron acceptor followed by resolution of the different acyl-CoA esters by means of HPLC as described in Andresen et al. [48] and Madsen et al. [49]. The gene encoding 2-methylbutyryl-CoA dehydrogenase, named *ACADSB* is located on chromosome 10q26.13 and contains 11 exons. At present many different mutations have been reported in SBCAD deficient patients (see [48–51] for review).
- *2-Methyl-3-hydroxybutyryl-CoA dehydrogenase deficiency (MHBD) (MIM 300438)*. One of the most puzzling and mysterious disorders of branched chain amino acid oxidation is 2-methyl-3-hydroxybutyryl-CoA dehydrogenase deficiency which was first described by Zschocke et al. in a boy who experienced a transient metabolic derangement in the neonatal period and subsequently manifested mild developmental delay. The finding of increased urinary excretion of 2-methyl-3-hydroxybutyrate and tiglylglycine prompted the diagnosis beta-ketothiolase deficiency. However, the subsequent course of the disease in this boy was highly unusual and did not mimic that described for patients with beta-ketothiolase deficiency. Indeed, although psychomotor development was only moderately delayed at 1 year of age, the boy showed progressive psychomotor regression with loss of motor skills, marked restlessness, choreoathetoid movements, loss of vision, and epilepsy.

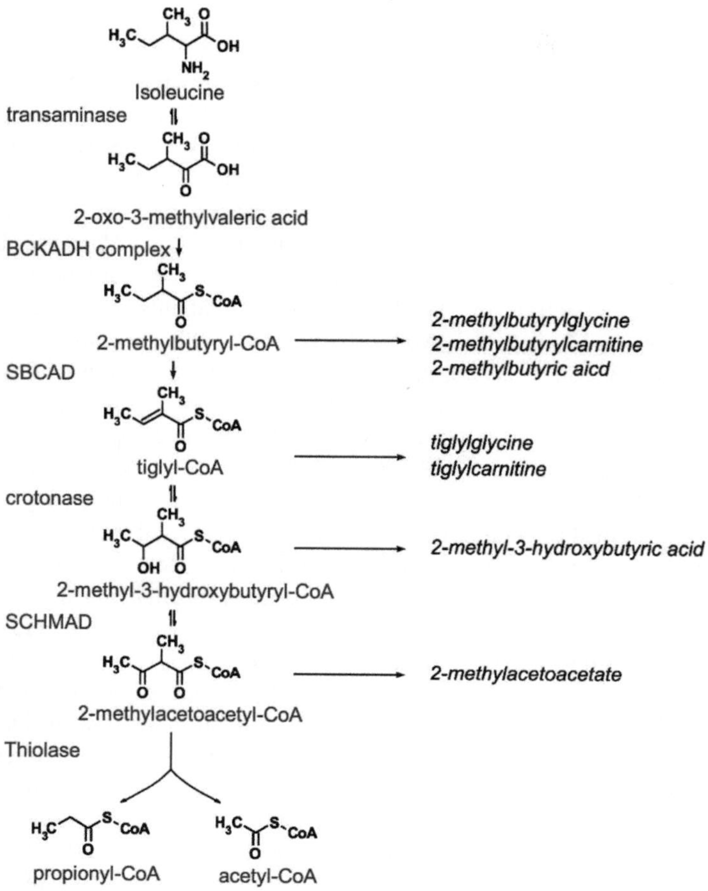

Fig. 11.4 Catabolic pathway of isoleucine. The structures and names of the intermediates are shown in the *center* with the name of each individual enzyme on the *left* and the main metabolites which accumulate as a direct consequence of the enzyme block on the *right*. Abbreviations used (see text): SBCAD = short/branched chain acyl-CoA dehydrogenase, SCHMAD = HADH2 = HSD17B10 = 2-methyl-3-hydroxybutyryl-CoA dehydrogenase

Triggered by this unusual disease course, the diagnosis of beta-ketothiolase deficiency was reconsidered and proper enzyme studies were done which revealed normal beta-ketothiolase activity but deficient 2-methyl-3-hydroxybutyryl-CoA dehydrogenase activity [52]. The molecular basis of this defect was resolved in our own laboratory in 2003 [53]. The gene involved was initially named HADH2, is located on the short arm of the X chromosome (Xp11.2), and is now officially termed HSD17B10.

Since the first description by Zschocke et al. in 2000, many additional patients have been described. Indeed, in the last comprehensive review on the topic by Zschocke in 2012 [54] 19 families with pathogenic mutations in the HSD17B10 gene have been reported. According to Zschocke, 2012 different forms of HSD17B10 deficiency can be discriminated. These include: (1) a severe form starting in the neonatal period with little neurological development, severe progressive cardiomyopathy, and early death; (2) the more frequent, classical form characterized by a more or less normal development in the first 6–18 months of life followed by a progressive neurodegenerative course with retinopathy and cardiomyopathy leading to death at 2–4 years of age; and (3) a juvenile form described in only few patients. Interestingly, heterozygous females may also show symptoms ranging from severe with psychomotor delay, intellectual disability, and a sensorineural hearing impairment to asymptomatic (see [54] for more detailed discussion and references).

Through the years it has become clear that there is no correlation between residual enzyme activity and phenotype. In fact, patients have been described with the severe form of HSD17B10 deficiency with early death with only partially reduced enzyme activity amounting to 30 % of mean control [55–57]. Although not definitively resolved yet, the fact that the HSD17B10 enzyme protein also harbors a second functionality may well be of major significance in this respect. Indeed, work by Rauschenberger et al. [57] has shown that HSD17B10 is crucial for mitochondrial integrity and cell survival. Interestingly, earlier work by Holzmann et al. [58] had already shown that MRPP2 which later turned out to be the same protein as HSD17B10, plays an indispensable role in mitochondrial tRNA processing. It remains to be demonstrated whether the true pathophysiological mechanism in this disease is indeed the disturbance in tRNA processing as caused by the deficiency of MRPP2/HSD17B10 subsequently leading to mitochondrial dysfunction and cell death. HSD17B10 deficient patients show increased urinary excretion of tiglylglycine and 2-methyl-3-hydroxybutyric acid. Other metabolic abnormalities include: elevated CSF and/or blood lactate, and increased urinary excretion of 3-hydroxyisobutyric acid, 2-ethylhydracrylic acid, and 3-methylglutaconic acid.

We have developed a sensitive and robust enzyme assay allowing the activity of the enzyme to be determined accurately in many different cell types including lymphocytes and fibroblasts [53]. With respect to the molecular basis of HSD17B10 deficiency many different mutations have been described with one frequent mutation as discussed in Zschocke [54].

- *Beta-ketothiolase deficiency (MIM 203750).* Beta-ketothiolase deficiency was first described by Daum et al. in 1971 [59] as a new disorder of isoleucine metabolism characterized by the excretion of 2-methyl-3-hydroxybutyric acid, 2-methylacetoacetic acid, tiglylglycine, and butanone in all four patients identified except one in whom tiglylglycine was normal. Subsequent studies in fibroblasts from one of the patients established that oxidation of l-[U-14C]isoleucine was indeed deficient. Subsequently, a number of additional patients with this condition were reported. In 1979 Robinson et al. [60] published their detailed studies aimed to resolve the true enzyme defect in this condition and concluded that it was the mitochondrial potassium-dependent, acetoacetyl-CoA specific thiolase which was deficient. Delineation of the precise enzyme defect was hampered by the lack of the presumed true substrate of the enzyme, i.e., 2-methylacetoacetyl-CoA which was only resolved in 1983 when Middleton and Bartlett [61] reported the successful synthesis of 2-methylacetoacetyl-CoA. Since then many patients with true 2-methylacetoacetyl-CoA specific thiolase deficiency often referred to as beta-ketothiolase deficiency, have been described. Taken together, the main clinical manifestations of beta-ketothiolase deficiency include intermittent episodes of acute ketoacidosis which typically occur in a previously healthy child. Age of onset is usually between 6 and 24 months as triggered by a febrile illness although clinical heterogeneity is considerable. Vomiting and dehydration may occur which may progress to lethargy and coma if not properly treated. Some patients have died in fact or have suffered severe neurological sequelae as a result of prolonged episodes of ketoacidosis. Beta-ketothiolase deficiency may also go unnoticed as deduced from family studies once a patient has been identified in a family.
- Patients suffering from beta-ketothiolase deficiency may show a range of metabolic abnormalities especially during crises. These include: massive amounts of ketone bodies, 2-methylacetoacetic acid, 2-methyl-3-hydroxybutyric acid, and tiglylglycine.
- In candidate patients enzyme studies should be done. Since the enzyme is expressed in multiple cell types including white blood cells like lymphocytes, the true enzyme defect can be ascertained quickly. Enzyme studies should be done with the true substrate, i.e., 2-methylacetoacetyl-CoA rather than with acetoacetyl-CoA simply because acetoacetyl-CoA is also reactive with other thiolases in contrast to 2-methylacetoacetyl-CoA which is fully specific. Indeed, in cells from patients with mutations in the ACAT1 gene allowing no protein to be synthesized, there is no enzyme activity to be detected with 2-methylacetoacetyl-CoA as substrate which is not true if acetoacetyl-CoA is used as substrate. Molecular analysis of the ACAT1 gene notably by Fukao and coworkers has identified a great number of different mutations (see for instance [62]).

References

1. Roe CR, Cederbaum SD, Roe DS, Mardach R, Galindo A, Sweetman L. Isolated isobutyryl-CoA dehydrogenase deficiency: an unrecognized defect in human valine metabolism. Mol Genet Metab. 1998;65:264–71.
2. Koeberl DD, Young SP, Gregersen NS, et al. Rare disorders of metabolism with elevated butyryl- and isobutyryl-carnitine detected by tandem mass spectrometry newborn screening. Pediatr Res. 2003;54:219–23.
3. Sass JO, Sander S, Zschocke J. Isobutyryl-CoA dehydrogenase deficiency: isobutyrylglycinuria and ACAD8 gene mutations in two infants. J Inherit Metab Dis. 2004;27:741–5.
4. Pedersen CB, Bischoff C, Christensen E, et al. Variations in IBD (ACAD8) in children with elevated C4-carnitine detected by tandem mass spectrometry newborn screening. Pediatr Res. 2006;60:315–20.
5. Oglesbee D, He M, Majumder N, et al. Development of a newborn screening follow-up algorithm for the diagnosis of isobutyryl-CoA dehydrogenase deficiency. Genet Med. 2007;9:108–16.
6. Brown GK, Hunt SM, Scholem R, et al. Beta-hydroxyisobutyryl coenzyme A deacylase deficiency: a defect in valine metabolism associated with physical malformations. Pediatrics. 1982;70:532–8.
7. Loupatty FJ, Clayton PT, Ruiter JPN, et al. Mutations in the gene encoding 3-hydroxyisobutyryl-CoA hydrolase results in progressive infantile neurodegeneration. Am J Hum Genet. 2007;80:195–9.
8. Loupatty FJ, van der Steen A, IJlst L, et al. Clinical, biochemical, and molecular findings in three patients with 3-hydroxyisobutyric aciduria. Mol Genet Metab. 2006;87:243–8.
9. Pollitt RJ, Green A, Smith R. Excessive excretion of beta-alanine and of 3-hydroxypropionic, R- and S-3-aminoisobutyric, R- and S-3-hydroxyisobutyric and S-2-(hydroxymethyl)butyric acids probably due to a defect in the metabolism of the corresponding malonic semialdehydes. J Inherit Metab Dis. 1985;8:75–9.
10. Gray RG, Pollitt RJ, Webley J. Methylmalonic semialdehyde dehydrogenase deficiency: demonstration of defective valine and beta-alanine metabolism and reduced malonic semialdehyde dehydrogenase activity in cultured fibroblasts. Biochem Med Metab Biol. 1987;38:121–4.
11. Goodwin GW, Rougraff PM, Davis EJ, Harris RA. Purification and characterization of methylmalonate-semialdehyde dehydrogenase from rat liver. Identity to malonate-semialdehyde dehydrogenase. J Biol Chem. 1989; 264:14965–71.
12. Kedishvili NY, Popov KM, Rougraff PM, Zhao Y, Crabb DW, Harris RA. CoA-dependent methylmalonate-semialdehyde dehydrogenase, a unique member of the aldehyde dehydrogenase superfamily. cDNA cloning, evolutionary relationships, and tissue distribution. J Biol Chem. 1992;267:19724–9.
13. Chambliss KL, Gray RG, Rylance G, Pollitt RJ, Gibson KM. Molecular characterization of methylmalonate semialdehyde dehydrogenase deficiency. J Inherit Metab Dis. 2000;23:497–504.
14. Sass JO, Walter M, Shield JP, et al. 3-Hydroxyisobutyrate aciduria and mutations in the ALDH6A1 gene coding for methylmalonate semialdehyde dehydrogenase. J Inherit Metab Dis. 2012;35:437–42.
15. Shield JP, Gough R, Allen J, Newbury-Ecob R. 3-Hydroxyisobutyric aciduria: phenotypic heterogeneity within a single family. Clin Dysmorphol. 2001;10:189–91.
16. Tanaka K, Budd MA, Efron ML, Isselbacher KJ. Isovaleric acidemia: a new genetic defect of leucine metabolism. Proc Natl Acad Sci U S A. 1966;56:236–42.
17. Vockley J, Ensenauer R. Isovaleric acidemia: new aspects of genetic and phenotypic heterogeneity. Am J Med Genet C Semin Med Genet. 2006;142C:95–103.
18. Tokatli A, Coskun T, Ozalp I. Isovaleric acidemia. Clinical presentation of 6 cases. Turk J Pediatr. 1998;40:111–9.
19. Tanaka K. Isovaleric acidemia: personal history, clinical survey and study of the molecular basis. Prog Clin Biol Res. 1990;321:273–90.
20. Ensenauer R, Vockley J, Willard JM, et al. A common mutation is associated with a mild, potentially asymptomatic phenotype in patients with isovaleric acidemia diagnosed by newborn screening. Am J Hum Genet. 2004;75: 1136–42.
21. Gregersen N, Kolvraa S, Mortensen PB. Acyl-CoA: glycine N-acyltransferase: in vitro studies on the glycine conjugation of straight- and branched-chained acyl-CoA esters in human liver. Biochem Med Metab Biol. 1986;35:210–8.
22. Roe CR, Millington DS, Maltby DA, Kahler SG, Bohan TP. L-Carnitine therapy in isovaleric acidemia. J Clin Invest. 1984;74:2290–5.
23. Tanaka K, Isselbacher KJ. The isolation and identification of N-isovalerylglycine from urine of patients with isovaleric acidemia. J Biol Chem. 1967;242:2966–72.
24. Cohn RM, Yudkoff M, Rothman R, Segal S. Isovaleric acidemia: use of glycine therapy in neonates. N Engl J Med. 1978;299:996–9.
25. Yudkoff M, Cohn RM, Puschak R, Rothman R, Segal S. Glycine therapy in isovaleric acidemia. J Pediatr. 1978; 92:813–7.
26. Eminoglu FT, Ozcelik AA, Okur I, et al. 3-Methylcrotonyl-CoA carboxylase deficiency: phenotypic variability in a family. J Child Neurol. 2009;24:478–81.

27. Visser G, Suormala T, Smit GP, et al. 3-Methylcrotonyl-CoA carboxylase deficiency in an infant with cardiomyopathy, in her brother with developmental delay and in their asymptomatic father. Eur J Pediatr. 2000;159:901–4.
28. Grunert SC, Stucki M, Morscher RJ, et al. 3-Methylcrotonyl-CoA carboxylase deficiency: clinical, biochemical, enzymatic and molecular studies in 88 individuals. Orphanet J Rare Dis. 2012;7:31.
29. Stadler SC, Polanetz R, Maier EM, et al. Newborn screening for 3-methylcrotonyl-CoA carboxylase deficiency: population heterogeneity of MCCA and MCCB mutations and impact on risk assessment. Hum Mutat. 2006; 27:748–59.
30. Uematsu M, Sakamoto O, Sugawara N, et al. Novel mutations in five Japanese patients with 3-methylcrotonyl-CoA carboxylase deficiency. J Hum Genet. 2007;52:1040–3.
31. Dantas MF, Suormala T, Randolph A, et al. 3-Methylcrotonyl-CoA carboxylase deficiency: mutation analysis in 28 probands, 9 symptomatic and 19 detected by newborn screening. Hum Mutat. 2005;26:164.
32. Ihara K, Kuromaru R, Inoue Y, et al. An asymptomatic infant with isolated 3-methylcrotonyl-coenzyme: a carboxylase deficiency detected by newborn screening for maple syrup urine disease. Eur J Pediatr. 1997;156:713–5.
33. Wortmann SB, Kluijtmans LA, Rodenburg RJ, et al. 3-Methylglutaconic aciduria—lessons from 50 genes and 977 patients. J Inherit Metab Dis. 2013;36(6):913–21.
34. Barth PG, Scholte HR, Berden JA, et al. An X-linked mitochondrial disease affecting cardiac muscle, skeletal muscle and neutrophil leucocytes. J Neurol Sci. 1983;62:327–55.
35. Clarke SL, Bowron A, Gonzalez IL, et al. Barth syndrome. Orphanet J Rare Dis. 2013;8:23.
36. Wortmann SB, Vaz FM, Gardeitchik T, et al. Mutations in the phospholipid remodeling gene SERAC1 impair mitochondrial function and intracellular cholesterol trafficking and cause dystonia and deafness. Nat Genet. 2012;44:797–802.
37. Costeff H, Gadoth N, Apter N, Prialnic M, Savir H. A familial syndrome of infantile optic atrophy, movement disorder, and spastic paraplegia. Neurology. 1989;39:595–7.
38. Davey KM, Parboosingh JS, McLeod DR, et al. Mutation of DNAJC19, a human homologue of yeast inner mitochondrial membrane co-chaperones, causes DCMA syndrome, a novel autosomal recessive Barth syndrome-like condition. J Med Genet. 2006;43:385–93.
39. Honzik T, Tesarova M, Mayr JA, et al. Mitochondrial encephalocardio-myopathy with early neonatal onset due to TMEM70 mutation. Arch Dis Child. 2010;95:296–301.
40. Duran M, Beemer FA, Tibosch AS, Bruinvis L, Ketting D, Wadman SK. Inherited 3-methylglutaconic aciduria in two brothers—another defect of leucine metabolism. J Pediatr. 1982;101:551–4.
41. Wortmann SB, Kremer BH, Graham A, et al. 3-Methylglutaconic aciduria type I redefined A syndrome with late-onset leukoencephalopathy. Neurology. 2010;75:1079–83.
42. Loupatty FJ, Ruiter JPN, IJlst L, Duran M, Wanders RJA. Direct nonisotopic assay of 3-methylglutaconyl-CoA hydratase in cultured human skin fibroblasts to specifically identify patients with 3-methylglutaconic aciduria type I. Clin Chem. 2004;50:1447–50.
43. IJlst L, Loupatty FJ, Ruiter JPN, Duran M, Lehnert W, Wanders RJA. 3-Methylglutaconic aciduria type I is caused by mutations in AUH. Am J Hum Genet. 2002;71:1463–6.
44. Ly TB, Peters V, Gibson KM, et al. Mutations in the AUH gene cause 3-methylglutaconic aciduria type I. Hum Mutat. 2003;21:401–7.
45. Faull KF, Bolton PD, Halpern B, Hammond J, Danks DM. The urinary organic acid profile associated with 3-hydroxy-3-methylglutaric aciduria. Clin Chim Acta. 1976;73:553–9.
46. Wanders RJA, Zoeters PH, Schutgens RBH, et al. Rapid diagnosis of 3-hydroxy-3-methylglutaryl-coenzyme A lyase deficiency via enzyme activity measurements in leukocytes or platelets using a simple spectrophotometric method. Clin Chim Acta. 1990;189:327–34.
47. Gibson KM, Burlingame TG, Hogema B, et al. 2-Methylbutyryl-coenzyme A dehydrogenase deficiency: a new inborn error of L-isoleucine metabolism. Pediatr Res. 2000;47:830–3.
48. Andresen BS, Christensen E, Corydon TJ, et al. Isolated 2-methylbutyrylglycinuria caused by short/branched-chain acyl-CoA dehydrogenase deficiency: identification of a new enzyme defect, resolution of its molecular basis, and evidence for distinct acyl-CoA dehydrogenases in isoleucine and valine metabolism. Am J Hum Genet. 2000; 67:1095–103.
49. Madsen PP, Kibaek M, Roca X, et al. Short/branched-chain acyl-CoA dehydrogenase deficiency due to an IVS3+3A>G mutation that causes exon skipping. Hum Genet. 2006;118:680–90.
50. Sass JO, Ensenauer R, Roschinger W, et al. 2-Methylbutyryl-coenzyme A dehydrogenase deficiency: functional and molecular studies on a defect in isoleucine catabolism. Mol Genet Metab. 2008;93:30–5.
51. Alfardan J, Mohsen AW, Copeland S, et al. Characterization of new ACADSB gene sequence mutations and clinical implications in patients with 2-methylbutyrylglycinuria identified by newborn screening. Mol Genet Metab. 2010;100:333–8.
52. Zschocke J, Ruiter JPN, Brand J, et al. Progressive infantile neurodegeneration caused by 2-methyl-3-hydroxybutyryl-CoA dehydrogenase deficiency: a novel inborn error of branched-chain fatty acid and isoleucine metabolism. Pediatr Res. 2000;48:852–5.

53. Ofman R, Ruiter JPN, Feenstra M, et al. 2-Methyl-3-hydroxybutyryl-CoA dehydrogenase deficiency is caused by mutations in the HADH2 gene. Am J Hum Genet. 2003;72:1300–7.
54. Zschocke J. HSD10 disease: clinical consequences of mutations in the HSD17B10 gene. J Inherit Metab Dis. 2012;35:81–9.
55. Garcia-Villoria J, Navarro-Sastre A, Fons C, et al. Study of patients and carriers with 2-methyl-3-hydroxybutyryl-CoA dehydrogenase (MHBD) deficiency: difficulties in the diagnosis. Clin Biochem. 2009;42:27–33.
56. Perez-Cerda C, Garcia-Villoria J, Ofman R, et al. 2-Methyl-3-hydroxybutyryl-CoA dehydrogenase (MHBD) deficiency: an X-linked inborn error of isoleucine metabolism that may mimic a mitochondrial disease. Pediatr Res. 2005;58:488–91.
57. Rauschenberger K, Scholer K, Sass JO, et al. A non-enzymatic function of 17beta-hydroxysteroid dehydrogenase type 10 is required for mitochondrial integrity and cell survival. EMBO Mol Med. 2010;2:51–62.
58. Holzmann J, Frank P, Loffler E, Bennett KL, Gerner C, Rossmanith W. RNase P without RNA: identification and functional reconstitution of the human mitochondrial tRNA processing enzyme. Cell. 2008;135:462–74.
59. Daum RS, Lamm PH, Mamer OA, Scriver CR. A "new" disorder of isoleucine catabolism. Lancet. 1971;2: 1289–90.
60. Robinson BH, Sherwood WG, Taylor J, Balfe JW, Mamer OA. Acetoacetyl CoA thiolase deficiency: a cause of severe ketoacidosis in infancy simulating salicylism. J Pediatr. 1979;95:228–33.
61. Middleton B, Bartlett K. The synthesis and characterisation of 2-methylacetoacetyl coenzyme A and its use in the identification of the site of the defect in 2-methylacetoacetic and 2-methyl-3-hydroxybutyric aciduria. Clin Chim Acta. 1983;128:291–305.
62. Fukao T, Yamaguchi S, Orii T, Hashimoto T. Molecular basis of beta-ketothiolase deficiency: mutations and polymorphisms in the human mitochondrial acetoacetyl-coenzyme A thiolase gene. Hum Mutat. 1995;5:113–20.

Chapter 12
Branched Chain Amino Acids and Maple Syrup Urine Disease

Kevin Carpenter

Key Points

- Maple syrup urine disease is an autosomal recessive condition affecting metabolism of valine, leucine, and isoleucine, caused by mutations in four genes encoding the proteins which make up the branched chain α-keto acid decarboxylase enzyme complex.
- The disorder may be classified according to clinical and biochemical presentation.
- Classical MSUD is the most common form and untreated neonates will begin to develop symptoms by 48 h including "maple syrup" odor, particularly in the cerumen, lethargy, poor feeding, cerebral edema, encephalopathy, and coma.
- Milder forms of the disorder may present later with intermittent, intermediate, and thiamine responsive classifications described.
- Newborn screening programs which use tandem mass spectrometry to measure leucine isomers will detect classical MSUD but milder forms may be missed.
- Presence of alloisoleucine in plasma is considered pathognomonic for the condition.
- Patients are treated by restriction of dietary leucine, supplementation with BCAA free medical foods, valine, and isoleucine as indicated by biochemical monitoring.
- Acute metabolic decompensation may require extracorporeal removal of leucine to prevent life threatening cerebral edema.
- Neurological outcome appears to be related to the degree and duration of hyperleucinemia.

Keywords Inborn error • Maple syrup urine disease • Alloisoleucine • Newborn screening • Branched chain amino acids

Abbreviations

BCAA	Branched chain amino acids
BCAT	Branched chain amino acid aminotransferase
BCKD	Branched chain α-keto acid decarboxylase/dehydrogenase

K. Carpenter, Ph.D. (✉)
NSW Biochemical Genetics Service, The Children's Hospital at Westmead, University of Sydney, Hawkesbury Road, Westmead, NSW 2145, Australia
e-mail: Kevin.carpenter@health.nsw.gov.au

R. Rajendram et al. (eds.), *Branched Chain Amino Acids in Clinical Nutrition: Volume 1*, Nutrition and Health, DOI 10.1007/978-1-4939-1923-9_12,

BCKDC	Branched chain α-keto acid decarboxylase complex
BIA	Bacterial inhibition assay
CT	Computerized tomography
GABA	Gamma-aminobutyric acid
KIC	2-ketoisocaproic acid
KIV	2-ketoisovaleric acid
KMV	2-keto-3-methylvaleric acid
LNAA	Large neutral amino acids
MR	Magnetic resonance
MSUD	Maple syrup urine disease
NMDA	*N*-methyl-D-aspartate

Introduction

In 1954 Menkes et al. [1] described a new syndrome affecting four siblings with a progressive infantile cerebral dysfunction, commencing in the first week of life and resulting in death within 3 months of age. The patient's urine had a distinctive smell resembling burnt sugar or maple syrup. Three years later, Westall et al. coined the term Maple Syrup Urine Disease (MSUD) and found the branched chain amino acids (BCAA), valine, leucine, and isoleucine were markedly elevated in these patients [2]. Menkes subsequently identified the corresponding branched chain keto acids in the urine, suggesting the defect lay in the decarboxylase step of the catabolic pathway [3]. MSUD is thus an organic acid disorder although some of the clinical features result from accumulation of the amino acid precursors. Overall incidence is around 1:185,000 but is much higher in certain populations (see Sect. "Genetics" below).

Enzymology

The first two steps in the catabolic pathway are shared by all three branched chain amino acids (Fig. 12.1). The first step is a reversible transamination catalyzed by branched chain amino acid aminotransferase (BCAT) to produce the keto acids. The second step is an oxidative decarboxylation catalyzed by the branched chain α-keto acid decarboxylase complex (BCKDC). These two enzymes are now known to associate as a supramolecular complex effecting efficient channelling of substrates and control over the initial step of BCAA oxidation [4].

The branched chain α-keto acid decarboxylase enzyme complex comprises the catalytic components; branched chain α-keto acid decarboxylase/dehydrogenase, E1 (EC 1.2.4.4), an α2β2 heterotetramer, dihydrolipoyltransacylase, E2, (EC 2.3.1.168), and dihydrolipoamide dehydrogenase E3, (EC 1.8.1.4). There are also a BCKD kinase and BCKD phosphatase which have a regulatory role effected through phosphorylation of the E1 component which results in complete inactivation. Inactivating mutations of BCKD kinase have recently been associated with low levels of BCAA, autism, epilepsy, and intellectual disability that may respond to dietary treatment [5].

The BCKD complex is organized around a core of 24 E2 subunits forming a cuboid structure, which is surrounded by multiple copies of the E1 and E3 proteins, the BCKD kinase, and the BCKD phosphatase through a subunit-binding domain [6].

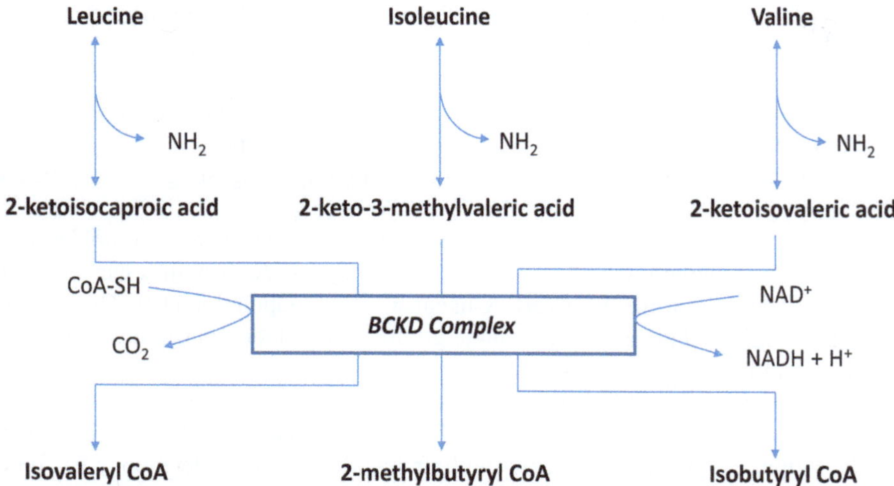

Fig. 12.1 Early steps in the catabolic pathways for branched chain amino acids. The first step is a reversible transamination catalyzed by branched chain amino acid aminotransferase (BCAT) to produce the keto acids. The second step is an oxidative decarboxylation catalyzed by the branched chain α-keto acid decarboxylase complex (BCKDC). These two enzymes associate as a supramolecular complex effecting efficient channelling of substrates and control over the initial step of BCAA oxidation

Table 12.1 Summary of MSUD mutations reported in the literature to March 2013 [8, 9]

Summary of reported MSUD mutations					
Gene	Missense/ nonsense	Small deletions/ insertions	Large deletions	Other (splicing, regulatory)	Total
BCKHDA	47	10	2	6	65
BCKHDB	48	16	0	2	66
DBT	27	11	8	8	54
DLD	12	3	0	2	17

Genetics

Two genes, BKCDHA, chromosomal locus 19q13.2 and BKCDHB, 6q14.1, encode the E1α and E1β subunits with the E2 subunit encoded by the gene DBT, 1p21.2. Homozygous or compound heterozygous mutations on the same allele in any of these genes can lead to reduced activity of the BCKD complex and the biochemical and clinical features of MSUD.

The E3 subunit, encoded by DLD, is also a component of pyruvate dehydrogenase complex, α-ketoglutarate dehydrogenase complex and the glycine cleavage system and mutations in this gene have a different and usually more severe phenotype.

Maple syrup urine disease can be classified based on mutation groups. Mutations in the BKCDHDA and BKCDHB genes result in type IA and IB respectively. Type II MSUD results from mutations in the E2 subunit and missense mutations in this gene mostly result in thiamine responsive or intermittent clinical phenotypes [7]. Type III MSUD is more often referred to as dihydrolipoamide dehydrogenase deficiency, reflecting the multiple pathways involved.

There is great heterogeneity in the mutations underlying MSUD. In early 2013 there were 202 mutations described; 65 in type IA, 66 in type IB, 54 in type II, and 17 in type III [8, 9] (Tables 12.1).

Type IA mutations are mostly missense with small numbers of splicing, frameshift and nonsense mutations. They are predominantly associated with a classical MSUD phenotype and appear panethnic in origin. The commonest mutation, c.1312T>A, is responsible for the very high disease frequency amongst Old Order Mennonites (1:176) due to a founder effect [10]. However this mutation has also been found in non-Mennonite patients not all of whom share common Mennonite haplotypes, suggesting the occurrence of the defect in these families is due to either pre-Mennonite or de novo events [11].

A second founder effect type IA mutation occurs in Portuguese Gypsies, c.117delC, but this mutation has also been identified in another distinct population group, suggesting this may be a mutation hotspot [12].

Type IB mutations also largely result in a classical phenotype but with some milder variants [6]. There is one founder mutation, c.832G>A, found in Ashkenazi Jews with a carrier frequency of approximately 1:113 found in the New York Ashkenazi Jewish population [13]. The remaining type IB mutations are distributed across several ethnic groups and include nonsense and missense mutations but with a significant number of small deletions.

Type II mutations are also widely distributed amongst ethnic groups. Two significant deletions, one of 4.7 kb in the Paiwan aboriginal tribe in Taiwan [14], and the other 4.1 kb with a carrier frequency of 1:100 in the Filipino population [15] represent founder effects. A 2-bp deletion, c.75_76delAT, has been reported in a number of different populations [16]. Almost 40 % of type II mutations are associated with an intermittent or thiamine responsive phenotype [6].

Type III mutations reported in the DLD gene are predominantly missense with c.685G>T being the most frequent and found in the Ashkenazi Jewish population worldwide [17]. Several other mutations have been reported in this group with most other type III mutations representing private mutations.

Clinical Presentations

Classical MSUD

The condition described by Menkes in 1954 is now known to represent the severe end of the clinical spectrum, typically with less than 2 % residual enzyme activity in fibroblasts. Around 80 % of MSUD patients fall into this category [18] and are characterized by irritability and poor feeding with onset typically within the first 48 h of life (Tables 12.2). Untreated, they develop cerebral edema which results in encephalopathy, alternating hypertonia and hypotonia, opisthotonus, lethargy, and coma, within the first 2 weeks [19]. Irreversible brain damage will occur without detoxification and initiation of dietary leucine restriction.

Intermediate MSUD

Schulman et al. reported a new variant of MSUD with a milder phenotype [20]. This child had an unremarkable neonatal course and had several mild illnesses without acute symptoms. However, she exhibited marked developmental delay and was unable to sit alone at 16 months of age. Her physicians noted she had a smell of maple syrup and her blood showed increased branched chain amino acids. Residual enzyme activity (in leucocytes) ranged from 15 to 25 % of normal for each of the three branched chain amino acid substrates.

Untreated, intermediate MSUD patients show consistently abnormal plasma branch chain amino acids but may escape acute metabolic crises except during severe catabolism. However, the chronic increases in leucine and keto acids result in developmental delays and because these patients are unlikely to present clinically and start treatment early, developmental outcomes are often worse than neonatal presenting cases [21].

Table 12.2 Summary of clinical and biochemical phenotypes in MSUD

Clinical presentations of maple syrup urine disease			
Type	Biochemical features	Clinical features	Age of presentation
Classical	Increased plasma BCAA with loss of normal ratios	Poor feeding, lethargy Cerebral edema	Neonatal
	Presence of alloisoleucine in plasma	Hypertonia/hypotonia Opisthotonus	
	Increased urine BCKA on organic acids Ketosis	Coma Maple syrup odor	
Intermittent (including thiamine responsive)	Increased BCAA and alloisoleucine in plasma *but may not be detected on newborn screening*	Failure to thrive Developmental delay	Variable
	Increased urine BCKA on organic acids	Encephalopathy during intercurrent illness	
Intermediate	Plasma BCAA may be normal when well or on protein restriction	Episodic ataxia, lethargy, and confusion, associated with intercurrent illness	Variable
	During catabolic illness, pattern similar to classical	Normal development unless metabolic decompensation causes acute brain injury	

Intermittent MSUD

This variant form of MSUD was first described by Morris et al. [22]. The patient developed episodes of ataxia, lethargy, and semi-coma following otitis media at the age of 16 months. Symptoms recurred several times over a 12-month period and each time a smell of maple syrup and increased excretion of keto acids was noted. Previously her growth and development were normal and between episodes her biochemistry was normal. Several other patients have been described with normal development but episodic presentation of MSUD typically associated with intercurrent illness [19]. Residual enzyme activity is in the 5–20 % range but unlike the intermediate form, BCAA levels are normal when well.

Thiamine Responsive MSUD

This variant was first described by Scriver et al. [23] who reported a patient with the intermediate form of the disease who's metabolic phenotype improved when given pharmacological doses of thiamine. This patient was reviewed at 15 years of age and there was no definitive long-term response to thiamine in vivo [24]. Several other patients with varying degrees of biochemical response to thiamine have been described but none have been treated with thiamine alone and the phenotype is similar to intermediate MSUD [19].

Diagnosis

Newborn Screening

Maple syrup urine disease was one of the first disorders for which a newborn blood spot screening test was developed. A bacterial inhibition assay (BIA) for increased leucine, similar in principle to the PKU test, was developed by Children's Hospital Buffalo in 1964 and its use was reviewed in 1974 by which time over 9.5 million newborns had been screened and 43 confirmed cases detected [25]. The introduction of tandem mass spectrometry based newborn screening in the late 1990s allowed quantitation of total leucine species (leucine, isoleucine and alloisoleucine) as well as valine and phenylalanine and afforded improved discrimination over BIA method when amino acid values were used in combination [total leucines:phenylalanine ratio] [26]. False positive rates can be improved further by second tier testing on the initial dried blood spot to separate leucine isomers [27].

Newborn screening has allowed early treatment and less severe biochemical disturbances compared to unscreened clinically presenting cases [28]. Median day of life for confirmation of diagnosis was 7 days in the screened population compared to 12 days in the clinically presenting cohort. Plasma leucine concentration at commencement of treatment was also lower; mean 2,100 μmol/L in cases identified on screening, 2,873 μmol/L in clinically presenting cases. However, newborn screening may fail to identify intermediate and intermittent forms of the condition [21] so MSUD still needs to be considered in appropriate clinical situations later in life.

Plasma Amino Acids

Abnormalities in plasma amino acid concentrations in classical MSUD include marked elevations of valine, leucine, and isoleucine with a characteristic loss of the normal ratios between valine–leucine (normally 2:1) and leucine–isoleucine (normally 2:1). Leucine may reach concentrations of 5 mmol/L with valine and isoleucine as high as 1 mmol/L. There is also a strong reciprocal relationship between the concentrations of leucine and alanine and leucine and glutamine [19]. In addition, patients with MSUD produce alloisoleucine in vivo from isoleucine by a mechanism which appears to involve condensation with the aldehyde moiety of the pyridoxyl phosphate–aminotransferase enzyme complex to form an aldimine Schiff base, followed by enzyme-mediated isomerization, and formation of alloisoleucine by reversal of the transamination step [29]. Although formation of alloisoleucine occurs in healthy individuals, plasma levels are very low and typically undetectable. In MSUD there is both increased level of isoleucine resulting in enhanced formation and impaired metabolism of alloisoleucine and the presence of this amino acid above the cutoff value of 5 μmol/L is the most specific and most sensitive diagnostic marker for all forms of MSUD [30].

Urine Organic Acids

As expected the characteristic organic acids excreted in MSUD are 2-ketoisocaproic acid (KIC) from leucine, 2-keto-3-methylvaleric acid (KMV) from isoleucine, and 2-ketoisovaleric acid (KIV) from valine. The hydroxyl derivatives 2-hydroxyisovaleric, 2-hydroxyisocaproic, and 2-hydroxy-3-methylvaleric acids, produced by the reduction of their respective α-keto acids, also accumulate [31]. These metabolites can be detected using the dinitrophenylhydrazine (DNPH) test by mixing equal volumes of urine and DNPH reagent (1 g/L DNPH in 2 M HCl) and observing for yellow-white precipitate after 10 min. This test is still used by some centers to allow parents to identify impending decompensation and initiate changes to therapy. During metabolic decompensation patients are also ketotic and often have elevated lactate. Urine test strips to identify ketonuria may be used as a marker of metabolic instability if access to more formal testing is unavailable or delayed. Although not required to confirm a diagnosis in a patient with elevated alloisoleucine, organic acid analysis is useful in identifying intermediate MSUD patients as part of a differential diagnosis of developmental delay.

Pathophysiology

The neurotoxic mechanisms involved in MSUD are still being elucidated. Leucine and KIC are considered to be the primary neurotoxins since acute symptoms are closely correlated with plasma concentrations [6]. During acute decompensation, MR and CT imaging shows diffuse generalized brain edema. In addition, a localized, more severe edema which involves the deep cerebellar white matter, the dorsal part of the brainstem, the cerebral peduncles, and the dorsal limb of the internal capsule is seen [32]. The cerebral edema may lead to herniation and death [33]. Jan et al. [34] used diffusion weighted imaging and demonstrated reversible marked restriction of proton diffusion compatible with cytotoxic or intramyelinic sheath edema in the brainstem, basal ganglia, thalami, cerebellar and periventricular white matter, and the cerebral cortex. This suggests increased intracellular and decreased extracellular fluid resulting from sodium pump failure. There is evidence that KIC acts as an uncoupler of oxidative phosphorylation and as a metabolic inhibitor possibly through its inhibitory effect on α-ketoglutarate dehydrogenase activity, while leucine acts as a metabolic inhibitor [35]. Branched chain amino acids and corresponding keto acids significantly inhibited Na(+), K(+)-ATPase activity in synaptic plasma membranes in rat cerebral cortex [36].

Hyperleucinemia also induces significant reductions in large neutral amino acids (LNAA) in brain [37, 38]. Since leucine and the other LNAA penetrate the blood–brain barrier exclusively through the L-amino acid transporter and leucine has the highest affinity for this system, gross increases in leucine would compromise the influx of the other amino acids. This may lead to an acute depletion of the neurotransmitters dopamine and serotonin due to restricted availability of tyrosine and tryptophan. This may account for some of the early changes, such as dystonia, seen during decompensation events. More chronic effects on protein synthesis may contribute to brain-specific growth restriction.

Leucine, isoleucine, KIV, and KMV also reduce glutamate uptake by the synaptic vesicles by up to 60 % [39]. The presence of high amounts of glutamate in the synaptic cleft may lead to excitotoxicity which has been implicated in the neuropathology of acute and chronic brain disorders [40]. In the rat, KIV has been shown to elicit clonic convulsions in a dose responsive manner which are attenuated by muscimol and MK-801, suggesting GABAergic and glutamatergic NMDA mechanisms are involved [41].

The pathophysiology of the chronic effects of MSUD on brain development probably includes the energy deprivation and neurotransmitter and protein synthesis failures described above. Oxidative stress is also implicated in the pathophysiology with BCKA inducing marked reduction of the non-enzymatic antioxidant defences and reduced activity of superoxide dismutase and glutathione peroxidase [42]. Plasma from MSUD patients shows thiobarbituric acid-reactive species are increased and inversely correlated with methionine and tryptophan levels [43]. Supplementing with L-carnitine may have an antioxidant effect, protecting against lipid peroxidation [44] and antioxidant treatment (*N*-acetylcysteine plus deferoxamine) prevented DNA damage in an animal model for MSUD [45].

Treatment

Dietary Therapy

The possible beneficial effects of a diet low in BCAA was proposed as early as 1959 by Mackenzie and Woolf [46] and this remains the cornerstone of management in MSUD today. Commercial BCAA free formula and protein substitutes are used to prevent deficiencies of essential amino acids and allow adequate growth. Because the levels of isoleucine and valine in natural protein are lower than leucine, restriction of natural protein may lead to low levels of other BCAA, even when leucine is high,

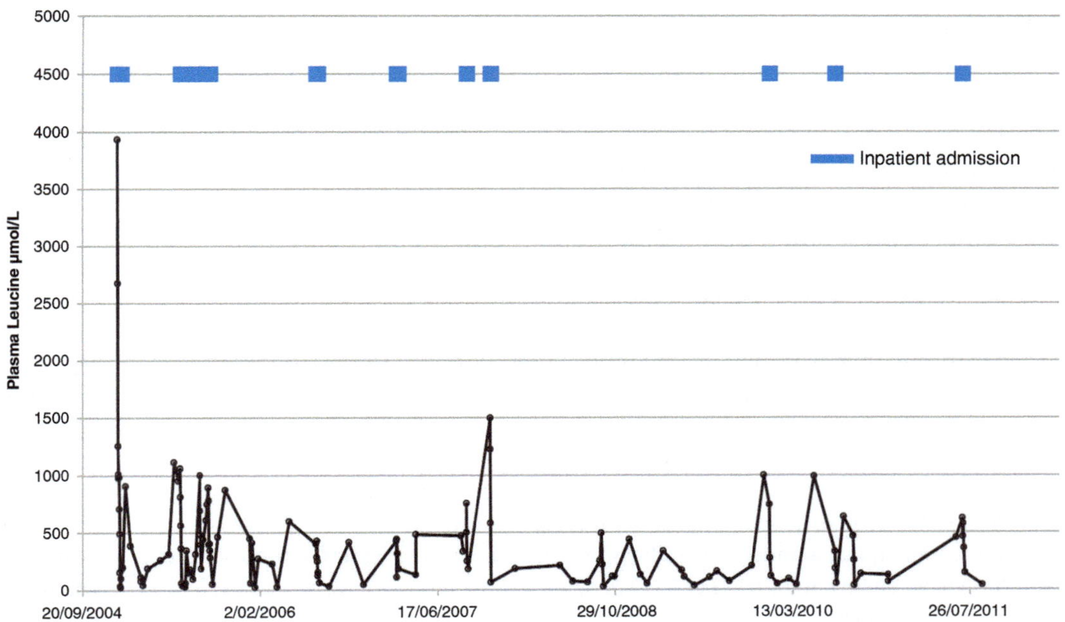

Fig. 12.2 Plasma leucine measurements over a 5-year period in a classical MSUD patient detected on newborn screening and admitted for CVVH on day 7 of life. Note repeated admissions triggered by intercurrent infections requiring parenteral nutrition

and valine and isoleucine may need to be supplemented to effect protein synthesis [47]. In contrast to leucine, elevated levels of these amino acids appear to be well tolerated [6].

Monitoring of blood or plasma levels of BCAA is an essential part of management and the use of blood spot cards mailed to a reference center has been shown to be effective for patients living in rural and regional areas where there are no local laboratory services [48]. The aim of dietary therapy is to achieve age appropriate tolerance of leucine and maintain stable plasma branched chain amino acid concentrations and ratios [19]. During intercurrent illness an enteral "sick day formula" with no leucine and enriched with calories, isoleucine, valine, and BCAA-free amino acids is routinely used but hospital admission and more aggressive therapy are sometimes inevitable (Fig. 12.2). Parenteral nutrition with BCAA free amino acid mixtures, if available, and use of intralipid to promote anabolism is used when oral intake is being poorly tolerated.

Extracorporeal Removal Therapies

In the acute encephalopathy of classical MSUD, it may be necessary to induce rapid removal of toxic metabolites to control cerebral edema and reduce the risk of herniation. There is also evidence that long-term intellectual outcome may be related to the length of exposure and degree of hyperleucinemia in the neonatal period [49]. Peritoneal dialysis was shown to increase the clearance of BCAA over sevenfold compared to urinary clearance and was successfully used in an 8-day-old patient [50]. Hemodialysis is also highly effective in reducing leucine levels [51] and continuous venovenous hemofiltration has been successfully used in our institution for many years [52]. In all instances the extracorporeal removal therapies are used in combination with nutritional therapy and promotion of anabolism. Some centers have found a combination of enteral and parenteral treatment results in leucine clearance rates consistently higher than those reported for dialysis or hemoperfusion alone [47].

Other Therapeutic Options

A trial of pharmacological doses of thiamine will reveal the small proportion of patients with the thiamine responsive form of the disorder. Although responsive patients will show increased leucine tolerance on thiamine, dietary restriction is still required [19].

Sodium phenylbutyrate has been proposed as a treatment to reduce plasma levels of branch chain amino and keto acids in intermediate MSUD [53]. The phenylbutyrate appears to prevent phosphorylation of E1 by inhibition of the BCKDC kinase, increasing BCKDC overall activity.

Reduction of leucine accumulation in the brain by competition for the L-amino acid transporter at the blood–brain barrier has been demonstrated in the mouse model using norleucine, an isomer of leucine and isoleucine [54]. The authors also reported a reduction in accumulation in KIC, suggesting norleucine may also inhibit transamination of leucine to KIC.

Liver transplantation has been used to correct MSUD. Strauss et al. [55] reviewed ten cases of orthotopic liver transplantation in classical MSUD patients. All patients were alive after a median follow-up period of 14 months and plasma amino acid concentration ratios were less variable and remained appropriately regulated through periods of dietary fluctuation, fasting and illness although alloisoleucine remained detectable at low levels in the plasma. The risks of surgery and immune suppression were not found to be different from those for other pediatric transplant patients and this was considered a potentially cost-effective alternative to lifelong dietary therapy and the risks of recurrent metabolic decompensation. However, recent data shows psychiatric morbidity remains significant in MSUD patients who have undergone liver transplant [56].

Hepatocyte transplantation has been investigated in a mouse model of intermediate MSUD and has shown improved leucine tolerance and extended survival [57]. The same researchers have recently described direct intrahepatic injections of human amnion epithelial cells performed eight times over the first 35 days of life in their mouse model. Branched chain α-keto acid dehydrogenase enzyme activity was significantly increased in transplanted livers [58]. The authors suggest treatment at the neonatal stage is clinically relevant for MSUD and may offer a donor cell engraftment advantage.

Long-Term Outcome

Despite improvements in early diagnosis and treatment, unimpaired outcome of patients with classic or very severe variant MSUD is rare. A 2007 study of 22 adults with severe MSUD found low educational levels, a low participation in the workforce and most could not live autonomously, did not have a steady partnership and had no children [59].

Mental health outcomes in patients with classic MSUD have been studied by Muelly et al. [56]. They reported on a cohort of 37 patients, aged 5–35 years, with the majority of subjects Old Order Mennonite and homozygous for the BCKDHA c.1312T>A mutation. Twenty-six were on conventional dietary therapy and 11 had undergone liver transplantation. Results were compared with 26 age matched control subjects.

Incidence of depression, anxiety, inattention, and impulsivity were all higher in MSUD patients and had a cumulative lifetime incidence of 83 % by age of 36 years. Neonatal encephalopathy was a strong predictor of mood disorders but depression and anxiety did not correlate strongly with indices of lifetime metabolic control.

IQ was lower in MSUD patients treated with diet (81 ± 19) or transplant (90 ± 15), compared to controls (106 ± 15). Measures of intelligence were inversely correlated to average lifetime plasma leucine and positively correlated with the frequency of amino acid monitoring.

Quantitative proton magnetic resonance spectroscopy revealed low glutamate in all brain regions of MSUD patients, and deficits of NAA and creatine in the anterior cingulate cortex and parietal white matter. The authors suggest depletion of glutamate may have consequences for cognition and behavior, whilst cortical NAA and creatine deficiency may simply reflect reduced neuronal density.

Simon et al. reviewed 16 individuals with non-classical MSUD aged between 6 and 30 years [18]. Eight patients had an early diagnosis either through newborn screening or early clinical presentation. Only one of these were considered to have significant mental impairment but she was able to live an independent life and the remainder were attending normal schools or were employed.

Six subjects presented with encephalopathy triggered by catabolic stress consistent with intermittent phenotype between the ages of 5 months and 7 years. Five of this group were treated with either modest protein restriction or full MSUD diet. The remaining patient had no protein restriction. Two treated patients had further metabolic crises but all patients in this group had normal psychomotor development.

The remaining two patients presented at ages of 4 and 5 years. One had a history of ketoacidotic episodes, ataxia, and somnolence. Both had psychomotor delay, spasticity, and seizures with no response to diet.

Conclusions

Maple syrup urine disease represents a significant proportion of the patient cohort in most inherited metabolic disease centers around the world. Despite the advances in early detection through newborn screening, classical patients require intensive monitoring and aggressive treatment throughout life. Intermittent and intermediate forms are likely to be missed by newborn screening and can have poor developmental outcomes despite their milder disease.

Dietary treatment remains the mainstay of therapy but liver transplantation is an option for the severe classical cases.

References

1. Menkes JH, Hurst PL, Craig JM. A new syndrome: progressive familial infantile cerebral dysfunction associated with an unusual urinary substance. Pediatrics. 1954;14:462–7.
2. Westall RG, Dancis J, Miller S. Maple syrup urine disease. Am J Dis Child. 1957;94:2.
3. Menkes JH. Maple syrup disease; isolation and identification of organic acids in the urine. Pediatrics. 1959;23: 348–53.
4. Islam MM, Wallin R, Wynn RM, et al. A novel branched-chain amino acid metabolon. Protein-protein interactions in a supramolecular complex. J Biol Chem. 2007;282:11893–903.
5. Novarino G, El-Fishawy P, Kayserili H, et al. Mutations in BCKD-kinase lead to a potentially treatable form of autism with epilepsy. Science. 2012;338:394–7.
6. Chuang DT, Wynn RM, Shih VE. Maple syrup urine disease (branched-chain ketoaciduria). In: Valle D, Beaudet AL, Vogelstein B, Kinzler KW, Antonarakis S, Ballabio A, editors. Online metabolic and molecular bases of inherited disease. New York: McGraw-Hill; 2006.
7. Chuang JL, Wynn RM, Moss CC, et al. Structural and biochemical basis for novel mutations in homozygous Israeli maple syrup urine disease patients: a proposed mechanism for the thiamin-responsive phenotype. J Biol Chem. 2004;279:17792–800.
8. Jaafar N, Moleirinho A, Kerkeni E, et al. Molecular characterization of maple syrup urine disease patients from Tunisia. Gene. 2013;517:116–9.
9. The Human Gene mutation Database. 2012. www.hgmd.cf.ac.uk. Accessed 25 Mar 2013.
10. Mitsubuchi H, Matsuda I, Nobukuni Y, et al. Gene analysis of Mennonite maple syrup urine disease kindred using primer-specified restriction map modification. J Inherit Metab Dis. 1992;15:181–7.

11. Love-Gregory LD, Grasela J, Hillman RE, Phillips CL. Evidence of common ancestry for the maple syrup urine disease (MSUD) Y438N allele in non-Mennonite MSUD patients. Mol Genet Metab. 2002;75:79–90.
12. Quental S, Gusmao A, Rodriguez-Pombo P, et al. Revisiting MSUD in Portuguese Gypsies: evidence for a founder mutation and for a mutational hotspot within the BCKDHA gene. Ann Hum Genet. 2009; 73:298–303.
13. Edelmann L, Wasserstein MP, Kornreich R, Sansaricq C, Snyderman SE, Diaz GA. Maple syrup urine disease: identification and carrier-frequency determination of a novel founder mutation in the Ashkenazi Jewish population. Am J Hum Genet. 2001;69:863–8.
14. Chi CS, Tsai CR, Chen LH, et al. Maple syrup urine disease in the Austronesian aboriginal tribe Paiwan of Taiwan: a novel DBT (E2) gene 4.7 kb founder deletion caused by a nonhomologous recombination between LINE-1 and Alu and the carrier-frequency determination. Eur J Hum Genet. 2003;11:931–6.
15. Silao CL, Padilla CD, Matsuo M. A novel deletion creating a new terminal exon of the dihydrolipoyl transacylase gene is a founder mutation of Filipino maple syrup urine disease. Mol Genet Metab. 2004;81:100–4.
16. Fisher CW, Fisher CR, Chuang JL, Lau KS, Chuang DT, Cox RP. Occurrence of a 2-bp (AT) deletion allele and a nonsense (G-to-T) mutant allele at the E2 (DBT) locus of six patients with maple syrup urine disease: multiple-exon skipping as a secondary effect of the mutations. Am J Hum Genet. 1993;52:414–24.
17. Shaag A, Saada A, Berger I, et al. Molecular basis of lipoamide dehydrogenase deficiency in Ashkenazi Jews. Am J Med Genet. 1999;82:177–82.
18. Simon E, Flaschker N, Schadewaldt P, Langenbeck U, Wendel U. Variant maple syrup urine disease (MSUD)—the entire spectrum. J Inherit Metab Dis. 2006;29:716–24.
19. Strauss KA, Puffenberger EG, Morton DH. Maple syrup urine disease. In: Pagon RA, Bird TD, Dolan CR, Stephens K, Adam MP, editors. GeneReviews. Seattle, WA; 2006 Jan 30 [updated 2009 Dec 15].
20. Schulman JD, Lustberg TJ, Kennedy JL, Museles M, Seegmiller JE. A new variant of maple syrup urine disease (branched chain ketoaciduria). Clinical and biochemical evaluation. Am J Med. 1970;49:118–24.
21. Bhattacharya K, Khalili V, Wiley V, Carpenter K, Wilcken B. Newborn screening may fail to identify intermediate forms of maple syrup urine disease. J Inherit Metab Dis. 2006;29:586.
22. Morris MD, Lewis BD, Doolan PD, Harper HA. Clinical and biochemical observations on an apparently nonfatal variant of branched-chain ketoaciduria (maple syrup urine disease). Pediatrics. 1961;28:918–23.
23. Scriver CR, Mackenzie S, Clow CL, Delvin E. Thiamine-responsive maple-syrup-urine disease. Lancet. 1971;1:310–2.
24. Scriver CR, Clow CL, George H. So-called thiamin-responsive maple syrup urine disease: 15-year follow-up of the original patient. J Pediatr. 1985;107:763–5.
25. Naylor EW, Guthrie R. Newborn screening for maple syrup urine disease (branched-chain ketoaciduria). Pediatrics. 1978;61:262–6.
26. Chace DH, Hillman SL, Millington DS, Kahler SG, Roe CR, Naylor EW. Rapid diagnosis of maple syrup urine disease in blood spots from newborns by tandem mass spectrometry. Clin Chem. 1995;41:62–8.
27. Matern D, Tortorelli S, Oglesbee D, Gavrilov D, Rinaldo P. Reduction of the false-positive rate in newborn screening by implementation of MS/MS-based second-tier tests: the Mayo Clinic experience (2004–2007). J Inherit Metab Dis. 2007;30:585–92.
28. Simon E, Fingerhut R, Baumkotter J, Konstantopoulou V, Ratschmann R, Wendel U. Maple syrup urine disease: favourable effect of early diagnosis by newborn screening on the neonatal course of the disease. J Inherit Metab Dis. 2006;29:532–7.
29. Mamer OA. Initial catabolic steps of isoleucine, the R-pathway and the origin of alloisoleucine. J Chromatogr B Biomed Sci Appl. 2001;758:49–55.
30. Schadewaldt P, Bodner-Leidecker A, Hammen HW, Wendel U. Significance of L-alloisoleucine in plasma for diagnosis of maple syrup urine disease. Clin Chem. 1999;45:1734–40.
31. Treacy E, Clow CL, Reade TR, Chitayat D, Mamer OA, Scriver CR. Maple syrup urine disease: interrelations between branched-chain amino-, oxo- and hydroxyacids; implications for treatment; associations with CNS dysmyelination. J Inherit Metab Dis. 1992;15:121–35.
32. Brismar J, Aqeel A, Brismar G, Coates R, Gascon G, Ozand P. Maple syrup urine disease: findings on CT and MR scans of the brain in 10 infants. AJNR Am J Neuroradiol. 1990;11:1219–28.
33. Riviello Jr JJ, Rezvani I, DiGeorge AM, Foley CM. Cerebral edema causing death in children with maple syrup urine disease. J Pediatr. 1991;119:42–5.
34. Jan W, Zimmerman RA, Wang ZJ, Berry GT, Kaplan PB, Kaye EM. MR diffusion imaging and MR spectroscopy of maple syrup urine disease during acute metabolic decompensation. Neuroradiology. 2003;45:393–9.
35. Amaral AU, Leipnitz G, Fernandes CG, Seminotti B, Schuck PF, Wajner M. Alpha-ketoisocaproic acid and leucine provoke mitochondrial bioenergetic dysfunction in rat brain. Brain Res. 2010;1324:75–84.
36. Wajner A, Burger C, Dutra-Filho CS, Wajner M, de Souza Wyse AT, Wannmacher CM. Synaptic plasma membrane Na(+), K(+)-ATPase activity is significantly reduced by the alpha-keto acids accumulating in maple syrup urine disease in rat cerebral cortex. Metab Brain Dis. 2007;22:77–88.

37. Araujo P, Wassermann GF, Tallini K, et al. Reduction of large neutral amino acid levels in plasma and brain of hyperleucinemic rats. Neurochem Int. 2001;38:529–37.
38. Wajner M, Coelho DM, Barschak AG, et al. Reduction of large neutral amino acid concentrations in plasma and CSF of patients with maple syrup urine disease during crises. J Inherit Metab Dis. 2000;23:505–12.
39. Tavares RG, Santos CE, Tasca CI, Wajner M, Souza DO, Dutra-Filho CS. Inhibition of glutamate uptake into synaptic vesicles of rat brain by the metabolites accumulating in maple syrup urine disease. J Neurol Sci. 2000;181:44–9.
40. Lipton SA, Rosenberg PA. Excitatory amino acids as a final common pathway for neurologic disorders. N Engl J Med. 1994;330:613–22.
41. Coitinho AS, de Mello CF, Lima TT, de Bastiani J, Fighera MR, Wajner M. Pharmacological evidence that alpha-ketoisovaleric acid induces convulsions through GABAergic and glutamatergic mechanisms in rats. Brain Res. 2001;894:68–73.
42. Funchal C, Latini A, Jacques-Silva MC, et al. Morphological alterations and induction of oxidative stress in glial cells caused by the branched-chain alpha-keto acids accumulating in maple syrup urine disease. Neurochem Int. 2006;49:640–50.
43. Barschak AG, Sitta A, Deon M, et al. Amino acids levels and lipid peroxidation in maple syrup urine disease patients. Clin Biochem. 2009;42:462–6.
44. Mescka CP, Wayhs CA, Vanzin CS, et al. Protein and lipid damage in maple syrup urine disease patients: l-carnitine effect. Int J Dev Neurosci. 2013;31:21–4.
45. Scaini G, Jeremias IC, Morais MO, et al. DNA damage in an animal model of maple syrup urine disease. Mol Genet Metab. 2012;106:169–74.
46. Mackenzie DY, Woolf LI. Maple syrup urine disease; an inborn error of the metabolism of valine, leucine, and isoleucine associated with gross mental deficiency. Br Med J. 1959;1:90–1.
47. Morton DH, Strauss KA, Robinson DL, Puffenberger EG, Kelley RI. Diagnosis and treatment of maple syrup disease: a study of 36 patients. Pediatrics. 2002;109:999–1008.
48. Alodaib A, Carpenter K, Wiley V, Sim K, Christodoulou J, Wilcken B. An improved ultra performance liquid chromatography-tandem mass spectrometry method for the determination of alloisoleucine and branched chain amino acids in dried blood samples. Ann Clin Biochem. 2011;48:468–70.
49. Kaplan P, Mazur A, Field M, et al. Intellectual outcome in children with maple syrup urine disease. J Pediatr. 1991;119:46–50.
50. McMahon Y, MacDonnell Jr RC. Clearance of branched chain amino acids by peritoneal dialysis in maple syrup urine disease. Adv Perit Dial. 1990;6:31–4.
51. Puliyanda DP, Harmon WE, Peterschmitt MJ, Irons M, Somers MJ. Utility of hemodialysis in maple syrup urine disease. Pediatr Nephrol. 2002;17:239–42.
52. Falk MC, Knight JF, Roy LP, et al. Continuous venovenous haemofiltration in the acute treatment of inborn errors of metabolism. Pediatr Nephrol. 1994;8:330–3.
53. Brunetti-Pierri N, Lanpher B, Erez A, et al. Phenylbutyrate therapy for maple syrup urine disease. Hum Mol Genet. 2011;20:631–40.
54. Zinnanti WJ, Lazovic J. Interrupting the mechanisms of brain injury in a model of maple syrup urine disease encephalopathy. J Inherit Metab Dis. 2012;35:71–9.
55. Strauss KA, Mazariegos GV, Sindhi R, et al. Elective liver transplantation for the treatment of classical maple syrup urine disease. Am J Transplant. 2006;6:557–64.
56. Muelly ER, Moore GJ, Bunce SC, et al. Biochemical correlates of neuropsychiatric illness in maple syrup urine disease. J Clin Invest. 2013;123:1809–20.
57. Skvorak KJ, Hager EJ, Arning E, et al. Hepatocyte transplantation (HTx) corrects selected neurometabolic abnormalities in murine intermediate maple syrup urine disease (iMSUD). Biochim Biophys Acta. 2009;1792:1004–10.
58. Skvorak KJ, Dorko K, Marongiu F, et al. Placental stem cell correction of murine intermediate maple syrup urine disease. Hepatology. 2013;57:1017–23.
59. Simon E, Schwarz M, Wendel U. Social outcome in adults with maple syrup urine disease (MSUD). J Inherit Metab Dis. 2007;30:264.

Chapter 13
Mental Retardation and Isoleucine Metabolism

Song-Yu Yang, Xue-Ying He, Carl Dobkin, Charles Isaacs, and W. Ted Brown

Key Points

- Mutations of the *HSD17B10* gene result in mental retardation.
- Accumulation of isoleucine metabolites is a clinical marker of hydroxysteroid (17β) dehydrogenase 10 (HSD10) deficiency.
- Mental retardation, X-linked, choreoathetosis, and abnormal behavior (MRXS10) is a *HSD17B10* transcript splicing disorder.
- Designations *E*ndoplasmic *R*eticulum-*A*ssociated *amyloid-β peptide* *B*inding *protein* and *A*myloid-β peptide *B*inding *A*lcohol *D*ehydrogenase reflect some dubious features of hydroxysteroid (17β) dehydrogenase 10.
- Loss of enzymatic as well as non-enzymatic functions of hydroxysteroid (17β) dehydrogenase 10 affects human development.

Keywords *HSD17B10* gene • Mutation • HSD10 deficiency • Developmental disability • ERAB/ABAD • Neurodegeneration

Abbreviations

ABAD	Amyloid β-peptide-binding alcohol dehydrogenase
AD	Alzheimer's disease
APP	Amyloid precursor protein
ERAB	Endoplasmic reticulum-associated amyloid β-peptide-binding protein
HSD10	Hydroxysteroid (17β) dehydrogenase X
MR	Mental retardation
MRXS10	Mental retardation, X-linked, choreoathetosis, and abnormal behavior
TCA	Tricarboxylic acid cycle

S.-Y. Yang, M.D., Ph.D. (✉) • X.-Y. He, Ph.D. • C. Dobkin, Ph.D.
C. Isaacs, Ph.D. • W.T. Brown, M.D., Ph.D.
New York State Institute for Basic Research in Developmental Disabilities
1050 Forest Hill Road, Staten Island, NY 10314, USA
e-mail: songyu.yang@csi.cuny.edu; song-yu.yang@opwdd.ny.gov;
xue-ying.he@opwdd.ny.gov; carl.dobkin@opwdd.ny.gov;
charles.isaacs@opwdd.ny.gov; ted.brown@opwdd.ny.gov

R. Rajendram et al. (eds.), *Branched Chain Amino Acids in Clinical Nutrition: Volume 1*, Nutrition and Health, DOI 10.1007/978-1-4939-1923-9_13,

Introduction

Mental retardation (MR) affects 2–3 % of the general population [1]. MR is also known as intellectual developmental disorders (IDD), and defined by an intelligence quotient (IQ) below 70 as well as significant delay or lack of at least two areas of adaptive skills [2]. MR is present from childhood and usually represents abnormal development in the central nervous system such that patients have a global deficiency in cognitive function. Epidemiological data show that males are significantly more disposed to MR than females presumably because of the hemizygosity of the X chromosome in males. About 40 % of the protein-coding genes on chromosome X are expressed in the brain. Mutations on the X chromosome cause X-linked mental retardation which accounts for about 10 % of all MR [3].

The branched chain carbon skeleton of isoleucine cannot be synthesized in mammals so that the source of this essential amino acid is only from diet. On the other hand, the oxidation of isoleucine is first catalyzed by a branched chain transaminase and then a branched chain α-keto acid dehydrogenase complex to form 2-methylbutyryl-CoA, which is further degraded under the catalysis of a series of enzymes similar to those in fatty acid β-oxidation [4]. After dehydrogenation, hydration, and a second dehydrogenation of the isoleucine degradation intermediate, the resulting short branched chain coenzyme A derivative undergoes thiolytic cleavage to yield acetyl-CoA and propionyl-CoA. Mutations in individual enzymes that block a particular step of the isoleucine degradation pathway cause particular diseases (Fig. 13.1). For example, maple syrup urine disease is due to mutations in the branched chain α-ketoacid dehydrogenase (BCKD) complex, which results in accumulation of all three branched chain amino acids [5].[1] Next SBCAD deficiency also known as 2-methylbutyryl-CoA dehydrogenase deficiency has been reported to be a harmless metabolic variant [6]. MR is not usually associated with mitochondrial acetoacetyl-CoA (β-KT) thiolase deficiency [7]. In contrast, MR occurs when patients carry a mutation in the *HSD17B10* gene, encoding a vital protein named hydroxysteroid (17β) dehydrogenase X (HSD10) [8, 9]. The continuing oxidation of isoleucine degradation products, e.g., propionyl-CoA, to CO_2 and water as well as the related disorders, propionic acidemia and methylmalonic aciduria, is also briefly discussed below. More than half of patients with these two inherited metabolic disorders have MR.

Mutations of *HSD17B10* Resulting in Mental Retardation

A silent mutation at the *HSD17B10* gene, HSD10 (p.R192R), causes mild mental retardation, choreoathetosis, and abnormal behavior (MRXS10) (OMIM#300220) [10, 11], whereas nine missense mutations have been found to result in worse damage to cognitive function and this disease is known as HSD10 deficiency (OMIM#300438) [12–15]. A complete knockout of the *HSD17B10* gene is an embryonic lethal in *Drosophila* [16] and the mouse [17], and is probably lethal in humans as well as no living cases have been yet identified. HSD10 is an ancient housekeeping enzyme present in all tissues. It was reported [18, 19] that HSD10 levels vary significantly among different brain regions and in response to stress and hypoxia-ischemia.

Patients suffering with HSD10 deficiency may have microcephaly but no dysmorphism or organomegaly. Clinical manifestations include epilepsy, choreoathetoid movements and ophthalmologic disorders [12, 14, 15, 20–25]. Psychomotor retardation or regression is always associated with this inherited metabolic disease, and its onset often occurs at 6–18 months after birth [12, 14, 15, 20–25]. When cardiomyopathy is present, HSD10 deficiency patients usually die at an early age [14]. The brain magnetic resonance imaging is normal for some patients, but frontotemporal or generalized cortical

[1] Please also refer to a previous chapter specifically devoted to this disease.

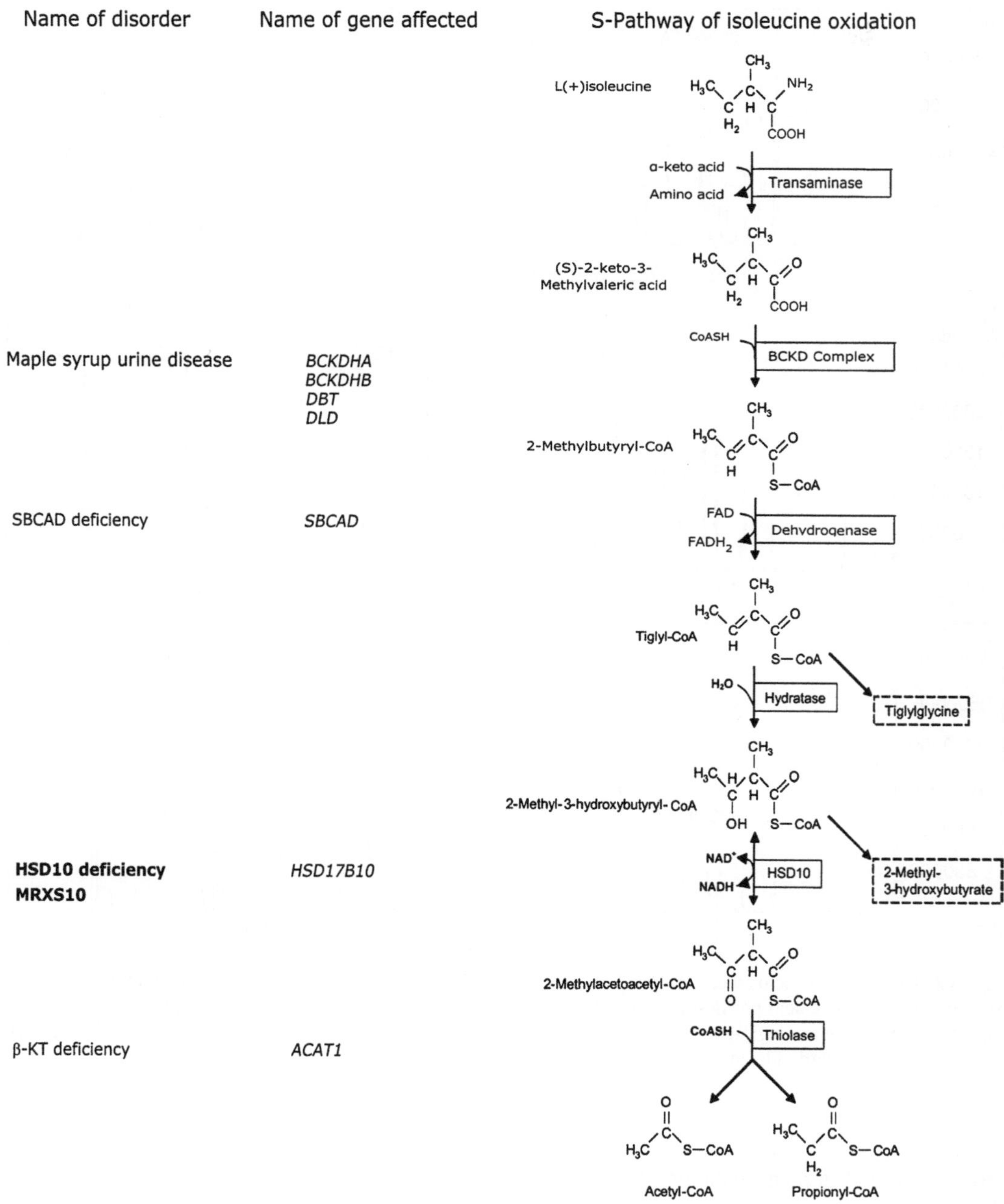

Fig. 13.1 Inborn errors in the isoleucine metabolism. *Dashed line boxes* indicate the compounds excreted in urine of HSD10 deficiency patients. A minor R pathway of isoleucine oxidation [26], where the breakdown of 2-ethylhydracrylic acid is very slow, is not shown here. *BCKD* branched chain α-keto acid dehydrogenase, *DBT* dihydrolipoamide branched chain transacylase, *DLD* dihydrolipoamide dehydrogenase, *SBCAD* short branched chain acyl-CoA dehydrogenase, *β-KT* β-ketothiolase, *ACAT1* acetyl-CoA acetyltransferase 1 (part of this figure was adapted from ref. [9] with permission from Elsevier)

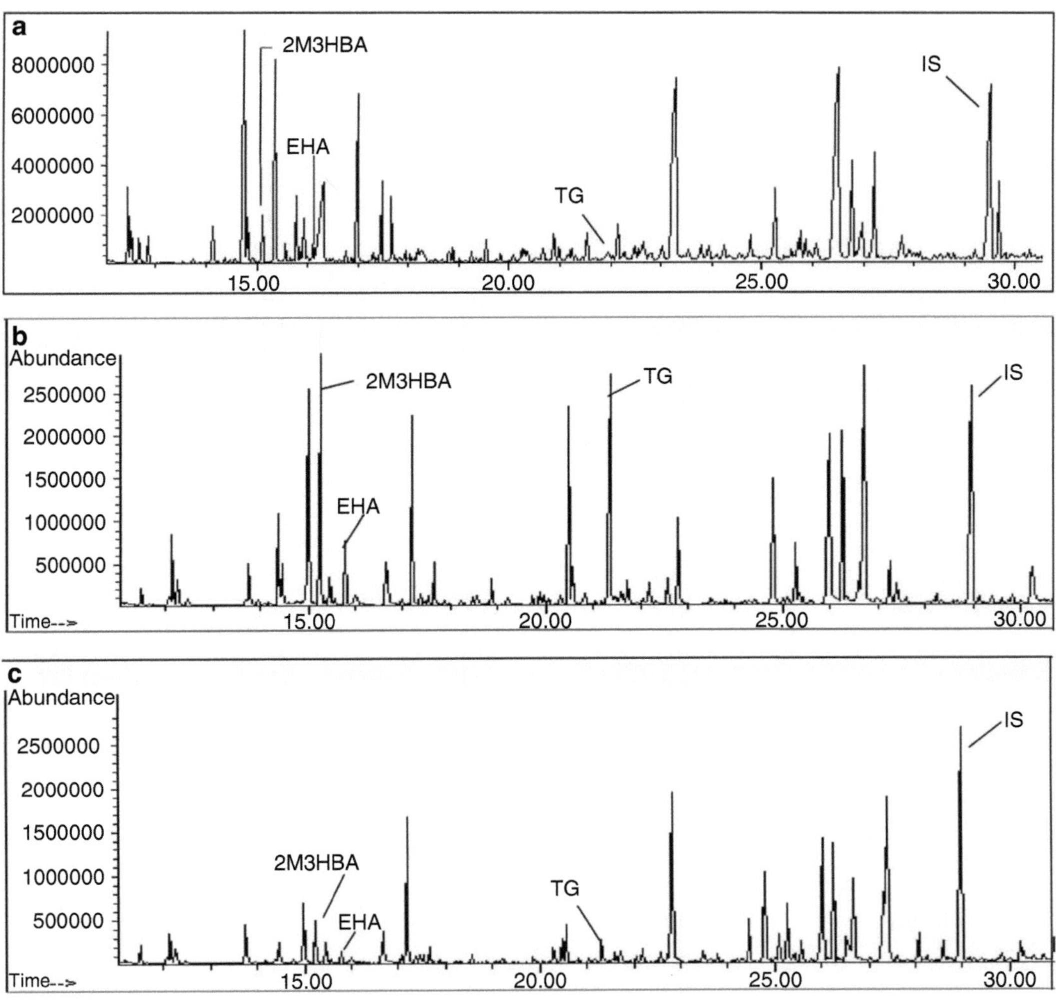

Fig. 13.2 Urine organic acids were analyzed as trimethylsilyl (TMS)-derivatives by gas chromatography-mass spectrometry. Urine organic acid profiles: (**a**) normal control; (**b**) HSD10 deficiency patient; (**c**) HSD10 deficiency patient under low protein dietary regimen. *2MHBA* 2-methyl-3-hydroxybutyrate, *EHA* 2-ethylhydracrylic acid, *TG* tiglylglycine, *IS* internal standard (adapted from ref. [14] with permission from Elsevier)

atrophy is found in many patients [22]. It is considered to be an infantile neurodegenerative disease [23], but no neuropathology is currently available. Elevated levels of isoleucine metabolites are consistently found in blood from patients with HSD10 deficiency but it has not been reported for the milder MRXS10 patients [10, 11]. It was proposed by one group [21] that these two disorders may be referred to as HSD10 disease. A blockade of the oxidation of isoleucine leads to the urinary excretion of 2-methyl-3-hydroxybutyrate, tiglylglycine, and 2-ethylhydracrylic acid (EHA) (Fig. 13.2). EHA is an intermediate in the R-pathway of isoleucine oxidation: (*S*)-2-keto-3-methylvaleric acid can be tautomerized to mixed enols, and then form (*R*)-2-keto-3-methylvaleric acid. Transamination of this isomeric keto acid can form L (+) alloisoleucine. On the other hand, dehydrogenation of (*R*)-2-keto-3-methylvaleric acid generates EHA, which can be rather slowly hydrolyzed to ethylmalonyl semialdehyde and then converted to a coenzyme A derivative again, namely, butyryl-CoA [26]. Therefore, a severe blockade at the downstream step in the S-pathway of isoleucine oxidation can lead to urinary excretion of EHA [14]. A urine organic acid profile is often used to diagnose β-KT deficiency or HSD10 deficiency. Enzymatic assays or mutations identified from DNA analysis can differentiate these two diseases [7, 15].

Possibility of Amyloid β (Aβ) Peptide Involvement

HSD10 was also known as Aβ-peptide-binding alcohol dehydrogenase (ABAD) since the loop D of HSD10 (residues 102–108) can bind to the Aβ peptide [27]. However, the previously reported alcohol dehydrogenase activities of ABAD/ERAB [19, 28] have been demonstrated to be inaccurate [29]. HSD10 is unable to catalyze the oxidation of straight-chain alcohols including ethanol at a spectrophotometrically measurable rate [29]. The solubility of (±)-2-octanol, (−)-2-octanol, and (+)-2-octanol is extremely limited, and n-decanol is *insoluble* in water. Whether alcohol dehydrogenase assays using medium-chain oily alcohols as substrates without solubilization could show the activity reported in ref. [28] (Table 13.1) is debatable. After HSD10 was found to be a mitochondrial enzyme [30, 31], ERAB was later replaced by a new term, ABAD (Aβ-peptide-binding alcohol dehydrogenase) [19, 28]. Since HSD10, an Aβ-binding protein, is involved in neurodegeneration in Alzheimer disease (AD) patients [32], it is of great interest to know whether Aβ peptide is also involved in MRXS10 [11] or HSD10 deficiency, particularly in infantile neurodegeneration [23]. It has been recently reported [33] that Aβ peptide is undetectable in the cerebrospinal fluid of an HSD10 deficiency patient. If this observation is confirmed, it means that a missense mutation on *HSD17B10* gene could alter the metabolism or trafficking of the amyloid precursor protein (APP) [34]. Since APP is a known risk factor for AD, studies on HSD10 deficiency may lead to new avenues for the treatment of AD.

Table 13.1 Alcohol dehydrogenase activities reported for ERAB/ABAD*

Substrate or alcohol	K_m (mM)	V_{max} (U/mg)	k_{cat}[a] (s^{-1})	Catalytic efficiency, k_{cat}/K_m ($M^{-1}s^{-1}$)
Reduction of *S*-acetoacetyl-CoA[b]				
S-Acetoacetyl-CoA	0.068±0.020	430±45	190	2.8×10^6
Oxidation of alcohol substrates[b]				
17β-Estradion	0.014±0.006	23±3	10	7.4×10^5
Methanol	No activity	No activity	No activity	No activity
Ethanol	1,210±260	2.2±0.4	1.0	0.82
Isopropanol	150±17	36±2	16	110
n-Propanol	272±62	4.2±0.5	1.9	6.9
n-Butanol	53±6	9.0±0.3	4.0	76
Isobutanol	56±16	8.0±0.7	3.6	64
n-Pentanol	18±5	6.9±0.4	3.1	170
(±)-2-Octanol	85±17	245±20	110	1,300
(+)-2-Octanol	84±16	102±8	46	540
(–)-2Octanol	43±9.0	133±23	60	1,400
n-Decanol	14±6.3	2.8±0.5	1.3	90

*This table was reproduced with permission from the American Society for Biochemistry and Molecular Biology. For experimental details please see ref. [28]

[a]Calculation based on 1 U representing one μmol of product formed per min and a molecular mass of the enzyme as 26,926 Da

[b]Experiments were performed by incubating ERAB/HADH II with a range of concentrations of the indicated substrates in the presence of NAD/NADH. Details of experimental procedures are described in the text

A Hotspot Responsible for the HSD10 Prevalent Mutation

About half of the cases of HSD10 deficiency are due to a missense C>T mutation in exon 4 of the *HSD17B10* gene. The resulting HSD10(p.R130C) loses most catalytic functions, and the males with this mutation have a much more severe clinical phenotype than those carrying p.V65A, p.L122V, or p.E249Q mutations [35]. In order to answer the question why the p.R130C mutation is more frequently found in HSD10 deficiency patients, Rauschenberger et al. [17] suggested that "most other mutations in the *HSD17B10* gene are not observed because they are incompatible with life".

Recently a mutational hotspot, mCpG, has been identified at the p.R130C position, +2259 nucleotides from the initiation codon ATG of the *HSD17B10* gene [35], by methylation analysis (Fig. 13.3). 5-Methylcytosine (5mC) is produced by post-synthetic modification of cytosine residues, and it is a hypermutable site because it can undergo spontaneous or enzyme catalyzed deamination to thymine. Deamination of unmethylated cytosine produces uracil which is recognized and removed efficiently by uracil DNA glycosylase. However, 5mC deaminates at a higher rate and the product of this reaction is a thymine. The repair of G-T mismatch is not highly efficient. Thus, the 5mC>T transition occurs about ten times more frequently than other transitions. For this reason a 5mC>T transition at this nucleotide will occur more frequently than other mutations in this gene. Finding this mutational hotspot also provides an explanation for the observation [21] that the p.R130C mutation occurs de novo in some HSD10 deficiency patients. HSD10 deficiency patients are not confined to a particular ethnicity although most previously reported cases were either Spanish or German descendants [15].

Controversy About the Residual Activity of Mutant HSD10

The clinical manifestations of patients with HSD10 deficiency are imperfectly correlated with the residual activity of mutated HSD10, but the prevalent mutation, p.R130C, always causes a significant loss of enzymatic activity and a severe phenotype [12–14, 25, 35]. Although it was claimed by

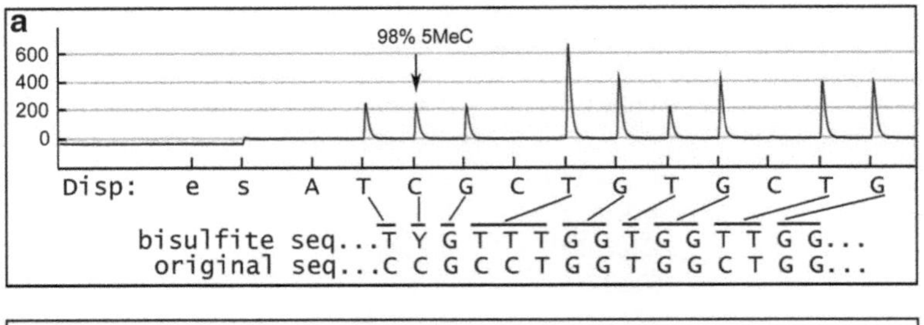

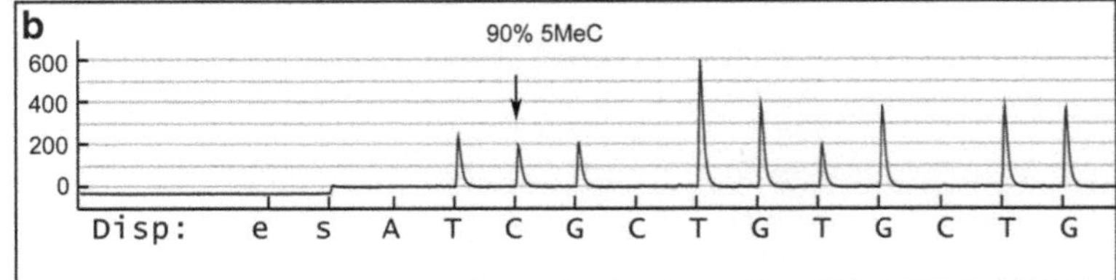

Fig. 13.3 Methylation analysis of the exon 4 of the *HSD17B10* gene. The bisulfite sequencing of the chromosome DNA from a normal male is displayed in (**a**); and that from a normal female in (**b**). The ordinate shows the relative light unit detected after dispensation of each nucleotide substrate. The abscissa shows the dispensation order ("e" and "s" are controls). The "bisulfite" line shows the predicted sequence after modification. "Y" represents C or T. Lines connecting the dispensed nucleotide to the bisulfite sequence indicate how the light signal represents the modified sequence. The height of the signal is proportional to the number of nucleotides at that position (adapted from ref. [35] with permission from Elsevier)

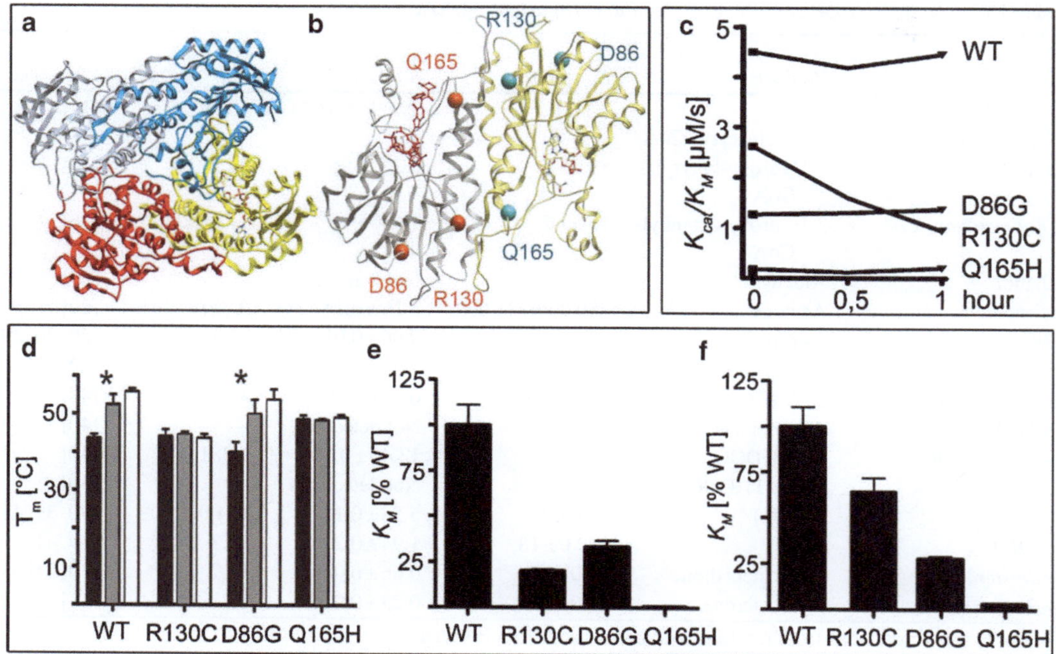

Fig. 13.4 Catalytic properties of HSD10 mutants as previously published in ref. [17] with permission for reuse from Wiley-VCH Verlag GmbH & Co. KGaA

Rauschenberger et al. [17] that HSD10(p.R130C) exhibits up to 64 % residual enzyme activity, detailed experimental data such as v versus [S] curves have not yet been presented to support this claim. Figure 13.4 was adapted from the Fig. 1 of ref. [17]. Part C indicates that the catalytic efficiency (k_{cat}/K_m) of the mutant enzyme, HSD10(p.R130C), was estimated to be about 0.6 of that of the wild type. However, the dimension of the ordinate should have been [$s^{-1}\cdot\mu M^{-1}$] rather than [µM/s] as shown. According to the Part F, the K_m value of this mutant enzyme was estimated to be about 0.6 of that of the wild type which means that the residual enzyme activity should be no greater than 36 % of the wild type HSD10 activity. As a result, the claim of "up to 64 %" was not supported by the figures in the report. If on the other hand, Part F does actually show k_{cat}/K_m values rather than K_m values, the inconsistency between the Part E, Part F, and the corresponding legends becomes even greater. It is difficult to interpret these results without more details.

HSD10 Is a Multifunctional Protein

Hydroxysteroid (17β) dehydrogenase X (HSD10) is a vital protein [8] encoded by the *HSD17B10* gene [9, 36], formerly termed the *HADH2* gene [37]. Human HSD10 was first cloned from brain [36]. Active sites of this enzyme can accommodate different substrates. This feature enables HSD10, an NAD^+-dependent dehydrogenase to be a "flexible" enzyme that is involved in the oxidation of isoleucine and branched chain fatty acids, the metabolism of sexual hormones and neuroactive steroids as well as aldehyde detoxification [8, 38] (Table 13.2). Since the role of this mitochondrial enzyme in the metabolism of steroid hormones and neuroactive steroids was recognized [18, 40, 41], it was renamed as type 10 17β-hydroxysteroid dehydrogenase [8, 31, 32]. This enzyme can also complex with other proteins to generate RNase P activity [42]. When HSD10 is bound to estrogen receptor alpha or amyloid β-peptides, various HSD10 functions are inhibited [43, 44].

Table 13.2 Catalytic properties of hydroxysteroid (17β) dehydrogenase X

Substrate	Product	K_m (μM)	K_{cat} (min^{-1})	K_{cat}/K_m (min^{-1} mM^{-1})	Reference
2-Methyl-3-hydroxybutyryl-CoA	2-Methyl acetoacetyl-CoA	7.1±1.1	398.96±37.74	56,191.5	[12]
Acetoacetyl-CoA	3-Hydroxy-butyryl-CoA	89±5.4	2,220±96	24,943.8	[36]
3-Ketooctanoyl-CoA	3-Hydroxy-octanoyl-CoA	15±1.7	1,680±90	112,000.0	[36]
Ethanol	Aldehyde	–	Not detectable	–	[29]
2-Propanol	Acetone	280,000±33,000	2.16±0.14	0.0077	[29]
17β-Estradiol	Estrone	43.0±2.1	0.66±0.01	15.3	[29, 30, 39]
Estrone	17β-Estradiol	—	Not detectable	—	[29, 30]
Allopregnanolone	5α-DHP	15.0±1.7	6.40±0.20	426.7	[40]
5α-DHP	Allopregnanolone	—	Not detectable	—	[40]
3α, 5α-THDOC	5α-DHDOC	9.7±1.3	13.40±0.70	1,381.0	[40]
5α-DHDOC	3α,5α-THDOC	—	Not detectable	—	[40]
Adiol	5α-DHT	34.0±2.4	5.58±0.17	164.0	[30, 39, 41]
5α-DHT	Adiol	112.0±18	1.94±0.21	17.3	[39, 41]
Androsterone	Androstenedione	45.0±9.3	0.66±0.08	14.7	[30, 41]
Androstenedione	Androsterone	44.0±2.5	0.23±0.01	5.2	[41]

This table was reproduced from ref. [38] with permission from Elsevier

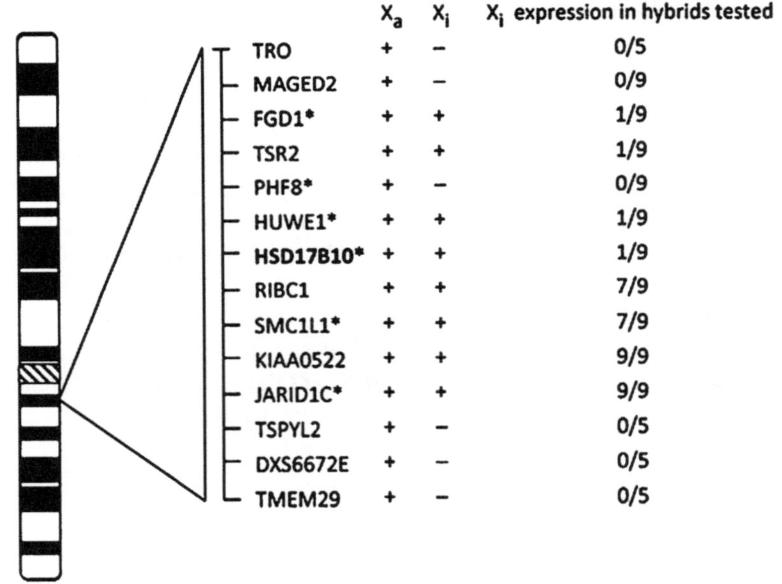

Fig. 13.5 Mental retardation genes at Xp11.22 are marked with an *asterisk*. Their mutation(s) or copy number variation would damage human global cognitive function. X_a and X_i represent active and inactive X-chromosome, respectively. The active or silent expression status of a particular gene is indicated by sign + and – accordingly. The data from expression in nine hybrids demonstrate that *HSD17B10* is a gene with variable escape from X-chromosome inactivation (adapted from ref. [45] with permission from the Nature Publishing Group)

The *HSD17B10* gene maps at Xp11.2, a region closely associated with X-linked mental retardation [45] (Fig. 13.5). Besides enzymatic catalytic activities, the capability of HSD10 to bind with other proteins and peptides also plays an important role in human health and disease [38]. Based on a rescue experiment using HSD10 knockdown in *Xenopus* embryos, human HSD10 was found to have a potential non-enzymatic function [17]. It is unlikely that orthologues of these two species would have an identical contour because their primary structures have only a 76 % identity. Rauschanberger et al. [17] inferred that HSD10 has a non-enzymatic function based on the observation that HSD10(p.R130C)

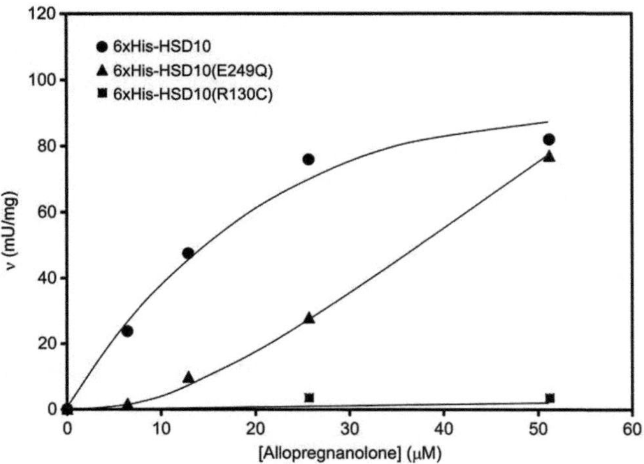

Fig. 13.6 Initial velocities of the oxidative reaction catalyzed by 6xHis-tagged HSD10 (*circle*), HSD10(p.E249Q) (*triangle*), or HSD10(p.R130C) (*square*) as a function of allopregnanolone concentrations (adapted from ref. [12] with permission from PNAS)

retained as much as 64 % of residual enzyme activity but was unable to rescue the knockdown *Xenopus* embryos. Nevertheless, the data underlying this claim were incomplete as mentioned above. The missense mutation p.R130C is a well-known potent mutation which diminishes the enzymatic functions of HSD10 (Fig. 13.6) [12, 13], or short chain 3-hydroxy-2-methylacyl-CoA dehydrogenase (SCHMAD) indispensable to the oxidation of isoleucine.[2]

Pathogenesis of MRXS10 and HSD10 Deficiency

A silent mutation in the *HSD17B10* gene (c.574C>A, p.R192R) affects only males and causes mild mental retardation (MRXS10) [11]. The *HSD17B10* gene generates at least three kinds of transcripts [9]. The longest one, isoform 1, specifies HSD10 (261 residues). The isoform 2 could be translated to a shorter polypeptide (251 residues) in brain with unknown function. Both isoform 2 and 3 are resulting from the alternative splicing [9]. This silent mutation causes a reduction of the amount of HSD10 mRNA isoform 1 to about one-third, and an increase of isoform 2 and 3 14-fold and 3.5-fold, respectively [9, 11]. As a result, the full length wild-type HSD10 was slightly reduced in patient cells [11]. However, the accumulation of isoleucine metabolites, if any, was not reported in such patients [10, 11]. In contrast, an accumulation of isoleucine metabolites is always detected in body fluids of patients suffering with HSD10 deficiency by acylcarnitine profile in blood and organic acid profile in urine [12, 14, 15, 20–25]. Disease-causing missense mutations in the *HSD17B10* gene diminish catalytic activities of HSD10 [12–15, 20, 22–25]. For example, p.R130C is a potent mutation that abolishes HSD10 enzymatic activities. Patients carrying this mutation present severe clinical manifestations [12–14, 25]. On the other hand, patients with p.E249Q or p.V65A mutations retain more HSD10 enzymatic activities and present with mild MR or merely learning disabilities [12, 15, 24]. In spite of the imperfect genotype-phenotype relationship in HSD10 deficiency, a blockade of the isoleucine oxidation pathway is the clinical marker of this disease [12, 15, 35]. The urine organic acid profile (see Fig. 13.2) is essential to the diagnosis of this disease. Therefore, it is reasonable to classify HSD10 deficiency as an inborn error of isoleucine metabolism even if MR cannot be attributed to the accumulation of isoleucine metabolites

[2] Luo MJ, Mao LF, Schulz H. Short-chain 3-hydroxy-2-methylacyl-CoA dehydrogenase from rat liver: purification and characterization of a novel enzyme of isoleucine metabolism. *Arch Biochem Biophys.* 1995;321:214–20.

[12]. It was reported [46] that 2-methyl-3-hydroxybutyrate and 2-methylacetoacetate inhibit the energy metabolism in the cerebral cortex of developing rats. The accumulation of isoleucine metabolites may be a contributing factor to MR, but is inadequate to cause MR by itself. MRXS10 appears not to be related to a significant disturbance of isoleucine metabolism. It is puzzling how an alternative splicing disorder of *HSD17B10* transcripts could result in mild MR. The pathogenesis of these two disorders causing MR is still unsolved. It has recently been reported [47] that a steroidal inhibitor of HSD10 is available. This compound may be useful to the study of imbalance of neurosteroid metabolism in HSD10 deficiency [12].

Disorders in the Oxidation of Isoleucine Degradation Products

Butyryl-CoA is produced by the minor R-pathway of the isoleucine degradation, and oxidized via the β-oxidation cycle to acetyl-CoA. As shown in Fig. 13.1, isoleucine is mainly broken down to acetyl-CoA and propionyl-CoA through the S-pathway. These coenzyme A derivatives are further oxidized to CO_2 and water via the tricarboxylic acid cycle (TCA) to supply energy, or can serve as building blocks for the synthesis of large molecules, e.g. cholesterol. In contrast to acetyl-CoA, propionyl-CoA must first be converted to succinyl-CoA for entering the TCA cycle. This is a relatively complicated process as described below.

The carboxylation of propionyl-CoA to D-methylmalonyl-CoA is catalyzed by the biotin-dependent propionyl-CoA carboxylase in an ATP consuming reaction. The D-isomer is then converted to the L-isomer under the catalysis of methylmalonyl-CoA racemase. Finally, the L-isomer is converted to succinyl-CoA under the catalysis of methylmalonyl-CoA mutase that requires adenosylcobalamine as coenzyme. The isoleucine catabolism generates ATP to supply energy for cells. However, mutations in this mutase cause methylmalonic aciduria, mut type (OMIM#251000), which is unresponsive to B12 therapy. In fact, some methylmalonic aciduria patients do have a positive response to B12 therapy because they have defects in the synthesis of the coenzyme rather than the mutase itself. Under these circumstances, mutations are found in the MMAA gene on chromosome 4q31 or the MMAB gene on 12q24. Clinical features of methylmalonic aciduria are very similar to propionic acidemia due to mutations in propionyl-CoA carboxylase (OMIM#606054). This carboxylase consists of six α-subunits encoded by the PCCA gene in chromosome 13q32.3 and six β-subunits encoded by the PCCB gene in 3q22.3. Dozens of different mutations in these genes have been identified in patients with these autosomal recessive disorders. Elevated levels of isoleucine metabolites such as tiglylglycine and 2-methyl-3-hydroxybutyrate (see Fig. 13.1) were found in a propionic acidemia patient's urine organic acid profile in addition to 3-hydroxypropionic acid and methylcitric acid [48]. A short-chain (C4) or odd-numbered (C3) acyl-CoA pool can have diverse precursors such as valine and odd-numbered fatty acids in addition to the isoleucine. The fate of these short-chain coenzyme A derivatives has been briefly discussed above for a more complete view of isoleucine metabolism.

The clinical manifestations of propionic acidemia often include episodic vomiting, lethargy, ketoacidosis, and hyperammonemia. The incidence of propionic acidemia is high in some tribes of Saudi Arabia probably due to a founder effect [49]. Asymptomatic patients suffering with propionic acidemia or methylmalonic aciduria are much rarer than those with SBCAD deficiency or acetoacetyl-CoA thiolase deficiency. The diagnosis of propionic acidemia and methylmalonic aciduria relies on enzyme assays and identification of mutations in related genes. More than half of patients with these inherited metabolic disorders suffer from MR. Most such patients attend special education programs [50].

Treatment and Prognosis

The therapeutic treatment of such metabolic disorders is a low protein high-energy dietary regimen with carnitine supplementation [15, 24, 48]. Such treatment does reduce the accumulation of branched chain amino acid metabolites in blood and urine (see Fig. 13.2), but it cannot ameliorate the patients' psychomotor regression. In order to increase mitochondrial energy metabolism, a cocktail of vitamins and cofactors such as coenzyme Q_{10} and lipoic acid was tested, but this treatment had no measurable benefits to HSD10 deficiency patients [21]. At present there is no effective therapy for HSD10 deficiency [12, 15, 21]. Since HSD10 deficiency may sometimes mimic a mitochondrial disease [20], it was thought that valproic acid should not be prescribed to treat the epilepsy. However, in cases where the patients' mitochondrial respiratory chain enzyme activities are normal, cautious administration of valproic acid has been reported to be very useful for the control of refractory epilepsy [15].

For the neonatal-onset form of propionic acidemia or methylmalonic aciduria, emergency treatment is needed that is based on the avoidance of nitrogen intake and inhibition of endogenous catabolism by supplying parenteral glucose and insulin. Acidosis is corrected by rehydration and alkali therapy, and the administration of carbamylglutamate can improve hyperammonemia. Liver transplantation does not correct the long-term neurological deterioration as it is unable to prevent the massive accumulation of toxic metabolites in other tissues such as the central nervous system. Therapeutic doses of vitamin B12 can be used to effectively treat true B12 responsive methylmalonic aciduria, and those patients have the best prognosis [48].

As a result of newborn screening, early diagnosis and initiation of treatment have been achieved. More patients with propionic acidemia or methylmalonic aciduria survive neonatal onset and this reduces the mortality of such inherited metabolic disorders. However, the advances in our understanding of the etiology have not yet improved the patients' intellectual development. The prognosis of HSD10 deficiency is largely dependent upon the kind of mutation found in the *HSD17B10* gene and the sex of the patient. A potent mutation p.R130C leads to severe MR or even death [12–14, 25]. Relatively benign ones, e.g., p.E249Q and p.V65A or a silent mutation p. R192R result in mild MR or learning disabilities [10, 12, 15, 24]. Female patients uniformly present milder clinical manifestations than male patients carrying the same mutations, because of mosaicism and a variable escape of the *HSD17B10* gene from X-chromosome inactivation (see Fig. 13.5) [45].

Conclusions

Several isoleucine metabolic disorders can result in MR. Although the newborn screening has reduced mortality of patients, the incidence of MR has not yet been significantly reduced. Among these disorders, males carrying a mutation in the *HSD17B10* gene all suffer from MR, and thus HSD10 deficiency is discussed in detail. HSD10 is a unique mitochondrial hydroxysteroid (17β) dehydrogenase, and its alcohol dehydrogenase activity is marginal. Trivial names of HSD10, in particular, endoplasmic reticulum-associated amyloid β-peptide binding protein (ERAB) and amyloid β-peptide-binding alcohol dehydrogenase (ABAD), reflect some dubious features of HSD10. Elimination of these distractions will allow efforts to be focused on the pathogenesis of MR due to mutations in the *HSD17B10* gene. It is likely that HSD10 has yet uncharacterized functions. Both enzymatic and non-enzymatic functions of HSD10 may be involved in MR. Such new concepts may lead to the design of more effective treatments for this inborn error of isoleucine metabolism.

Acknowledgment This work was supported by the New York State Office of People with Developmental Disabilities. We thank Dr. David Miller for helpful discussions.

Conflicts of Interest
Authors declare no conflicts of interest.

References

1. Ropers HH, Hamel BCJ. X-linked mental retardation. Nat Rev Genet. 2005;6:46–57.
2. Carulla LS, Reed GM, Vaez-Azizi LM, et al. Intellectual developmental disorders: towards a new name, definition and framework for "mental retardation/intellectual disability" in ICD-11. World Psychiatry. 2011;10:175–80.
3. Stevenson RE. Advances in X-linked mental retardation. Curr Opin Pediatr. 2005;17:720–4.
4. Yang SY, He XY. Molecular mechanisms of the fatty acid β-oxidation enzyme catalysis. In: Quant PA, Eaton S, editors. Current views of fatty acid oxidation and ketogenesis. New York, NY: Kluwer; 1999. p. 133–44.
5. Strauss KA, Morton DH. Branched-chain ketoacyl dehydrogenase deficiency: maple syrup urine disease. Curr Treat Options Neurol. 2004;5:329–41.
6. Sass JO, Ensenauer R, Roschinger W, et al. Methylbutyryl-coenzyme A dehydrogenase deficiency: functional and molecular studies on a defect in isoleucine catabolism. Mol Genet Metab. 2008;93:30–5.
7. Fukao T, Scriver CR, Kondo N. The clinical phenotype and outcome of mitochondrial acetoacetyl-CoA thiolase deficiency (β-ketothiolase or T2 deficiency) in 26 enzymatically proved and mutation-defined patients. Mol Genet Metab. 2001;72:109–14.
8. Yang SY, He XY, Schulz H. Multiple functions of type 10 17beta-hydroxysteroid dehydrogenase. Trends Endocrinol Metab. 2005;16:167–75.
9. Yang SY, He XY, Miller D. HSD17B10: a gene involved in cognitive function through metabolism of isoleucine and neuroactive steroids. Mol Genet Metab. 2007;92:36–42.
10. Reyniers E, Van Bogaert P, Peeters N, et al. A new neurological syndrome with mental retardation, choreoathetosis, and abnormal behavior maps to chromosome Xp11. Am J Hum Genet. 1999;65:1406–12.
11. Lenski C, Kooy RF, Reyniers E, et al. The reduced expression of the HADH2 protein causes X-linked mental retardation, choreoathetosis, and abnormal behavior. Am J Hum Genet. 2007;80:372–7.
12. Yang SY, He XY, Olpin SE, et al. Mental retardation linked to mutations in the *HSD17B10* gene interefering with neurosteroid and isoleucine metabolism. Proc Natl Acad Sci U S A. 2009;106:14820–4.
13. Ofman R, Ruiter JPN, Feenstra M, et al. Wanders, R.J. 2-Methyl-3-hydroxybutyryl-CoA dehydrogenase deficiency is caused by mutations in the *HADH2* gene. Am J Hum Genet. 2003;72:1300–7.
14. García-Villoria J, Navarro-Sastre A, Fons C, et al. A study of patients and carriers with 2-methyl-3-hydroxybutyryl-CoA dehydrogenase (MHBD) deficiency: difficulties in the diagnosis. Clin Biochem. 2009;42:27–33.
15. Seaver LH, He XY, Abe K, et al. A novel mutation in the *HSD17B10* gene of a 10-year-old boy with refractory epilepsy, choreoathetosis and learning disability. PLoS One. 2011;6:e27348.
16. Torroja L, Ortuño-Sahagún D, Ferrús A, et al. *scully*, an essential gene of Drosophila, is homologous to mammalian mitochondrial type II L-3-hydroxyacyl-CoA dehydrogenase/amyloid-beta peptide-binding protein. J Cell Biol. 1998;18:1009–17.
17. Rauschenberger K, Schöler K, Sass JO, et al. A non-enzymatic function of 17beta-hydroxysteroid dehydrogenase type 10 is required for mitochondrial integrity and cell survival. EMBO Mol Med. 2010;2:51–62.
18. He XY, Wegiel J, Yang SY. Intracellular oxidation of allopregnanolone by human brain type 10 17beta-hydroxysteroid dehydrogenase. Brain Res. 2005;1040:29–35.
19. Yan SD, Zhu Y, Stern ED, et al. Amyloid β-peptide-binding alcohol dehydrogenase is a component of the cellular response to nutritional stress. J Biol Chem. 2000;275:27100–9.
20. Perez-Cerda C, García-Villoria J, Ofman R. 2-Methyl-3-hydroxybutyryl-CoA dehydrogenase (MHBD) deficiency: an X-linked inborn error of isoleucine metabolism that may mimic a mitochondrial disease. Pediatr Res. 2005;58:488–91.
21. Zschocke J. HSD10 disease: clinical consequences of mutations in the HSD17B10 gene. J Inherit Metab Dis. 2012;35:81–9.
22. Cazorla MR, Verdú A, Pérez-Cerdá C, et al. Neuroimage findings in 2-methyl-3-hydroxybutyryl-CoA dehydrogenase deficiency. Pediatr Neurol. 2007;36:264–7.
23. Zschocke J, Ruiter JP, Brand J, et al. Progressive infantile neurodegeneration caused by 2-methyl-3-hydroxybutyryl-CoA dehydrogenase deficiency: a novel inborn error of branched-chain fatty acid and isoleucine metabolism. Pediatr Res. 2000;48:852–5.
24. Olpin SE, Pollitt RJ, McMenamin J. 2-methyl-3-hydroxybutyryl-CoA dehydrogenase deficiency in a 23-year-old man. J Inherit Metab Dis. 2002;25:477–82.
25. Sutton VR, O'Brien WE, Clark GD, et al. 3-Hydroxy-2-methylbutyryl-CoA dehydrogenase deficiency. J Inherit Metab Dis. 2003;26:69–71.
26. Mamer OA, Tjoa SS, Scriver CR, et al. Demonstration of a new isoleucine catabolic pathway yielding an R series of metabolites. Biochem J. 1976;160:417–26.
27. Marques AT, Antunes A, Fernandes PA, et al. Comparative evolutionary genomics of the HADH2 gene encoding Abeta-binding alcohol dehydrogenase/17beta-hydroxy-steroid dehydrogenase type 10 (ABAD/HSD10). BMC Genomics. 2006;7:201–21.

28. Yan SD, Shi Y, Zhu A, et al. Role of ERAB/L-3-hydroxyacyl-coenzyme A dehydrogenase type II activity in Abeta-induced cytotoxicity. J Biol Chem. 1999;274:2145–56.
29. He XY, Yang YZ, Schulz H, et al. Intrinsic alcohol dehydrogenase and hydroxysteroid dehydrogenase activities of human mitochondrial short chain L-3-hydroxyacyl-CoA dehydrogenase. Biochem J. 2000;345:139–43.
30. He XY, Merz G, Mehta P, et al. Human brain short chain L-3-hydroxyacyl coenzyme A dehydrogenase is a single-domain multifunctional enzyme. Characterization of a novel 17β-hydroxysteroid dehydrogenase. J Biol Chem. 1999;274:15014–9.
31. He XY, Merz G, Yang YZ, et al. Characterization and localization of human type 10 17β-hydroxysteroid dehydrogenase. Eur J Biochem. 2001;268:4899–907.
32. Yang SY, He XY. Role of type 10 17beta-hydroxysteroid dehydrogenase in the pathogenesis of Alzheimer's disease. In: Tolnay M, Probst A, editors. Neuropathology and genetics of dementia. New York, NY: Kluwer; 2001. p. 101–10.
33. Ortez C, Villar C, Fons C, et al. Undetectable levels of CSF amyloid-β peptide in a patient with 17β-hydroxysteroid dehydrogenase deficiency. J Alzheimers Dis. 2011;27:253–7.
34. He XY, Miller D, Yang SY. Possible alteration of amyloid-β protein precursor metabolism or trafficking in a 17β-hydroxysteroid dehydrogenase X deficiency patient. *J Alzheimer's Dis*. http://j-alz.com/letterseditor/index.html#December2011
35. Yang SY, Dobkin C, He XY, et al. A 5-methylcytosine hotspot responsible for the prevalent *HSD17B10* mutation. Gene. 2013;515:380–4.
36. He XY, Schulz H, Yang SY. A human brain L-3-hydroxyacyl coenzyme A dehydrogenase is identical with an amyloid β-peptide binding protein involved in Alzheimer's disease. J Biol Chem. 1998;273:10741–6.
37. Korman SH, Yang SY. HSD17B10 replaces HADH2 as the approved designation for the gene mutated in 2-methyl-3-hydroxybutyryl-CoA dehydrogenase deficiency. Mol Genet Metab. 2007;91:115.
38. Yang SY, He XY, Miller D. Hydroxysteroid (17β) dehydrogenase X in human health and disease. Mol Cell Endocrinol. 2011;343:1–6.
39. Shafqat N, Marschall H, Filling C, et al. Expended substrate screenings of human and *Drosophila* type 10 17β-hydroxysteroid dehydrogenase reveal multiple specificities in bile acid and steroid hormone metabolism. Characterization of multifunctional 3α/7α/7β/20β/21-hydroxysteroid dehydrogenase. Biochem J. 2003;376:49–60.
40. He XY, Wegiel J, Yang YZ, et al. Type 10 17beta-hydroxysteroid dehydrogenase catalyzing the oxidation of steroid modulators of γ-aminobutyric acid type A receptors. Mol Cell Endocrinol. 2005;229:111–7.
41. He XY, Merz G, Yang YZ, et al. Function of human short chain L-3-hydroxyacyl-CoA dehydrogenase in androgen metabolism. Biochim Biophys Acta. 2000;1484:267–77.
42. Holzmann J, Frank P, Löffler E, et al. RNase P without RNA: identification and functional reconstitution of the human mitochondrial tRNA processing enzyme. Cell. 2008;135:462–74.
43. Oppermann UC, Salim S, Tjernberg LO, et al. Binding of amyloid β-peptide to mitochondrial hydroxyacyl-CoA dehydrogenase (ERAB): regulation of an SDR enzyme activity with implication for apoptosis in Alzheimer's disease. FEBS Lett. 1999;451:238–42.
44. Jazbutyte V, Kehl F, Neyses L, et al. Estrogen receptor alpha interacts with 17beta-hydroxysteroid dehydrogenase type 10 in mitochondria. Biochem Biophys Res Commun. 2009;384:450–4.
45. He XY, Dobkin C, Yang SY. Does the HSD17B10 gene escape from X-inactivation? Eur J Hum Genet. 2011;19:123–4.
46. Rossa RB, Schuck PF, de Assis DR, et al. Inhibition of energy metabolism by 2-methylacetoacetate and 2-methyl-3-hydroxybutyrate in cerebral cortex of developing rats. J Inherit Metab Dis. 2005;28:501–15.
47. Ayan D, Maltais R, Poirier D. Identification of a 17β-hydroxysateoid dehydrogenase type 10 steroidal inhibitor: a tool to investigate the role of type 10 in Alzheimer's disease and prostate cancer. ChemMedChem. 2012;7:1181–4.
48. Deodato F, Boenzi S, Santorelli FM, et al. Methylmalonic and propionic aciduria. Am J Med Genet. 2006;142C: 104–12.
49. Al Essa M, Rahbeeni Z, Jumaah S, et al. Infectious complications of propionic acidemia in Saudi Arabia. Clin Genet. 1998;54:90–4.
50. Pena L, Franks J, Chapman KA, et al. Natural history of propionic acidemia. Mol Genet Metab. 2012;105:5–9.

Chapter 14
Anorexia and Valine-Deficient Diets

Tetsuya Takimoto, Chie Furuta, Hitoshi Murakami, and Makoto Bannai

Key Points

- Valine is an essential nutrient for body maintenance in mammals.
- Several characteristic phenotypes are observed when valine is deficient in the diet, the most distinguished of which is anorexia due to decreased food intake.
- Anorexic behavior is controlled by the hypothalamus and is due to metabolic pathway changes.
- The hypothalamus regulates food intake by sensing valine deficiency, inducing somatostatin secretion and decreasing ghrelin sensitivity in rats.
- Metabolic acidosis caused by a valine-deficient diet induces anorexia via excessive production of carboxylic acids and acetoacetate through metabolic pathway changes.

Keywords Food intake • Valine deficiency • BCAA metabolism • Hypothalamus • Acidosis • Bioinformatic analysis

Abbreviations

AGRP	Agouti-related protein
CCK	Cholecystokinin
CGRP	Calcitonin gene-related peptide
EAA	Essential amino acids
GHS	Growth hormone secretagogue
NPY	Neuropeptide Y
POMC	Pro-opiomelanocortin
PVN	Paraventricular nucleus
NEAA	Nonessential amino acids
BCAA	Branched chain amino acid
BCAT	Branched chain aminotransferase

T. Takimoto, Ph.D. • C. Furuta, D.V.M., Ph.D. • H. Murakami, Ph.D. • M. Bannai, Ph.D. (✉)
Institute for Innovation, Ajinomoto, Co. Inc.,
1-1 Suzuki-cho, Kawasaki-ku, Kawasaki-shi, Kanagawa, 210-8681, Japan
e-mail: tetsuya_takimoto@ajinomoto.com; chie_furuta@ajinomoto.com; chie.furuta@gmail.com; hitoshi_murakami@ajinomoto.com; makoto_bannai@ajinomoto.com

R. Rajendram et al. (eds.), *Branched Chain Amino Acids in Clinical Nutrition: Volume 1*, Nutrition and Health, DOI 10.1007/978-1-4939-1923-9_14,

PHGDH	3-Phosphoglycerate dehydrogenase
SDS	Serine dehydratase
APC	Anterior piriform cortex
BCKDH	Branched chain α-keto acid dehydrogenase
BDK	BCKDH kinase
KEGG	Kyoto Encyclopedia of Genes and Genomes
DAVID	Database for Annotation, Visualization and Integrated Discovery
ALDH	Aldehyde dehydrogenase
AOX	Aldehyde oxidase
ACAT	Acetyl-CoA acetyltransferase

Introduction

Hypothalamic Regulation of Food Intake

The hypothalamic signaling pathways that regulate food intake have been studied in detail. The arcuate nucleus of the hypothalamus contains neuropeptide Y (NPY)/agouti-related protein (AGRP) and pro-opiomelanocortin (POMC)/cocaine- and amphetamine-regulated transcript (CART) neurons, and the activation of NPY/AGRP or POMC/CART neurons stimulates or inhibits feeding behavior, respectively. Ghrelin was identified as a hormone that is primarily secreted from the stomach to stimulate food intake [1], mainly via gastric vagal afferents [2], and ghrelin also serves as an endogenous ligand for growth hormone secretagogue (GHS) receptors [3]. In the hypothalamus, GHS receptor mRNA is co-expressed with NPY/AGRP and POMC/CART neuron mRNA [4], and ghrelin intraventricular administration increases NPY and AGRP mRNA expression in the arcuate nucleus [5]. In a multiple-gene knockout experiment, deletion of the genes encoding NPY and AGRP completely abolished the orexigenic effect of ghrelin [6]. Altogether, the current evidence suggests that ghrelin signaling stimulates afferent vagus nerve activity from the stomach as well as NPY/AGRP neurons, which results in feeding, and anorexia and bulimia therefore seem to involve disturbed regulation of these NPY/AGRP neurons.

Anorexia Induced by Dietary Nutrients

Food contains many types of nutrients, such as protein, fat, carbohydrate, and micronutrients, and each of these nutrients and their metabolites have the potential to serve as nutrient intake signals; therefore, the act of feeding delivers a massive amount of information to the body. Such nutrient information is integrated in the brain, which can then send a signal to instruct the body to eat more or to stop eating. Thus, dietary nutrients are an important regulator of food intake.

Some reports have indicated that dietary nutrient imbalance is associated with altered food intake [7, 8]. Zinc deficiency is well known to induce anorexia in animals and humans [9]. The effects of zinc deficiency include dwarfism, hypogonadism and poor appetite, and reduction of food intake begins within 3–5 days of deficiency in rats. One of the potent mechanisms of zinc deficiency-induced anorexia is impaired NPY release from the paraventricular nucleus (PVN). Furthermore, the gene expression levels of other neuropeptides that have inhibitory feeding effects, such as cholecystokinin (CCK), calcitonin gene-related peptide (CGRP) and serotonin are increased in zinc deficiency [9].

However, the mechanism for decreased food intake resulting from zinc deficiency is complicated and remains poorly characterized.

Other nutrients are also associated with food intake regulation. For example, glucose levels in the liver are associated with decreased food intake [10], and most gut hormones released following carbohydrate intake inhibit feeding [11]. Dietary fatty acids, such as medium-chain triacylglycerol, also inhibit food intake, but long-chain triacylglycerol does not. One potential mechanism for food intake inhibition is fat oxidation in liver, which influences food intake [8], as medium-chain triacylglycerol is easily oxidized. In addition, diets with low protein quality or quantity have been shown to decrease rat food intake, which involves the activation of the central histamine and histamine receptor (H1) activation [12]. Furthermore, suppressed food intake can be recovered following an injection of histaminergic antagonists, which suggests that histaminergic activity in the brain reduces food intake. Protein is catabolized into amino acids in organs, and amino acids can also be important regulators of food intake; for example, deficiency of essential amino acids (EAA) is recognized to decrease food intake, and reports of anorexia with EAA deficiency date back to the early 1900s [13].

In this chapter, we review EAA deficiency-induced anorexia and the potential mechanisms responsible from a metabolic point of view. Additionally, we investigate the hypothesis of valine deficiency-induced anorexia using bioinformatics methods.

Metabolism Under Essential Amino Acid Deficiency

During the course of mammalian evolution, EAAs were obtained from dietary nutrients rather than synthesized by the body, and mammals require dietary EAAs to meet optimal requirements for survival. The general function of EAAs is the synthesis of proteins and precursors for synthesizing other nonessential amino acids (NEAAs), such as alanine and glutamine. Muscle is the largest protein reservoir and turns over protein to amino acids and vice versa. Therefore, muscle is able to acutely sense and identify deficiencies in amino acids. Several typical phenotypes of EAA deficiency include body weight loss due to anorectic behavior, neural symptoms that induce seizure susceptibility [13] and disturbances in amino acid metabolism involving gene transcription changes in amino acid transporters [14]. These stark phenotypes clearly illustrate the importance of EAAs for appropriate body maintenance.

BCAA Metabolism

Along with leucine and isoleucine, valine is classified as a branched chain amino acid (BCAA). BCAAs are abundant in most animal-based foods and account for approximately 15–20 % of total dietary protein [15]. Unlike other EAAs, BCAAs possess a unique metabolic characteristic because they are metabolized by mitochondrial branched chain aminotransferase (BCAT2 or BCATm; EC 2.6.1.42). However, expression of this enzyme is relatively low in the liver compared to the muscle, which means that BCAAs absorbed from the intestine to the circulation bypass initial liver metabolism [16]. Conversely, the other 17 amino acids are initially metabolized by the liver, which results in a sharp elevation of plasma BCAAs after feeding. This elevation in BCAAs is thought to send a unique signal to the peripheral nervous system and the brain regarding the protein or amino acid content of the absorbed nutrients. Thus, circulating BCAAs are known to act as signals that regulate protein synthesis and proteolysis; in addition, they are responsive to the inhibitory actions of insulin and have been associated with satiety control in the central nervous system.

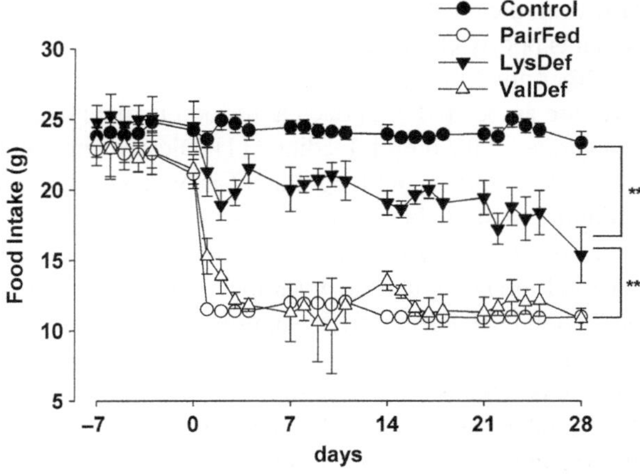

Fig. 14.1 Changes in food intake of rats fed experimental diets ($n=6$). Data are presented as the means ± SEM. The significant differences of daily food intake between experimental groups are shown as $^{**}P<0.01$ (Nagao et al. [17] with permission)

Metabolism Under Dietary Valine Deficiency

As indicated above, EAA dietary deficiency induces several clear phenotypes. However, valine is reported to induce more severe phenotypes compared to other single EAA deficiencies, such as lysine (Fig. 14.1). This phenomenon occurs because lysine is strongly conserved due to its high storage and slow metabolism [17], whereas BCAAs, especially valine, act as a signal for dietary intake, and their levels are acutely and sensitively affected by nutrient deficiency. Any EAA deficiency causes fluctuation in a number of different plasma amino acid levels, but in experimentally induced valine-deficient states, the common feature is elevated plasma serine and threonine levels [17, 18], which is the result of 3-phosphoglycerate dehydrogenase (PHGDH; EC1.1.1.95) upregulation and serine dehydratase (SDS; EC4.2.1.13) downregulation in the liver. PHGDH plays a critical role in de novo serine biosynthesis in the liver, and SDS catalyzes the conversions of serine to pyruvate and threonine to α-ketobutyrate. Because the reservoir of serine and threonine in the muscle or kidney is relatively small, the majority of serine and threonine is derived from the liver. The reason for the elevation of these two amino acids is thought to have evolved from strategies related to body protection and conservation. Because storing amino radicals as serine and threonine is less toxic than ammonia accumulation due to urea cycle metabolism [19], the body is protected from the toxic effect of urea when protein proteolysis is enhanced by valine deficiency. In addition, the conversion of glutamate to serine requires less energy than the conversion to urea.

Regulation of Food Intake Under Dietary EAA Deficiency

Previously, Goto et al. reported that a valine-deficient-diet-induced anorexia and that this process involved the production of acyl ghrelin and des-acyl ghrelin [20]. However, even when these authors injected acyl ghrelin intraperitoneally, anorexia was not improved (Fig. 14.2). These results suggest that valine deficiency induces ghrelin resistance and impairs hypothalamic feeding regulation. Additionally, Nakahara et al. reported that somatostatin in the hypothalamus is involved in mechanisms responsible for anorexia [21]. Interestingly, reduced food intake was dose-dependently restored by valine supplementation [21]. In another study, a chemosensor for detecting amino acid imbalance was discovered in the rat anterior piriform cortex (APC) [22], and rats with APC lesions failed to reject threonine deficient diets [22]. These results demonstrate that there are strong sensing mechanisms for EAA deficiency in the brain.

Contribution of Metabolic Changes to Dietary Valine Deficiency-Induced Anorexia

Because the contribution of metabolic changes induced by a valine-deficient diet to anorexia remains unclear, we sought to analyze Web-based human data to develop a hypothesis for the anorexic mechanism of valine deficiency. As previously described, BCAAs are first deaminized by BCAT and decarboxylated by the branched chain α-keto acid dehydrogenase (BCKDH add EC 1.2.4.4) complex, and these steps are common among all three BCAA members (Fig. 14.3a) [23]. Our strategy focused on individual differences in BCAA metabolism pathways, which originate with isobutyryl-CoA (Fig. 14.3b), isovaleryl-CoA (Fig. 14.3c), and 2-methylbutanoyl-CoA (Fig. 14.3d), which are metabolites of valine, leucine, and isoleucine, respectively. When rats were fed a valine-deficient diet, plasma valine, as well as leucine and isoleucine, concentrations decreased significantly (Table 14.1) [17]. These results suggest that a valine-deficient diet activates leucine and isoleucine metabolism involving with the increase of the availability of BCAT and BCKDH for both leucine and isoleucine. These metabolic changes under valine-deficient condition would induce low levels of isobutyryl-CoA (from valine, Fig. 14.3b; because of lack of valine) and high levels of isovaleryl-CoA (from leucine, Fig. 14.3c) and 2-methylbutanoyl-CoA (from isoleucine, Fig. 14.3d; because of the high availability of BCAT and BCKDH). This is the presupposition of our anorexia hypothesis with valine-deficient diet.

First, we analyzed metabolic pathways using the Kyoto Encyclopedia of Genes and Genomes (KEGG, http://www.genome.jp/kegg/) to identify specific metabolic pathways for each BCAA. As shown in Table 14.2, we identified 9, 11, and 6 genes that encoded enzymes specific for valine, leucine, and isoleucine catabolic pathways, respectively. Because most of these genes seemed to be common to other metabolic pathways (for example, aldehyde dehydrogenase pathway), we tried to extract common pathways by using the Database for Annotation, Visualization and Integrated Discovery (DAVID, http://david.abcc.ncifcrf.gov/) as a bioinformatics resource [24]. Table 14.3 shows the results of the DAVID analysis. The pathways representing "valine, leucine, and isoleucine degradation," "propanoate metabolism," and "fatty acid metabolism" were common metabolic pathways among the BCAAs; however, with the exception of "valine, leucine, and isoleucine degradation," the actual enzymatic reactions are quite different. For example, in the valine metabolism-specific pathway, "propanoate metabolism" was composed of "2-propyn-1-al to propynoate," "malonate-semialdehyde to/from beta-Alanine," and "3-hydroxy-propionyl-CoA to/from 3-hydroxy-propanoate," in the leucine metabolism-specific pathway, it consisted of

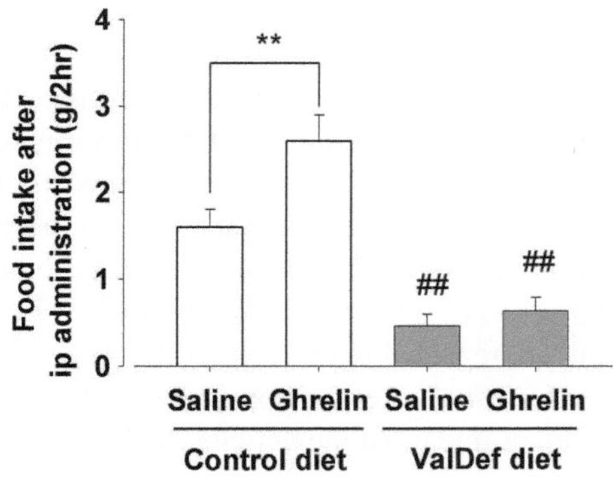

Fig. 14.2 An orexigenic effect of peripheral ghrelin was not observed in valine-deficient rats. Food intake was measured after intraperitoneal administration of ghrelin in rats fed control or valine-deficient diets for 5 days ($n = 12$–13). Data are presented as the means ± SEM. $^{**}P < 0.01$, $^{\#\#}P < 0.01$ for control diet versus valine-deficient diet (Goto et al. [20], with permission)

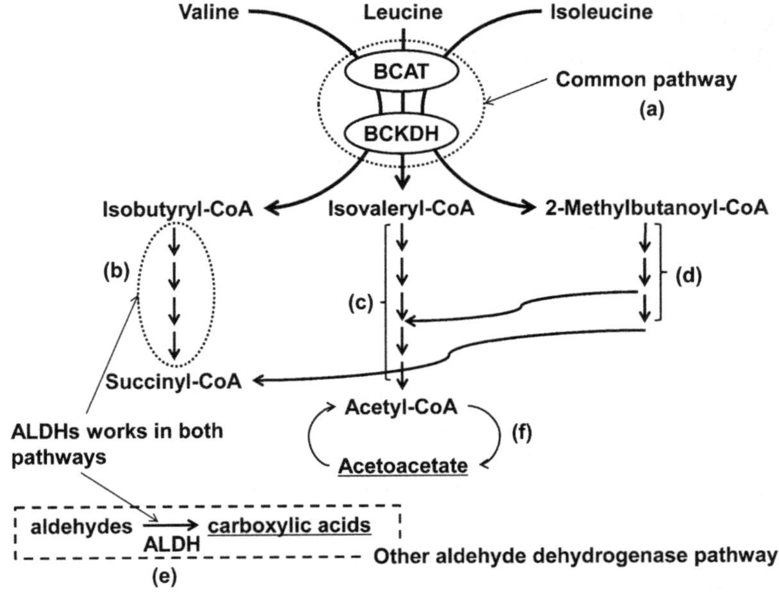

Fig. 14.3 Schematic illustration of the metabolic mechanisms of valine-deficient diet-induced anorexia

Table 14.1 Plasma BCAA concentrations of rats fed each diet for 28 days (modified from Nagao et al. [22] with permission)

μmol/L	Control	Pairfed	ValDef
Val	190.8±9.8	164.1±9.7	80.5±16.2
Leu	139.9±6.7	128.6±11.0	91.1±9.2
Ile	82.3±4.0	75.8±4.9	57.1±6.4

"acetoacetyl-CoA to/from acetyl-CoA," "acetoacetyl-CoA to HMG-CoA," and "HMG-CoA to acetoacetate," and in the isoleucine metabolism-specific pathway, it consisted of "Propionyl-CoA to Methylmalonyl-CoA." Because there were a greater number of valine and leucine metabolism-specific pathways compared to isoleucine (Table 14.3), we chose to focus on valine and leucine metabolism to describe the mechanism of valine-deficient diet induced anorexia.

Valine metabolism-specific genes encode different types of aldehyde dehydrogenases (ALDHs) that also play roles in other aldehyde dehydrogenase pathways (for example, alcohol metabolism, Table 14.3; Fig. 14.3b, e). However, leucine metabolism-specific genes do not encode ALDHs but encode enzymes that catalyze acetoacetyl-CoA to/from acetyl-CoA (Table 14.3; Fig. 14.3f). Altogether, we hypothesized that in valine-deficient situations, ALDHs involved in valine metabolism would mainly work for other aldehyde dehydrogenase pathways, resulting in increased carboxylic acid levels (Fig. 14.3b, e), and that the leucine metabolic pathway would induce acetoacetate accumulation (Fig. 14.3f). Excess acetate accumulation lowers blood pH and induces acidosis, which is known to be a major trigger of anorexia [25]. Moreover, ketone bodies can act as a satiety signal [26] and are reported to inhibit feeding behavior, which could lead to anorexia. Thus, the accumulation of acetoacetate, a ketone body, would likely contribute to anorexia, and we believe that these metabolic changes due to valine deficiency could be part of a novel mechanism underlying anorexia.

Table 14.2 Gene sets and substrate/product of each enzyme related to valine, leucine and isoleucine

	Enzyme	Gene symbol	EC	Substrate	Product	
Val metabolite	3-hydroxyisobutyryl-CoA hydrolase	HIBCH	3.1.2.4	(*S*)-3-hydroxyisobutyryl-CoA	(*S*)-3-hydroxyisobutyrate	
	3-hydroxyisobutyrate dehydrogenase	HIBADH	1.1.1.31	(*S*)-3-hydroxyisobutyrate	(*S*)-methylmalonate semialdehyde	
	4-aminobutyrate aminotransferase	ABAT	2.6.1.22	(*S*)-methylmalonate semialdehyde	(*S*)-3-amino-isobutyrate	
	Aldehyde dehydrogenase 2 family (mitochondrial)	ALDH2	1.2.1.3	(*S*)-methylmalonate semialdehyde	Methylmalonate	
	Aldehyde dehydrogenase 1 family, member B1	ALDH1B1	1.2.1.3	(*S*)-methylmalonate semialdehyde	Methylmalonate	
	Aldehyde dehydrogenase 9 family, member A1	ALDH9A1	1.2.1.3	(*S*)-methylmalonate semialdehyde	Methylmalonate	
	Aldehyde dehydrogenase 3 family, member A2	ALDH3A2	1.2.1.3	(*S*)-methylmalonate semialdehyde	Methylmalonate	
	Aldehyde dehydrogenase 7 family, member A1	ALDH7A1	1.2.1.3	(*S*)-methylmalonate semialdehyde	Methylmalonate	
	Aldehyde oxidase 1	AOX1	1.2.3.1	(*S*)-methylmalonate semialdehyde	Methylmalonate	
Leu metabolite	Isovaleryl-CoA dehydrogenase	IVD	1.3.8.4	Isovaleryl-CoA	3-methylcrotonyl-CoA	
	Methylcrotonoyl-CoA carboxylase 1	MCCC1	6.4.1.4	3-methylcrotonyl-CoA	3-methylglutaconyl-CoA	
	Methylcrotonoyl-CoA carboxylase 2	MCCC2	6.4.1.4	3-methylcrotonyl-CoA	3-methylglutaconyl-CoA	
	AU RNA binding protein/enoyl-CoA hydratase	AUH	4.2.1.18	3-methylglutaconyl-CoA	HMG-CoA	
	3-hydroxymethyl-3-methylglutaryl-CoA lyase	HMGCL	4.1.3.4	HMG-CoA	Acetoacetate	
	3-hydroxy-3-methylglutaryl-CoA synthase 1	HMGCS1	2.3.3.10	Acetoacetyl-CoA	HMG-CoA	
	3-hydroxy-3-methylglutaryl-CoA synthase 2	HMGCS2	2.3.3.10	Acetoacetyl-CoA	HMG-CoA	
	3-oxoacid CoA transferase 1	OXCT1	2.8.3.5	Acetoacetate	Acetoacetyl-CoA	
	3-oxoacid CoA transferase 2	OXCT2	2.8.3.5	Acetoacetate	Acetoacetyl-CoA	
	Acetyl-CoA acetyltransferase 1	ACAT1	2.3.1.9	Acetoacetyl-CoA	Acetyl-CoA	
	Acetyl-CoA acetyltransferase 2	ACAT2	2.3.1.9	Acetoacetyl-CoA	Acetyl-CoA	
Ile metabolite	Hydroxysteroid (17-beta) dehydrogenase 10	HSD17B10	1.1.1.178	(*S*)-3-hydroxy-2-methylbutyryl-CoA	2-methylacetoacetyl-CoA	
	Acetyl-CoA acyltransferase 1	ACAA1	2.3.1.16	2-methylacetoacetyl-CoA	Propanoyl-CoA	Acetyl-CoA
	Acetyl-CoA acyltransferase 2	ACAA2	2.3.1.16	2-methylacetoacetyl-CoA	Propanoyl-CoA	Acetyl-CoA
	Hydroxyacyl-CoA dehydrogenase	HADHB	2.3.1.16	2-methylacetoacetyl-CoA	Propanoyl-CoA	Acetyl-CoA
	Propionyl CoA carboxylase, Alpha polypeptide	PCCA	6.4.1.3	Propionyl-CoA	(S)-methylmalonyl-CoA	
	Propionyl CoA carboxylase, Beta polypeptide	PCCB	6.4.1.3	Propionyl-CoA	(S)-methylmalonyl-CoA	

Table 14.3 Valine, leucine, and isoleucine metabolism: specific gene-related metabolic pathways

Valine	Related metabolism	Leucine	Related metabolism	Isoleucine
Valine, leucine, and isoleucine degradation		Valine, leucine, and isoleucine degradation		Valine, leucine, and isoleucine degradation
Propanoate metabolism	2-propyn-1-al to Propynoate Malonate-semialdehyde to/from Beta-alanine 3-hydroxy-propionyl-CoA to/from 3-hydroxy-propanoate	Propanoate metabolism	Acetoacetyl-CoA to/from Acetyl-CoA Acetoacetyl-CoA to HMG-CoA HMG-CoA to Acetoacetate	Propanoate metabolism
Fatty acid metabolism	Aldehyde to fatty acid	Fatty acid metabolism	Acetoacetyl-CoA to/from Acetyl-CoA	Fatty acid metabolism
Butanoate metabolism	Succinate-semialdehyde to/from 4-aminobutanoate	Butanoate metabolism	Acetoacetyl-CoA to/from Acetyl-CoA	
Pyruvate metabolism	Acetaldehyde to/from Acetate	Pyruvate metabolism	Acetoacetyl-CoA to/from Acetyl-CoA	
Tryptophan metabolism	5-hydroxyindle-acetaldehyde to 5-hydroxy-indoleacetate Indole-3-acetaldehyde to Indoleacetate	Tryptophan metabolism	Acetoacetyl-CoA to/from Acetyl-CoA	
Lysine degradation	4-trimethyl-ammoniobutanal to 4-trimethyl-ammoniobutanoate	Lysine degradation	Acetoacetyl-CoA to/from Acetyl-CoA	
		Synthesis and degradation of ketone bodies	Acetoacetyl-CoA to/from Acetyl-CoA Acetoacetyl-CoA to HMG-CoA HMG-CoA to Acetoacetate	
		Terpenoid backbone biosynthesis	Acetoacetyl-CoA to/from Acetyl-CoA Acetoacetyl-CoA to HMG-CoA HMG-CoA to Acetoacetate	
Beta-alanine metabolism	Malonate-semialdehyde to/from Beta-alanine Beta-aminopropion-aldehyde to Beta-alanine			
Ascorbate and aldarate metabolism	D-Glucuronolactone to/from D-glucarate			
Histidine metabolism	Imidazole acetaldehyde to Imidazole-4-acetate			
Glycerolipid metabolism	D-glyceraldehyde to/from D-glycerate			
Glycolysis/Gluconeogenesis	Acetaldehyde to/from Acetate			
Arginine and proline metabolism	4-acetamidobutanal to 4-acetamidobutanoate			

Conclusions

Essential amino acids are indispensable nutrients for mammalian body maintenance. When they are missing from the diet, many typical phenotypes such as anorexia, body weight loss, and impairment of neural functions are reported. In this chapter, we discuss the mechanisms related to anorexia induced by valine deficiency. Food intake is controlled mainly through the hypothalamus and is strongly affected by dietary valine deficiency, which induces ghrelin resistance and increases somatostatin expression in hypothalamus, and both of these factors act through the endocrine system to induce anorexia. Additionally, from a metabolic point of view, valine deficiency significantly decreases plasma valine, leucine, and isoleucine levels. These physiological changes would inactivate the valine metabolic pathway, resulting in excess amount of ALDHs. When these ALDHs would work in other metabolic pathways, acidosis will arise in blood. Additionally, leucine pathway would be activated and accumulate acetoacetate, resulting in ketosis. These should contribute to valine-deficient-diet-induced anorexia.

Conflict of Interest and Funding Disclosure Author disclosures: T. T., C. F., H. M., and M. B. are employees of Ajinomoto Co., Inc.

References

1. Nakazato M, Murakami N, Date Y, et al. A role for ghrelin in the central regulation of feeding. Nature. 2001;409(6817):194–8.
2. Date Y, Murakami N, Toshinai K, et al. The role of the gastric afferent vagal nerve in ghrelin-induced feeding and growth hormone secretion in rats. Gastroenterology. 2002;123(4):1120–8.
3. Sun Y, Wang P, Zheng H, Smith RG. Ghrelin stimulation of growth hormone release and appetite is mediated through the growth hormone secretagogue receptor. Proc Natl Acad Sci U S A. 2004;101(13):4679–84.
4. Willesen MG, Kristensen P, Romer J. Co-localization of growth hormone secretagogue receptor and NPY mRNA in the arcuate nucleus of the rat. Neuroendocrinology. 1999;70(5):306–16.
5. Kamegai J, Tamura H, Shimizu T, Ishii S, Sugihara H, Wakabayashi I. Chronic central infusion of ghrelin increases hypothalamic neuropeptide Y and Agouti-related protein mRNA levels and body weight in rats. Diabetes. 2001;50(11):2438–43.
6. Chen HY, Trumbauer ME, Chen AS, et al. Orexigenic action of peripheral ghrelin is mediated by neuropeptide Y and agouti-related protein. Endocrinology. 2004;145(6):2607–12.
7. Johansen JE, Fetissov S, Fischer H, Arvidsson S, Hokfelt T, Schalling M. Approaches to anorexia in rodents: focus on the anx/anx mouse. Eur J Pharmacol. 2003;480(1–3):171–6.
8. Ooyama K, Kojima K, Aoyama T, Takeuchi H. Decrease of food intake in rats after ingestion of medium-chain triacylglycerol. J Nutr Sci Vitaminol. 2009;55(5):423–7.
9. Jing MY, Sun JY, Weng XY. Insights on zinc regulation of food intake and macronutrient selection. Biol Trace Elem Res. 2007;115(2):187–94.
10. Tordoff MG, Friedman MI. Hepatic portal glucose infusions decrease food intake and increase food preference. Am J Physiol. 1986;251(1 Pt 2):R192–6.
11. Sclafani A, Ackroff K. Role of gut nutrient sensing in stimulating appetite and conditioning food preferences. Am J Physiol Regul Integr Comp Physiol. 2012;302(10):R1119–33.
12. Mercer LP, Kelley DS, Haq A, Humphries LL. Dietary induced anorexia: a review of involvement of the histaminergic system. J Am Coll Nutr. 1996;15(3):223–30.
13. Gietzen DW, Rogers QR. Nutritional homeostasis and indispensable amino acid sensing: a new solution to an old puzzle. Trends Neurosci. 2006;29(2):91–9.
14. Kilberg MS, Pan YX, Chen H, Leung-Pineda V. Nutritional control of gene expression: how mammalian cells respond to amino acid limitation. Annu Rev Nutr. 2005;25:59–85.
15. Layman DK. The role of leucine in weight loss diets and glucose homeostasis. J Nutr. 2003;133(1):261S–7.
16. Hutson SM. Subcellular distribution of branched-chain aminotransferase activity in rat tissues. J Nutr. 1988;118(12):1475–81.

17. Nagao K, Bannai M, Seki S, Mori M, Takahashi M. Adaptational modification of serine and threonine metabolism in the liver to essential amino acid deficiency in rats. Amino Acids. 2009;36(3):555–62.
18. Shikata N, Maki Y, Noguchi Y, et al. Multi-layered network structure of amino acid (AA) metabolism characterized by each essential AA-deficient condition. Amino Acids. 2007;33(1):113–21.
19. Schutz Y. Protein turnover, ureagenesis and gluconeogenesis. Int J Vitam Nutr Res. 2011;81(2–3):101–7.
20. Goto S, Nagao K, Bannai M, et al. Anorexia in rats caused by a valine-deficient diet is not ameliorated by systemic ghrelin treatment. Neuroscience. 2010;166(1):333–40.
21. Nakahara K, Takata S, Ishii A, et al. Somatostatin is involved in anorexia in mice fed a valine-deficient diet. Amino Acids. 2012;42(4):1397–404.
22. Rudell JB, Rechs AJ, Kelman TJ, Ross-Inta CM, Hao S, Gietzen DW. The anterior piriform cortex is sufficient for detecting depletion of an indispensable amino acid, showing independent cortical sensory function. J Neurosci. 2011;31(5):1583–90.
23. Harper AE, Miller RH, Block KP. Branched-chain amino acid metabolism. Annu Rev Nutr. 1984;4:409–54.
24. da Huang W, Sherman BT, Lempicki RA. Systematic and integrative analysis of large gene lists using DAVID bioinformatics resources. Nat Protoc. 2009;4(1):44–57.
25. Winston AP. The clinical biochemistry of anorexia nervosa. Ann Clin Biochem. 2012;49(Pt 2):132–43.
26. Hawkins RA, Williamson DH, Krebs HA. Ketone-body utilization by adult and suckling rat brain in vivo. Biochem J. 1971;122(1):13–8.

Part III
Experimental Models of Growth and Disease States: Role of Branched Chain Amino Acids

Chapter 15
Leucine and Fetal Growth

Julio Tirapegui, Daiana Vianna, Gabriela Fullin Resende Teodoro, and Lucas Carminatti Pantaleão

Key Points

- Poor maternal nutrition during pregnancy in turn implies a risk of poor nutritional availability to the fetus.
- One of the most important factors that determine fetal growth is nutrient availability to the fetus.
- Poor maternal nutrition has however been shown to be one of the major causal determinants of intrauterine growth restriction.
- The treatment to alleviate the complications of intrauterine growth restriction is supplementation with proteins or leucine.
- Leucine not only is important to stimulate protein synthesis but also serves as an energy source for the fetus and is a regulator of embryo development.

Keywords BCAA • Intrauterine growth restriction • Leucine • mTOR • Protein synthesis

Abbreviations

AGA	Gestational age
ATGL	Triglyceride lipase
AKT	Protein Kinase B
BCAA	Branched chain amino acids
C/EBP	CCAAT/enhancer-biding protein
4E-BP1	Eukaryotic initiation factor 4E-binding protein 1
eIF4E	Eukaryotic initiation factor 4E
eIF4G	Eukaryotic initiation factor 4G

J. Tirapegui, Ph.D. (✉)
Department of Food and Experimental Nutrition, Faculty of Pharmaceutical Sciences,
University of São Paulo, Avenida Professor Lineu Prestes, 580, bloco 14, São Paulo, SP 05508-000, Brazil
e-mail: tirapegu@usp.br

D. Vianna, M.Sc. • G.F.R. Teodoro, M.Sc. • L.C. Pantaleão, M.Sc.
University of São Paulo, Avenida Professor Lineu Prestes, 580, São Paulo, SP 05508-000, Brazil
e-mail: daiana.vianna@gmail.com; daianavianna@usp.br; gabrielafrt@gmail.com;
lucascp@usp.br; luquetasp@yahoo.com.br

R. Rajendram et al. (eds.), *Branched Chain Amino Acids in Clinical Nutrition: Volume 1*, Nutrition and Health, DOI 10.1007/978-1-4939-1923-9_15,

eIF2α	Eukaryotic initiation factor 2α
eIF4F	Ribosomal complex
HSL	Hormone sensitive lipase
HNF-1	Hepatocyte nuclear factor 1
HNF-2	Hepatocyte nuclear factor 2
IGF	Insulin-like growth factor
IGF-1	Insulin-like growth factor 1
IGF-2	Insulin-like growth factor 2
IUGR	Intrauterine growth restriction
LAT1 (SLC7A5)	The system L type transporter 1
LAT2 (SLC7A8)	The system L type transporter 2
MRF4	Myogenic regulatory factors 4
mRNA	Messenger ribonucleic acid
mTOR	Mammalian target of rapamycin
Myf-5	Myogenic factor 5
SANT1 (SLC38A1)	The system Na-linked neutral amino acid transporter 1
SANT2 (SLC38A2)	The system Na-linked neutral amino acid transporter 2
SANT4 (SLC38A4)	The system Na-linked neutral amino acid transporter 4
S6K1	70-kDa ribosomal protein S6 kinase 1

Introduction

Maternal nutrition is directly related to the growth and development of the fetus. However, epidemiological studies have demonstrated the consequences of poor maternal nutrition on the development of diseases in later life, such as obesity, cardiovascular diseases associated with a high incidence of myocardial infarction, diabetes, and dyslipidemias [1, 2]. On the basis of these observations, Barker [3] formulated the hypothesis of the fetal origin of diseases, which proposes that fetuses exposed to a limited supply of nutrients adapt to this situation by altering their metabolic processes. Fetal programming can therefore be defined as an adaptive process to an adverse intrauterine environment which redefines developmental processes, guaranteeing fetal survival.

In this respect, it has been observed that dietary modifications during pregnancy, such as protein restriction, promote specific changes in the functional activity and expression of placental transporters, particularly amino acid transporters, which are directly related to the regulation of fetal growth in response to the altered supply of nutrients by the placenta [4]. It is a known fact that embryonic cells respond directly to amino acid deficiency by altering the expression of a variety of genes that regulate growth, differentiation, and apoptosis [5]. In addition to maternal nutrition, there are other factors associated with fetal growth, including genetic factors and the production of hormones such as insulin, insulin-like growth factors (IGFs), ghrelin, and leptin [6].

In this chapter, we discuss aspects of the amino acid leucine and fetal growth and report the data of experimental studies. In addition, we highlight the molecular mechanisms underlying the leucine. We would like to draw the attention of the reader to the fact that this study seems to be a promising research field in nutrition and health, which is crucial for the development of public health strategies and programs designed to prevent chronic non-communicable diseases in adult life.

The Importance of Experimental Models

Animal studies are important for the understanding of fetal nutrition since current knowledge of this topic is derived from experiments investigating intrauterine growth restriction (IUGR), fetal nutrition, and supplementation with amino acids and other nutrients. Despite a large number of studies, it still remains unclear how the fetus responds to amino acid supplementation at physiological and molecular levels. Further studies using appropriate animal models are therefore needed.

However, before moving to the discussion of the results of different experimental studies on IUGR and leucine supplementation, we would like to stress the importance of basic knowledge of these topics for the understanding of research in this area.

Intrauterine Growth Restriction (IUGR)

One of the most important factors that determine fetal growth is nutrient availability to the fetus, which depends on placental transporter proteins. It is known that any change in the activity of placental nutrient transport alters fetal growth. For example, maternal protein malnutrition causes downregulation of placental amino acid transporters and contributes to the development of IUGR [7]. In addition, maternal protein restriction affects immunocompetence, body composition and the secretion of hormones that play a role in fetal growth [8]. The fetus is able to adapt to conditions of poor maternal nutrition, reducing its growth and development [2].

In agreement with this hypothesis, human and animal studies have shown that amino acid transport from the mother to the fetus and fetal amino acid metabolism are affected during IUGR [9]. Factors associated with IUGR include intrauterine infections, maternal illnesses, placental insufficiency, use of drugs and androgens, and congenital anomalies and some cases are idiopathic [9, 10] (Fig. 15.1). Moreover, pregnant women who had a prior pregnancy complicated by IUGR present a 50 % chance of manifesting this condition in the current pregnancy. The same is observed for a history of stillbirth since more than half of all stillbirths are associated with IUGR [11].

IUGR also causes various changes in islet cells, in the hypothalamic–pituitary–adrenal axis, and in the secretion of prolactin, progesterone, estradiol, and insulin, as well as in the glucose uptake by muscles, body fat content, and mitochondrial function. Furthermore, IUGR has been shown to induce

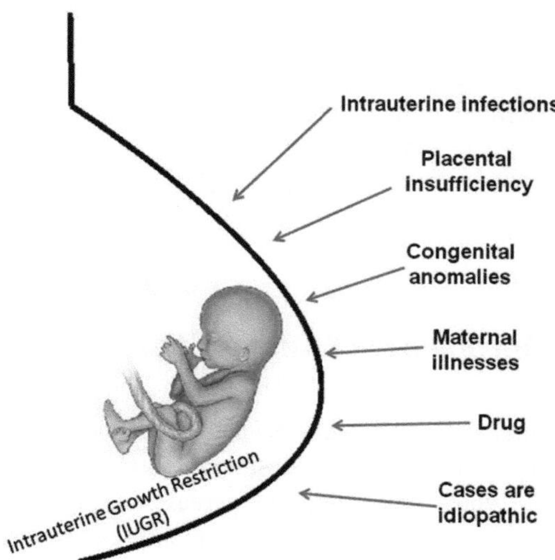

Fig. 15.1 Some risk factor for developing intrauterine growth restriction. Fetal intrauterine growth restriction (IUGR) is a frequently occurring and serious complication of pregnancy. The IUGR which is associated with perinatal mortality and morbidity, and developing non-communicable chronic diseases in later life. Unpublished

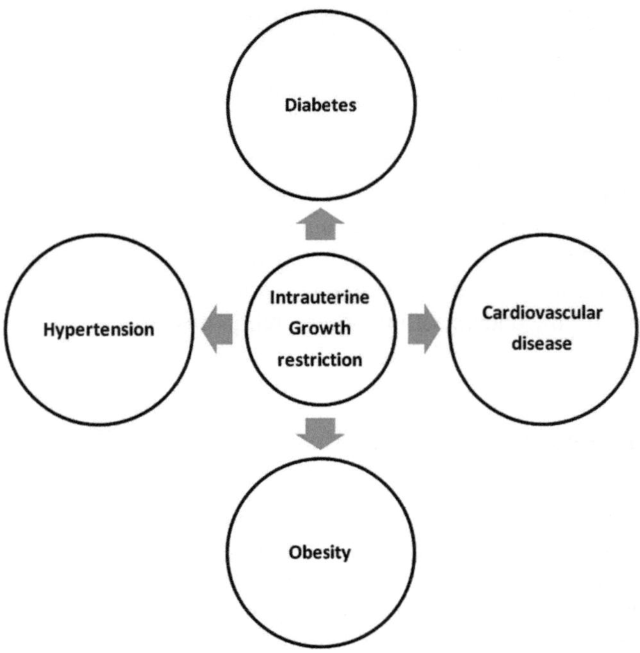

Fig. 15.2 Intrauterine growth restriction is associated with the development of diseases in later life. The intrauterine growth restriction is associated with the development of diabetes, cardiovascular disease, hypertension, and obesity later in life. Unpublished

oxidative stress. Changes in insulin and leptin signaling pathways, modifying lipid metabolism, have also been reported [8, 12]. Human and animal studies have demonstrated that IUGR increases the risk of developing cardiovascular diseases, diabetes, and obesity in adult life [1, 9, 12] (Fig. 15.2). Depending on the severity of IUGR, the infant presents a fivefold to tenfold risk of dying in utero [13].

In addition to the complications reported above, IUGR causes a reduction in organ growth and permanent changes in organ metabolism and/or structure. During fetal life, tissues and organs undergo critical periods of development that coincide with periods of rapid cell division. Acute or chronic changes in the fetal environment that occur during these periods cause a selective reduction of mitosis rates in certain vulnerable fetal tissues, reducing tissue hyperplasia and preserving other more important vital tissues, including the central nervous system [2, 3]. Retardation of fetal growth during critical periods of development can cause irreversible damage.

There is still no standard treatment to reverse the alterations caused by IUGR. However, evidence indicates that good prenatal care may reduce mortality, complications of IUGR, and prematurity [9]. Another treatment suggested to alleviate the complications of IUGR is supplementation with proteins or single amino acids, since both human and animal studies have shown that placental transport of essential amino acids such as leucine and threonine is reduced in IUGR pregnancies [14, 15]. Amino acids are important not only to stimulate protein synthesis but also as precursors of hormones, neurotransmitters, purine and pyrimidine nucleotides, and nitric oxide. In addition, amino acids serve as an energy source for the fetus and are regulators of embryo development [15].

Morphology of Adipose Tissue

Since IUGR has been associated with a long-term increase of adiposity and leptin concentrations [8, 16], in this section we discuss the importance of adipose tissue at the beginning of life.

Concern regarding the increase of adipose mass is justified by the fact that recent studies point to a new understanding of the function of adipose tissue. It is now known that adipose tissue does not

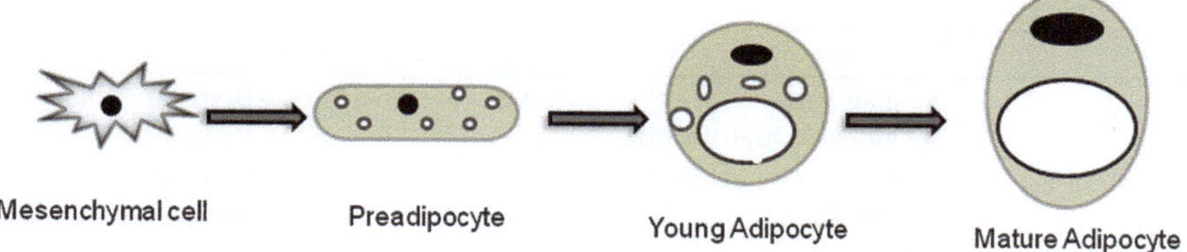

Fig. 15.3 Adipogenesis. The process of cell differentiation where preadipocytes become adipocytes. Unpublished

Table 15.1 Complications caused by intrauterine growth restriction

Tissue	Function	Complications
Placenta	Transports of amino acids	Amino acid transporters
Skeletal muscle	Glucose and triacylglycerol metabolism	Myocytes Satellite cell
Adipose	Lipid and carbohydrate metabolism	Adipose mass Obesity

only serve as a storage site for body fuels, but is also a dynamic organ that produces substances related to body equilibrium/homeostasis. These substances produced by adipocytes are called adipokines. Adipokines are involved in numerous processes, including the regulation of satiety and lipid and carbohydrate metabolism, and are also associated with growth [17].

The development of adipose tissue is characterized by the differentiation of precursor cells into mature adipocytes, a process known as adipogenesis. Morphological studies have shown that in human embryos this process starts before birth [18] (Fig. 15.3). Rapid expansion of adipose tissue is observed during the perinatal period and is a consequence of hypertrophic and hyperplastic processes. The potential to produce new mature cells continues even in adult life. Experimental studies indicate that the number of adipose cells tends to increase in response to diets rich in simple carbohydrates and saturated lipids. This phenomenon is related to a complex set of events that involve the proliferation and differentiation of pre-adipocytes [18].

The neonatal period is another important phase for the development of adipose mass in adults [19]. Increased expression of different IGF receptor isoforms is observed in adipose tissue of mammals during the first days after birth [19]. Prenatal nutritional restriction, followed by improved nutritional conditions after birth, is believed to increase the abundance of messenger ribonucleic acid (mRNA) encoding IGF receptors, thus potentiating the anabolic effect of IGF in adipose tissue and the consequent gain in adipose mass [18] (Table 15.1). This mechanism may explain the rapid weight gain characterized by greater adipose mass observed during catch-up growth.

Furthermore, it is known that excessive maternal energy intake directly affects the morphology of white adipose tissue, increasing the area of adipocytes in visceral adipose tissue already during the lactation period [20]. This phenomenon was described in recently weaned rats whose mothers received a diet rich in saturated fatty acids during gestation and lactation, but was not observed in pups of dams fed this diet only during gestation. In a similar study evaluating the effect of feeding a high-fat diet only during lactation, the magnitude of the increase in body adiposity and visceral fat pad mass was similar to that observed in animals of dams fed a high-fat diet during gestation and lactation. Both studies also reported the potential effect of a maternal postnatal high-fat diet on the morphology of adipose tissue at weaning [21] (Table 15.2). Taken together, these results suggest that both the fetal and the lactation periods are determinant for the establishment of the phenotype at the beginning of life, with maternal diet and nutritional status exerting a major influence on this process.

Table 15.2 Research in our group

Study	Population	Supplementation	Conclusion
Pantaleão et al. [21]	Weaning rats Submitted to high-fat diet	No	Body fat accumulation Reduced myofiber density
Donato et al. [37]	Adult rats submitted to food restriction	5 % Leucine	Body fat loss Protein Synthesis in liver
Vianna et al. [38]	Old rat	4 % Leucine	Body fat loss
Teodoro et al. [43]	Pregnant rats submitted to protein-restricted diet	4 % BCAA	Activation of the mTOR signaling pathway.

In conclusion, we emphasize the importance of considering the effects on the increase of adipose tissue mass since this increase is maintained during the postnatal period, contributing to a higher risk of obesity later in life associated with excess maternal energy intake.

Development of Muscle Tissue

Myogenesis is the development of embryonic muscle tissue. The predominant cells of this tissue, myocytes, are derived from mesenchymal cells that first differentiate into precursor cells, a process that requires the activity of different transcription factors. These factors include myogenic regulatory factors (MyoD, myogenin, Myf-5, and MRF4), which bind to specific DNA promoter regions and coordinate the differentiation of mesenchymal cells, maturation of myoblasts and the response to growth factors that regulate the myogenic process. These factors include IGFs which play a role in muscle tissue hyperplasia and hypertrophy [22, 23] (Fig. 15.4). In mammals, the hyperplastic phase occurs exclusively during fetal development, with the number of muscle fibers remaining constant after birth. However, there are studies reporting a slight increase after birth as a result of the maturation of preexisting myotubes [22]. During myogenesis, muscle fibers develop into two main types: fibers that form during the first stages of myoblast fusion, called primary fibers, and secondary cells that form during the second step of fetal myoblast differentiation. There is a third type of myoblasts that do not give origin to muscle fibers, but differentiate into a quiescent cell line found close to myofibrils, called satellite cells [22].

Since skeletal muscle tissue is the main site of glucose and triacylglycerol metabolism, the study of early muscle alterations has contributed to the understanding of the relationship between maternal dietary habits and different metabolic disorders in adult life. The effect of pregnancy diet involves mechanisms ranging from the molecular regulation of preferential differentiation of mesenchymal cells and myoblasts during fetal development to alterations in protein expression rates. Numerous enzymes and transcription and translation factors, as well as other elements that form an extensive intracellular network, participate in the modulation of these processes.

Fetal growth retardation programs the organism to develop a smaller proportion of muscle mass by reducing cell replication during gestation, with a consequent permanent reduction in the number of myocytes and satellite cells [23] (Table 15.1). In this respect, studies using animal models have shown that energy restriction (50 % of total intake) during gestation negatively affects muscle protein synthesis in the fetus manifested as a delay in muscle development and a decrease in the number of secondary myofibrils, which are the result of negative regulation of mammalian target of rapamycin (mTOR) signaling in muscle [24]. As a consequence of this hyperplastic dysfunction during muscle development, disproportionality between lean mass and adipose mass is observed during the body weight gain period in offspring of these dams [25]. Epidemiological evidence indicates that this reduced muscle mass in IUGR fetuses persists into adult life [26].

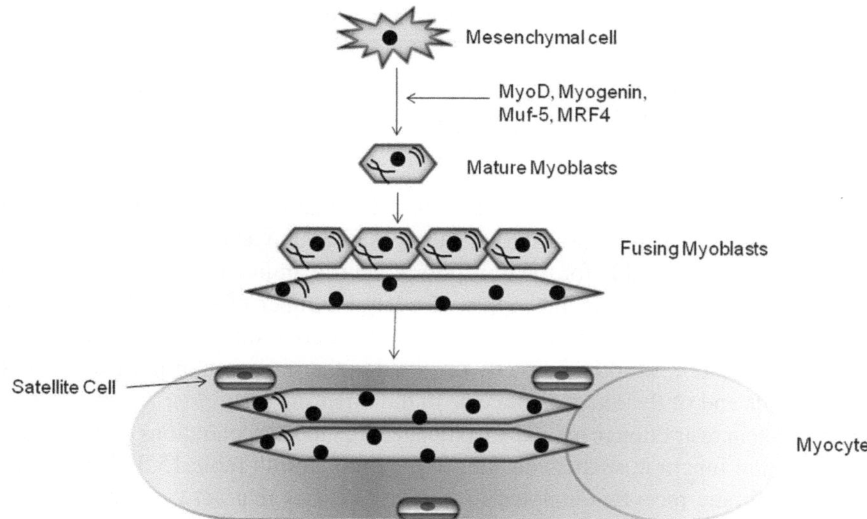

Fig. 15.4 Myogenesis. Myogenesis is the creation of muscle tissue from stem cells in the embryo. In this scheme shows the main cellular transitions that occur during myogenesis. The transcription factors MyoD, Myogen, Myf5, and MRF4 are crucial for determination. The myoblasts are derived from mesenchymal cell. The myoblasts in muscle can fuse to form myotubes in a myoblast–myoblast fusion event. Once formed myotubes initiate the formation of muscle tissue. Unpublished

Among the numerous factors that determine alterations in muscle development, mTOR activity, which mediates protein synthesis, growth, and mitosis, seems to be particularly important. Recent studies have shown that both diets rich in foods with high energy density and protein-deficient diets are responsible for the reduction of anabolic processes mediated by mTOR signaling [24].

Placental Transport of Amino Acids

During gestation, the placenta serves as an interface between maternal and fetal blood circulation, supplying oxygen and nutrients, removing metabolic waste, producing and secreting hormones, and regulating the uterine environment. The capacity of the placenta to supply nutrients to the fetus depends on a variety of factors, including placental size and morphology, blood flow, activity/expression of specific transporter proteins in the placental barrier, and the rate of nutrient production [27, 28]. The placenta is therefore of the utmost importance for fetal development.

It is known that fetal growth is determined by the supply of nutrients that are transported across the placenta and that any change in this process can cause alterations in fetal growth. The human placenta expresses more than 20 different amino acid transporters, including system A which is present in three isoforms: SNAT1 (*SLC38A1*), SNAT2 (*SLC38A2*), and SNAT4 (*SLC38A4*). This transport system permits the uptake of amino acids against their concentration gradient by simultaneously transporting sodium into the cell. System A transporter activity is reduced in the placenta of IUGR pregnancies [27, 28] (Table 15.1). Another amino acid transporter is the sodium-independent system L. This heterodimeric transporter consists of a light chain, large neutral amino acid transporter LAT1 (SLC7A5) or LAT2 (SLC7A8), and a heavy chain transmembrane protein, 4F2hc/CD98 (SLC3A2). System L transports branched chain amino acids (BCAA), phenylalanine, tyrosine, tryptophan, histidine, and methionine. As observed for system A, system L transporter activity is also reduced in IUGR pregnancies [28] (Table 15.1).

Table 15.3 Influence of protein-restricted diet on the developing fetus

- Reduced growth
- Reduced production of important hormones that regulate placental function
- Development of diseases in later life

Within this context, it has recently been demonstrated that severe protein restriction in rodents reduces the expression and activity of different subtypes of amino acid transport systems A and L in the placenta, which are responsible for the transport of neutral amino acids from the maternal to fetal side of the placental barrier and for the supply of a large number of essential amino acids such as leucine and tryptophane to the fetal circulation [29]. As a consequence, the deficiency and lower activity of these transporters would be directly involved in the reduction of plasma concentrations of essential amino acids and of the nitrogen balance in the offspring [30].

Furthermore, protein restriction in pregnant women reduces the concentrations of important hormones that regulate placental function, such as insulin, IGF-1, and leptin (Table 15.3). These hormones are recognized by membrane receptors coupled to tyrosine kinases that activate intracellular signaling pathways (mTOR and AKT) in the placenta. These signaling cascades, in turn, increase the expression and activity of the transporters cited above and, consequently, the supply of essential amino acids to the fetus. Therefore, a reduction in maternal anabolic hormone concentrations would decrease amino acid transport to the fetus in situations of protein restriction [30].

Roos et al. [31] demonstrated that mTOR is highly expressed in the human placental epithelium and that the inhibition of placental mTOR promotes a significant reduction in leucine uptake by the cell in placental villous samples. The authors suggested that the reduced cell growth mediated by inhibition of mTOR may be reinforced by decreased cellular leucine transport since intracellular leucine concentration is a potent activator of the mTOR signaling pathway. In addition, mTOR activity is reduced in IUGR pregnancies, with the observation of decreased expression of proteins involved in the mTOR signaling cascade in the placenta of fetuses with IUGR defined by a birth weight below the third percentile for term infants (37–41 weeks and 6 days).

Role of Leucine in Fetal Protein Synthesis

The better understanding of the mechanisms involved in the supply of amino acids for fetal metabolism across the placenta permitted to define that amino acids act as regulators of fetal and placental development by influencing metabolic pathways between the placenta and fetus. In addition, these amino acids directly participate in the regulation of protein synthesis in different fetal tissues and in the placenta by controlling the activity and/or expression of proteins involved in protein translation [7, 28]. In this respect, some investigators demonstrated that this stimulus mainly depends on BCAA, which include leucine, valine, and isoleucine. Leucine is particularly important since it stimulates protein synthesis in skeletal muscle, adipose tissue, hepatocytes, and other tissues [32, 33].

Experimental evidence indicates that leucine modulates translation initiation, which is fundamental for the control of protein synthesis by the cell. This modulation occurs through the phosphorylation and consequent activation of mTOR kinase. This protein serves as a nutrient sensor of cells, able to integrate environmental factors with the survival capacity of the organism [33].

It is known that mTOR is activated in the presence of mitogens such as insulin and some nutrients, especially leucine. When amino acid reserves, particularly leucine, are increased, mTOR activates other proteins in order to phosphorylate key components of the complex that translates mRNA into proteins [34]. Furthermore, mTOR acts on the signaling pathway that regulates cell growth and protein synthesis. In this process, phosphorylation of eukaryotic translation initiation factor 4E-binding protein 1 (4E-BP1) by mTOR leads to the inhibition of this protein and consequent dissociation of the

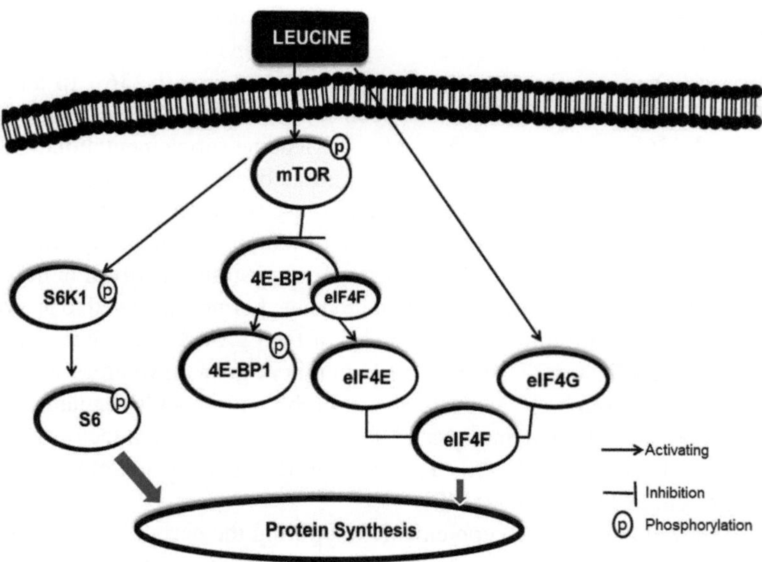

Fig. 15.5 Role of leucine in the regulation of protein synthesis. The mammalian target of rapamycin (mTOR) is activated in the presence of leucine. mTOR is involved in the activation of 70 kDa ribosomal protein S6 kinase (S6K1) and repression of the eukaryotic initiation factor 4E-biding protein 1 (4E-BP1). S6k1 phosphorylates of the ribosomal subunit protein s6. S6K1 controls several aspects of protein synthesis. The mTOR phosphorylation of 4E-BP1 disassociates it from eukaryotic initiation factor 4E (eIF4E). Leucine stimulates eukaryotic initiation factor 4G (eIF4G). The eIF4e and eIF4G form the ribosomal complex. This complex is vital for translation initiation of protein synthesis. Unpublished

4E-BP1 complex, favoring the formation of eukaryotic initiation factor 4E (eIF4E) which is essential for translation initiation. 4E-BP1 is a translation repressor, whereas eIF4E is involved in mRNA-ribosome binding [35].

Leucine also stimulates a second eukaryotic initiation factor, eIF4G, through an mTOR-independent pathway. Factors eIF4E and eIF4G form the ribosomal complex eIF4F, which regulates translation initiation by mediating the binding of mRNA to the 43S preinitiation complex. In addition, mTOR phosphorylates and activates 70-kDa ribosomal protein S6 kinase 1 (S6K1). Once activated, S6K1 phosphorylates and activates ribosomal protein S6. S6 is a component of the 40S ribosomal subunit and is therefore thought to be involved in regulating translation [35] (Fig. 15.5). Thus, leucine also stimulates the synthesis of complexes involved in protein translation, ultimately increasing the cellular capacity of protein synthesis [35]. However, the reduction or deprivation of essential amino acids can reduce S6K1 activity by 50–90 % [36].

Studies have demonstrated that protein-rich diets improve glycemic control and promote weight loss, in addition to maintaining lean mass. BCAA, particularly leucine, play an important role in the regulation of protein synthesis and energy metabolism in different tissues such as white adipose tissue, liver, and muscle [36]. According to this hypothesis, Donato et al. [37] found that leucine supplementation during a period of calorie restriction promoted body fat loss and improved protein synthesis in liver and muscle (Table 15.2). Recently, Vianna et al. [38] showed that dietary leucine supplementation of old rats attenuates body fat gain during aging (Table 15.2). These effects of leucine on adiposity are due to its action on insulin signaling, its capacity to increase the secretion of leptin and adiponectin in white adipose tissue, and its activation of the mTOR signaling pathway [39, 40]. The activation of mTORC1 signaling in adipocytes inhibits the expression of adipose triglyceride lipase (ATGL) and hormone sensitive lipase (HSL), thus suppressing lipolysis of triacylglycerol and diacylglycerol in adipose tissue. In addition, mTOR activation in adipocytes promotes de novo lipogenesis and triacylglycerol accumulation. The activation of mTOR also increases the secretion of adiponectin in 3T3-L1 adipocytes [40]. Adiponectin is directly associated with improved glucose uptake by increasing insulin sensitivity [41].

Some studies proposed that stimulation of protein synthesis by leucine in adipose tissue favors adipogenesis. In this respect, mTOR signaling seems to be involved in the differentiation of pre-adipocytes, adipose tissue morphogenesis, and hypertrophic growth [36]. In addition, 4E-BP1, one of the substrates modulated by mTOR, has been suggested to be a new regulator of adipogenesis [43]. Lynch et al. [36], studying the effects of acute leucine supplementation in different tissues of rats, observed that leucine supplementation stimulated protein synthesis in adipose tissue. These findings indicate that leucine plays a role in adipose tissue morphology, but further studies are needed to explain the true mechanisms of action of leucine in this body compartment.

In addition to improving protein synthesis by activating substrates involved in mTOR signaling, it is speculated that leucine stimulates mechanisms that modulate hormones related to growth such as IGFs and insulin. These hormones are important regulators of fetal growth, stimulating cell proliferation, differentiation, and metabolism. IGF-2 is involved in embryo growth, whereas IGF-1 predominates at the end of gestation. Although the mechanisms whereby amino acids regulate IGF transcription are not well understood, it has been demonstrated that several transcription factors such as hepatocyte nuclear factor-1 and -3 (HNF-1 and HNF-3) and CCAAT/enhancer-binding protein (C/EBP) are modulated in the presence of amino acids [42].

It is therefore believed that leucine supplementation during the perinatal period and during childhood can minimize the acute impacts of maternal protein malnutrition and IUGR. However, although the effects of leucine on protein synthesis are important, low-protein diets containing high concentrations of this amino acid have been shown to promote a reduction in the concentration of valine and isoleucine in the organism, a phenomenon called the "leucine paradox." This fact might be attributed to the allosteric regulation of enzymes involved in the metabolism of BCAA, increasing their oxidation.

Under conditions of protein restriction in which leucine is the only amino acid supplemented, the activity of enzymes involved in its catabolism is increased in response to the greater availability of this amino acid, whereas the other BCAA, valine and isoleucine are depleted [43]. Furthermore, the reduced growth of rats caused by a protein-restricted diet becomes more pronounced when the concentration of these last two amino acids declines. In addition to reduced growth, a decrease in food intake is observed in the case of supplementation with only one of the BCAA and dietary deficiency of at least one of these amino acids. Taken together, these findings indicate that, although leucine is the main amino acid responsible for mTOR activation, exclusive chronic supplementation with this BCAA is often impractical [43].

Studies of Branched Chain Amino Acids and Fetal Growth

As mentioned earlier, during pregnancy growth and development of the fetus entirely depend on maternal nutrition. Changes in maternal diet lead to alterations in fetal health as a result of adaptation of the fetus to the adverse intrauterine environment and the effects induced in this environment are strongly associated with the development of chronic diseases in adult life [2]. On the basis of this premise, studies have been conducted to define strategies that would prevent or reduce the deleterious effects of the hostile environment on the fetus.

There are various animal models designed to induce IUGR and to evaluate its consequences in different systems. One of the most frequently used models is maternal protein restriction, in which rats are fed a low-protein (5–10 %) diet during pregnancy [3]. Studies using this model suggest that maternal intake of a sufficient amount of protein is necessary for normal fetal growth [42].

Within this context, a study investigating rats fed a low-protein diet showed that these animals presented lower plasma concentrations of BCAA, lower levels of plasma urea nitrogen throughout gestation, and a higher rate of oxygen consumption during late gestation. Elevated levels of BCAA are associated with dietary protein intake and the release of amino acids during protein degradation in

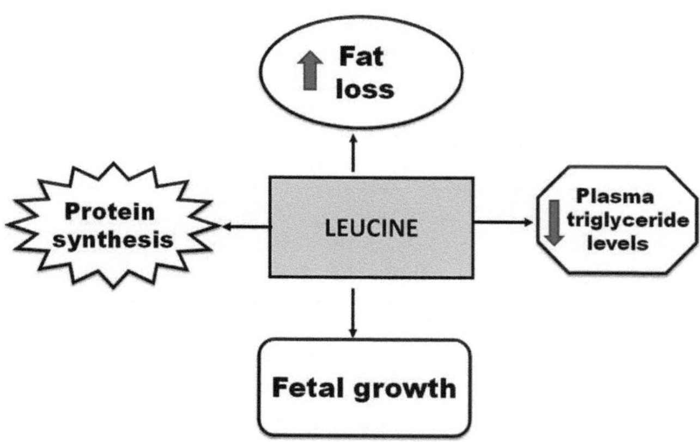

Fig. 15.6 Role of leucine. Leucine stimulates greater fat loss and reductions in plasma triglyceride levels. In addition, leucine activates the mTOR signaling pathway which controls protein synthesis and fetal growth. Unpublished

tissues. In addition, the authors observed increased phosphorylation of eIF2α in the liver, a finding suggesting reduced protein synthesis rates in fetuses with IUGR [44]. A study evaluating high-protein supplements administered during pregnancy reported a high frequency of low birth weight, an increased risk of small-for-gestational-age birth, and increased perinatal mortality [45]. In contrast, other studies found that amino acid supplementation reduces fetal acidosis, fetal hypoxia, and protein degradation and improves fetal protein accretion [26, 46]. Mogami et al. [42] observed that a high-protein diet, as well as BCAA-supplemented diets, improved parameters related to fetal growth (fetal weights and gene and protein expression of IGF-1 and IGF-2 in fetal liver), suggesting a pivotal role of BCAA in the amelioration of fetal growth restriction.

The regulatory action of amino acids in fetal and placental development has also been reported. This regulation occurs through the allosteric control of enzymes participating in metabolic pathways and physiological responses, which directly influence protein synthesis in different fetal and placental tissues by regulating the activity and/or expression of proteins involved in mRNA translation. The activity of mTOR is sensitive to the availability of amino acids, especially BCAA [7].

BCAA have been implicated in fetal development. In this respect, pregnant women with IUGR have been shown to present low plasma concentrations of BCAA in the umbilical artery and vein when compared to women with appropriate for gestational age fetuses [42]. Among these amino acids, leucine is particularly important since it is the main effector of mTOR signaling and stimulates protein synthesis in skeletal muscle and other tissues such as adipose tissue [36].

Leucine is an essential amino acid since it is not synthesized by the organism and must therefore be acquired through diet. This amino acid is found mainly in protein sources of animal origin and plays a crucial role in the regulation of protein synthesis and degradation [34, 36] (Fig. 15.6). Leucine is transported in the placenta by the L system, an amino acid transporter stimulated in the presence of neutral amino acids. As a consequence, high concentrations of neutral amino acids may stimulate the absorption of leucine [29]. However, a recent study showed that prolonged infusion of amino acids in sheep fetuses increased the concentration, transport and oxidation rate of leucine, findings indicating that the fetus used leucine for oxidative metabolism instead of protein synthesis. In other words, leucine stimulated the secretion of fetal glucagon and cortisol, changing the oxidative substrate used by the fetus for energy supply from glucose to the amino acid [45].

The fetal/maternal enrichment ratio for leucine is reduced in IUGR pregnancies [45] because of decreased placental blood flow, defects in more than one placental amino acid transporter, and conditions of fetal hypoxia and/or lactacidemia accompanying IUGR, reducing the transport of this amino acid across the placenta [47]. It would therefore be interesting to investigate the role of leucine in promoting fetal growth and in minimizing the consequences of IUGR. In this perspective, acute leucine supplementation has been shown to restore protein synthesis in skeletal muscle and visceral adipose tissue of neonatal pigs previously submitted to protein malnutrition (Table 15.4). Animals not

Table 15.4 Research in other groups

Study	Population	Leucine (%)	Conclusion
Murgas et al. [48]	Neonatal pigs	4	Increased of protein synthesis
Ventrucci et al. [49]	Pregnant rats	3	Higher muscle protein content
Ventrucci et al. [50]	Pregnant rats	3	Increased of protein synthesis

Table 15.5 Role of leucine in fetal growth

- Substrate and activation of protein synthesis
- Synthesis and secretion of hormones
- Body composition (e.g., fat loss and skeletal muscle increase)
- Regulator of protein, lipid, and energy metabolism

supplemented with this amino acid presented lower protein synthesis rates during the perinatal period, but supplementation resulted in muscle anabolism similar to that observed in animals born to well-nourished dams. This effect was attributed to positive regulation of the mTOR pathway as demonstrated by increased phosphorylation of mTOR, 4E-BP1, S6K1, and eIF4G and formation of the eIF4E-eIF4G complex [48].

Recently, Teodoro et al. [43] demonstrated the critical role of leucine combined with isoleucine and valine in the activation of the mTOR signaling pathway for the control of altered intrauterine growth induced by a maternal low-protein diet. The authors suggested that supplementation with BCAA partially reversed the growth deficit of neonatal rats born to dams subjected to maternal protein restriction by partially restoring body and intestinal weight and completely restoring heart, spleen, kidney and pancreas weight and body fat. These effects were due to the activation of mTOR and downstream proteins, such as 4E-BP1, and to lower phosphorylation of eIF2a (Table 15.2).

Studies of tumor-bearing rats have shown increased protein degradation, especially muscle protein, and fetal weight reduction. Ventrucci et al. [49] observed that dietary leucine supplementation (3 %) of tumor-bearing pregnant rats increased the absorption and availability of this amino acid in neoplastic cells and in the fetus. This fact contributed to fetal weight gain and higher muscle protein content (Table 15.4). In a subsequent study, the same group demonstrated that dietary leucine supplementation (3 %) of tumor-bearing pregnant rats increased the expression of protein kinase S6K1 and of eIF translation initiation factors. The authors suggested that a leucine-rich diet promotes an increase of protein synthesis in skeletal muscle of tumor-bearing pregnant rats, possibly through the activation of eIF factors and/or the S6K1 pathway, demonstrating the role of leucine in the translation initiation of protein synthesis [50] (Table 15.4). These studies show that cancer induces alterations in pregnant rats and that leucine supplementation may attenuate these complications and improve fetal development (Table 15.5).

Conclusions

Nutrient availability is the most important determinant of fetal growth and mainly depends on the transport capacity of the placenta. As a consequence, any alteration in amino acid transporters can cause IUGR. Increased protein intake might be beneficial for fetal growth, in particular the consumption of leucine which is a key amino acid in protein synthesis that can reduce the consequences of IUGR. However, further studies are needed to investigate whether the effects of leucine during the intrauterine period will effectively cause permanent changes in metabolism that predispose to or prevent diseases in adult life.

References

1. Barker DJP, Hales CN, Fall CHD, Osmond C, Phipps K, Clark PMS. Type 2 (non-insulin-dependent) diabetes mellitus, hypertension and hyperlipidaemia (syndrome X): relation to reduced fetal growth. Diabetologia. 1993;36:62–7.
2. Desai M, Gayle D, Babu J, Ross MG. Permanent reduction in heart and kidney organ growth in offspring of undernourished rat dams. Am J Obstet Gynecol. 2005;193:1224–32.
3. Barker DJ. Fetal nutrition and cardiovascular disease in later life. Br Med Bull. 1997;53:96–108.
4. Jansson T, Scholtbach V, Powell TL. Placental transport of leucine and lysine is reduced in intrauterine growth restriction. Pediatr Res. 1998;44:532–7.
5. Malandro MS, Beveridge MJ, Kilberg MS, Novak DA. Effect of low protein diet-induced intrauterine growth retardation on rat placental amino acid transport. Am J Physiol. 1996;271:C295–303.
6. Lesage J, Hahn D, Léonhardt M, Blondeau B, Bréant B, Dupouy JP. Maternal undernutrition during late gestation-induced intrauterine restriction in the rat is associated with impaired placental GLUT3 expression, but does not correlate with endogenous corticosterone levels. J Endocrinol. 2002;174:37–43.
7. Roos S, Powell TL, Jansson T. Placental mTOR links maternal nutrient availability to fetal growth. Biochem Soc Trans. 2009;37:295–8.
8. Chiesa C, Osborn JF, Haass C, Natale F, Spinelli M, Scapillati E, Spinelli A, Pacifico L. Ghrelin, leptin, IGF-1, IGFBP-3, and Insulin concentrations at birth: is there a relationship with fetal growth and neonatal anthropometry? Clin Chem. 2008;54:550–8.
9. Brown LD, Green AS, Limesand SW, Rozance PJ. Maternal amino acid supplementation for intrauterine growth restriction. Front Biosci (Schol Ed). 2011;3:428–44.
10. Figueras F, Gardosi J. Intrauterine growth restriction: new concepts in antenatal surveillance, diagnosis, and management. Am J Obstet Gynecol. 2011;204:288–300.
11. Villar J, Carroli G, Wojdyla D, et al. Preeclampsia, gestational hypertension and intrauterine growth restriction, related or independent conditions? Am J Obstet Gynecol. 2006;194:921–31.
12. Alexandre-Gouabau MC, Courant F, Le Gall G, Moyon T, Darmaun D, Parnet P, Coupé B, Antignac JP. Offspring metabolomic response to maternal protein restriction in a rat model of intrauterine growth restriction (IUGR). J Proteome Res. 2011;10:292–302.
13. Clausson B, Gardosi J, Francis A, Cnattingius S. Perinatal outcome in SGA births defined by customised versus population-based birthweight standards. BJOG. 2001;108:830–4.
14. Metcoff J, Cole TJ, Luff R. Fetal growth retardation induced by dietary imbalance of threonine and dispensable amino acids, with adequate energy and protein-equivalent intakes, in pregnant rats. J Nutr. 1981;111:1411–24.
15. Wu G, Pond WG, Ott T, Bazer FW. Maternal dietary protein deficiency decreases amino acid concentrations in fetal plasma and allantoic fluid of pigs. J Nutr. 1998;128:894–902.
16. Greenwood PL, Bell AW. Consequences of intra-uterine growth retardation for postnatal growth, metabolism and pathophysiology. Reprod Suppl. 2003;61:195–206.
17. Kimble RB, Matayoshi AB, Vannice JL, Kung VT, Williams C, Pacifici R. Simultaneous block of interleukin-1 and tumor necrosis factor is required to completely prevent bone loss in the early postovariectomy period. Endocrinology. 1995;136:3054–61.
18. Symonds ME, Pearce S, Bispham J, Gardner DS, Stephenson T. Timing of nutrient restriction and programming of fetal adipose tissue development. Proc Nutr Soc. 2004;63:397–403.
19. Bispham J, Clarke L, Symonds ME, Stephenson T. Postnatal ontogeny of insulin-like growth factor (IGF) and prolactin receptor (PRL-R) in ovine perirenal adipose tissue. J Physiol. 2003;65:547–65.
20. Bayol SA, Simbi BH, Stickland NC. A maternal cafeteria diet during gestation and lactation promotes adiposity and impairs skeletal muscle development and metabolism in rat offspring at weaning. J Physiol. 2005;567:951–61.
21. Pantaleão LC, Teodoro GF, Torres-Leal FL, et al. Maternal postnatal high-fat diet, rather than gestational diet, affects morphology and mTOR pathway in skeletal muscle of weaning rats. J Nutr Biochem. 2013. doi:10.1016/j.jnutbio.2012.10.009.
22. Du M, Yan X, Tong JF, Zhao J, Zhu MJ. Maternal obesity, inflammation, and fetal skeletal muscle development. Biol Reprod. 2010;82:4–12.
23. Lawrence T, Fowler V. Growth of farm animals. 2nd ed. CAB International: Wallingford; 2002.
24. Zhu MJ, Ford SP, Nathanielsz PW, Du M. Effect of maternal nutrient restriction in sheep on the development of fetal skeletal muscle. Biol Reprod. 2004;71:1968–73.
25. Eriksson JG, Forsen T, Jaddoe VWV, Osmond C, Barker DJP. Effects of size at birth and childhood growth on the insulin resistance syndrome in elderly individuals. Diabetologia. 2002;45:342–8.
26. Brown LD, Rozance PJ, Thorn SR, Friedman JE, Hay Jr WW. Acute supplementation of amino acids increases net protein accretion in IUGR fetal sheep. Am J Physiol Endocrinol Metab. 2012;303:352–64.
27. Shibata E, Hubel CA, Powers RW, et al. Placental system A amino acid transport is reduced in pregnancies with small for gestational age (SGA) infants but not in preeclampsia with SGA infants. Placenta. 2008;29:879–82.

28. Lager S, Powell TL. Regulation of nutrient transport across the placenta. J Pregnancy. 2012;2012:179827.
29. Glazier JD, Cetin I, Perugino G, Ronzoni S, Grey AM, Mahendran D, Marconi AM, Pardi G, Sibley CP. Association between the activity of the system A amino acid transporter in the microvillous plasma membrane of the human placenta and severity of fetal com-promise in intrauterine growth restriction. Pediatr Res. 1997;42:514–9.
30. Rosario FJ, Jansson N, Kanai Y, Parsad PD, Powell TL, Jansson T. Maternal protein restriction in the rat inhibits placental insulin, mTOR, and STAT3 signaling and down-regulates placental amino acid transporters. Endocrinology. 2011;152:1119–29.
31. Roos S, Jansson N, Palmberg I, Saljo K, Powell TL, Jansson T. Mammalian target of rapamycin in the human placenta regulates leucine transport and is down-regulated in restricted fetal growth. J Physiol. 2007;582:449–59.
32. Nicastro H, Da Luz CR, Chaves DF, Bechara LR, Voltarelli VA, Rogero MM, Lancha Jr AH. Does branched-chain amino acids supplementation modulate skeletal muscle remodeling through inflammation modulation? Possible mechanisms of action. J Nutr Metab. 2012;2012:136937.
33. Dodd KM, Tee AR. Leucine and mTORC1: a complex relationship. Am J Physiol Endocrinol Metab. 2012; 302:1329–42.
34. Kimball SR, Jefferson LS. Signaling pathways and molecular mechanisms through which branched-chain amino acids mediate translational control of protein synthesis. J Nutr. 2006;136:227–31.
35. Stipanuk MH. Leucine and protein synthesis: mTOR and beyond. Nutr Rev. 2007;65:122–9.
36. Lynch CJ, Fox HL, Vary TC, Jefferson LS, Kimball SR. Regulation of amino acid-sensitive TOR signaling by leucine analogues in adipocytes. J Cell Biochem. 2000;77:234–51.
37. Donato Jr J, Pedrosa RG, Cruzat VF, Pires ISO, Tirapegui J. Effects of leucine supplementation on the body composition and protein status of rats submitted to food restriction. Nutrition. 2006;22:520–7.
38. Vianna D, Teodoro GFR, Torres-Leal FL, Pantaleão LC, Donato Jr J, Tirapegui J. Long-term leucine supplementation reduces fat mass gain without changing body protein status of aging rats. Nutrition. 2012;28:182–9.
39. Lynch CJ, Gern B, Lioyd SM, Eicher R, Vary TC. Leucine in food mediates of the postprandial rise in plasma leptin concentrations. Am J Physiol Endocrinol Metab. 2006;291:621–30.
40. Torres-Leal FL, Fonseca-Alaniz MH, Teodoro GF, et al. Leucine supplementation improves adiponectin and total cholesterol concentrations despite the lack of changes in adiposity or glucose homeostasis in rats previously exposed to a high-fat diet. Nutr Metab. 2011;8:62.
41. Torres-Leal FL, Fonseca-Alaniz MH, Rogero MM, Tirapegui J. The role of inflamed adipose tissue in the insulin resistance. Cell Biochem Funct. 2010;28:623–31.
42. Mogami H, Yura S, Itoh H, et al. Isocaloric high-protein diet as well as branched-chain amino acids supplemented diet partially alleviates adverse consequences of maternal undernutrition on fetal growth. Growth Horm IGF Res. 2009;19:478–85.
43. Teodoro GF, Vianna D, Torres-Leal FL, et al. Leucine is essential for attenuating fetal growth restriction caused by a protein-restricted diet in rats. J Nutr. 2012;142:924–30.
44. Parimi PS, Cripe-Mamie C, Kalhan SC. Metabolic responses to protein restriction during pregnancy in rat and translation initiation factors in the mother and fetus. Pediatr Res. 2004;56:423–31.
45. Maliszewski AM, Gadhia MM, O'Meara MC, Thorn SR, Rozance PJ, Brown LD. Prolonged infusion of amino acids increases leucine oxidation in fetal sheep. Am J Physiol Endocrinol Metab. 2012;302:1483–92.
46. Rozance PJ, Crispo MM, Barry JS, O'Meara MC, Frost MS, Hansen KC, Hay Jr WW, Brown LD. Prolonged maternal amino acid infusion in late-gestation pregnant sheep increases fetal amino acid oxidation. Am J Physiol Endocrinol Metab. 2009;297:638–46.
47. Paolini CL, Marconi AM, Ronzoni S, et al. Placental transport of leucine, phenylalanine, glycine, and proline in intrauterine growth-restricted pregnancies. J Clin Endocrinol Metab. 2001;86:5427–32.
48. Murgas Torrazza R, Suryawan A, Gazzaneo MC, et al. Leucine supplementation of a low-protein meal increases skeletal muscle and visceral tissue protein synthesis in neonatal pigs by stimulating mTOR-dependent translation initiation. J Nutr. 2010;140:2145–52.
49. Ventrucci G, Mello MAR, Gomes Marcondes MCC. Effects of leucine supplemented diet on intestinal absorption in tumor bearing pregnant rats. BMC Cancer. 2002;2:7.
50. Ventrucci G, Mello MAR, Gomes Marcondes MCC. Leucine-rich diet alters the eukaryotic translation initiation factors expression in skeletal muscle of tumour-bearing rats. BMC Cancer. 2007;7:1–11.

Chapter 16
Enteral Leucine and Protein Synthesis in Skeletal and Cardiac Muscle

Agus Suryawan and Teresa A. Davis

Key Points

- The postprandial increase in protein synthesis supports protein anabolism.
- Leucine acts as an anabolic agent to enhance protein synthesis.
- The leucine signaling pathway to protein synthesis is still not well understood.
- Early explorations on the effects of leucine on protein synthesis were performed by in vitro study.
- Acute in vivo studies show leucine's effectiveness in stimulating protein synthesis.
- Studies on the long-term effect of enteral leucine supplementation are very limited.
- Further studies are needed to support whether the anabolic effect of leucine can be sustained in the long-term to maintain muscle mass or promote muscle accretion.

Keywords Skeletal muscle • Cardiac muscle • Protein synthesis • Leucine signaling • Amino acid transporters • Insulin signaling • mTORC1 • Translation initiation factors • Neonate • Sarcopenia

Abbreviations

4EBP1	Eukaryotic initiation factor 4E binding protein 1
BCAA	Branched chain amino acids
BCAT	Branched chain amino acid aminotransferase
BCKD	Branched chain α-keto acid dehydrogenase
EAA	Essential amino acids
eEF2	Elongation factor 2
eIF	Eukaryotic translation initiation factor
GβL	G Protein beta subunit-like
GTPase	Guanosine 5′-triphosphatase
LAT1	L-Type amino acid transporter 1
MAP4K3	Mitogen-activated protein kinase kinase kinase kinase-3
mTORC1	Mechanistic target of rapamycin complex 1

A. Suryawan, Ph.D. • T.A. Davis, Ph.D. (✉)
USDA/ARS Children's Nutrition Research Center, Baylor College of Medicine,
1100 Bates Street, Houston, TX 77030, USA
e-mail: suryawan@bcm.edu; tdavis@bcm.edu

R. Rajendram et al. (eds.), *Branched Chain Amino Acids in Clinical Nutrition: Volume 1*, Nutrition and Health, DOI 10.1007/978-1-4939-1923-9_16,

mTORC2	Mechanistic target of rapamycin complex 2
met-tRNA$_i$	Initiator methionyl-tRNA
PRAS40	Proline-rich Akt substrate of 40 kDa
Rag	Ras-related GTPase
Ragulator	Rag regulator
raptor	Regulatory-associated protein of mTOR
rictor	Rapamycin-insensitive companion of mTOR
SNAT2	Sodium-coupled neutral amino acid transporter 2
S6K1	p70 S6 kinase 1
TCA	Tricarboxylic acid
Vps34	Vacuolar protein sorting 34

Introduction

The ability of mammalian tissue to mount an increase in protein synthesis in response to a meal is fundamental for growth and repair throughout life. The contribution of the postprandial rise in insulin and amino acids is vital for this anabolic process. Although little attention has been given to other macronutrients, such as fats and carbohydrate, the postprandial rise in energy expenditure which starts immediately after a meal is crucial for the high efficiency of protein synthesis. Thus, the delicate coordination of these factors is necessary to ensure proper execution of optimal postprandial protein synthesis.

Skeletal muscle, which comprises approximately 40–50 % of total body mass, is a major site of whole body metabolism and a primary determinant of amino acid requirements [1]. Maintaining optimal muscle mass throughout life, from the neonatal period to old age, has been the subject of much study in recent years. Although the heart is much smaller in quantity than skeletal muscle, normal growth and protein turnover of cardiac muscle is crucial for heart function and whole body metabolism. Abnormal cardiac muscle overgrowth due to hypertrophy can lead to serious adverse health effects such as heart failure and vascular disease [2]. Taken together, the understanding of the regulation of postprandial protein synthesis is important, since this physiological process plays a crucial role in maintaining skeletal muscle as well as cardiac muscle mass.

The physiological increase in insulin and amino acid levels after a meal stimulates both skeletal and cardiac muscle protein synthesis [3, 4]. Anabolic effects of insulin in regulating overall protein metabolism and the mechanistic action of insulin toward protein synthesis is well characterized. On the other hand, the anabolic properties of amino acids, especially leucine, have just begun to surface. Although the notion that leucine acts not only as a substrate for protein synthesis but also as an anabolic agent that supports protein synthesis was discovered in vitro in 1975, a complete understanding of the role of leucine is still far from our reach. Promisingly, recent research focusing on the effect of leucine on protein synthesis has brought us closer to the understanding of how this unique amino acid affects human health.

In this review article, we discuss what is currently known about leucine's effect on skeletal and cardiac muscle protein synthesis and present some of our research findings in the highly metabolically sensitive animal model, the neonatal pig. The specific purpose of this review is to provide an overview of how leucine modulates skeletal muscle protein synthesis in normal and disease states. Studies on leucine's effect on cardiac muscle protein systems have focused primarily on certain conditions such as alcoholism and sepsis. In terms of leucine's mechanistic action, much of the information was obtained using in vitro experiments involving non-muscle cell lines. Thus, the applicability of the evidence on the mode of action by which leucine stimulates protein synthesis in skeletal and cardiac muscle must be established using in vivo animal models. In this chapter, we discuss whole animal experiments from our laboratory and others that have investigated the molecular mechanisms involved in the control of protein synthesis by leucine, with a particular focus on the effects of enteral leucine supplementation.

Leucine, a Unique Member of the Branched Chain Amino Acids (BCAA)

There are three members of the BCAA: leucine, isoleucine, and valine. As essential amino acids, these amino acids have important functions which include a primary role in protein structure and metabolism [5]. It is intriguing that the requirement for BCAA in humans comprise about 40–45 % of the total essential amino acids [6]. However, human deficiency of the BCAA usually does not occur due to a high percentage of these amino acids (up to 50 %) in the food supply [7]. Another unique property of the BCAA is that, unlike other amino acids, the majority of BCAA are metabolized in muscle and other nonsplanchnic tissues suggesting that BCAA metabolism is important in these tissues [8].

There are two steps in the metabolic pathway of BCAA [7]. The first step, which is reversible and catalyzed by branched chain amino acid aminotransferase (BCAT), is transamination of BCAA (leucine, isoleucine, and valine) to form their respective α-keto acids (α-ketoisocaproate, α-keto-β-methylvalerate, and α-ketoisovalerate). The second step is an irreversible catabolic process catalyzed by the branched chain α-keto acid dehydrogenase (BCKD) complex which involves oxidative decarboxylation of all of the branched chain keto acids. In the second step, the carbon skeletons of the BCAA are committed to be degraded for substrates of the tricarboxylic acid (TCA) cycle [7]. Consistent with the notion that BCAA are metabolized in nonsplanchnic tissues, the abundance and the activation of BCAT and BCKD are detectable in nonsplanchnic peripheral tissues from several species including rats, monkeys, and humans [9].

Hutson et al. [10] described the hallmarks of BCAA metabolism that set them apart from other essential amino acids. First, unlike the catabolic enzymes for other amino acids which are mainly located in the liver, the BCAA catabolic enzymes are widely distributed throughout the body resulting in inter- and intra-organ shuttling or exchange of BCAA metabolites. Second, BCAA also serve as major nitrogen donors for alanine and glutamine synthesis. Lastly, and the most important feature for this review is the role of leucine as an anabolic nutrient signal that participates in anabolic processes such as the stimulation of protein synthesis in various tissues.

To determine the possible role of BCAA in the regulation of muscle protein turnover, Buse and Reid [11] tested the effect of adding a mixture of three BCAA to the media containing glucose on the incorporation of [^{14}C]lysine into protein in isolated rat hemi-diaphragm muscles. They found that BCAA stimulated muscle protein synthesis in vitro. When they tested each member of the BCAA separately, only leucine stimulated muscle protein synthesis while valine had no effect and isoleucine inhibited protein synthesis. Experiments using actinomycin-D to block transcription and cycloheximide to block translation indicated that leucine stimulated protein synthesis and inhibited protein breakdown. Other in vitro studies using different skeletal muscle types and liver in rats resulted in similar conclusions [6].

To test whether Buse and Reid's findings [11] can be applied to the in vivo condition, we conducted a study using the neonatal pig as a model of the human infant to determine if leucine is the only member of BCAA that has the anabolic capacity to stimulate protein synthesis in various tissues [12]. Pigs were infused for 1 h with leucine, isoleucine, or valine to raise circulating concentrations to within the postprandial physiological levels and the fractional rate of protein synthesis was determined using the flooding dose technique. Figure 16.1 shows that leucine infusion enhanced protein synthesis in the longissimus dorsi muscle and heart by 28 % and 26 %, respectively. By contrast, valine and isoleucine infusion did not stimulate protein synthesis in either tissue. Thus, our in vivo study supported previous findings that, of the BCAA, only leucine demonstrates anabolic properties towards protein synthesis [11].

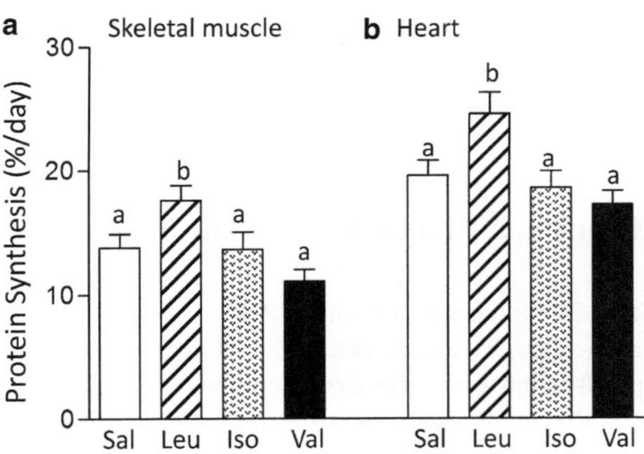

Fig. 16.1 Protein synthesis in skeletal (**a**) and heart (**b**) muscles of neonatal pigs after 60 min of infusion with saline or 400 μmol kg^{-1} h^{-1} of leucine, isoleucine, or valine. Values are means ± pooled SE; $n = 6–8$ per treatment. a, b: means with different letters differ at $P < 0.05$. (Adapted from Escobar et al. [12])

Cellular and Molecular Mechanistic Understanding of How Leucine Regulates Protein Synthesis

The molecular mechanism by which insulin regulates protein synthesis is largely known (for review, see Proud [13]). On the other hand, the signaling pathways responsible for the leucine-induced increase in protein synthesis have been discovered only recently and are not completely established. Interestingly, at a specific point downstream in the signaling pathway toward protein synthesis, insulin and leucine share a similar mechanism of action. Over the past few years, there has been a wealth of new information regarding the possible mechanism by which leucine acts to regulate protein synthesis. However, if we want to understand the role of leucine in the regulation of protein synthesis in skeletal and cardiac muscles in vivo, caution is needed in interpreting these new findings for several reasons. First, most of the studies were conducted using non-muscle cell culture systems. Considering that cell-specific biological processes occur, these findings in non-muscle cells may not be applicable to skeletal or cardiac muscle. Second, in the in vitro studies, the conditions used were mostly non-physiological, making it difficult to extrapolate to the in vivo situation. For example, in order to elucidate leucine's action, all other amino acids were depleted to nearly zero levels, an environment that does not occur in vivo. Lastly, most of the findings were not subjected to the reevaluation process such as retesting particular signaling pathways and signaling components using transgenic mice. Thus, without this assessment, the physiological role of these findings may be questionable.

Figure 16.2 shows the coordinated signaling by two independent pathways, one involving insulin-signaling leading to protein synthesis and the other for leucine-signaling towards protein synthesis. Although the relative contribution of each pathway is unknown, using metabolic clamp methods (hyperinsulinemic–euglycemic–euaminoacidemic clamp and euinsulinemic–euglycemic–hyperaminoacidemic clamp) we found that insulin and amino acids or leucine alone independently stimulate protein synthesis in skeletal muscle of the neonatal pig [12, 14]. In term of signaling actions, most or all molecular aspects involving the regulation of protein synthesis by insulin have been elucidated [13]. By contrast, the molecular mechanism by which leucine modulates protein synthesis has not been fully established and remains elusive. However, there are two points for which most investigators agree: first, there are distinct pathways by which insulin and leucine activate mechanistic target of rapamycin complex 1 (mTORC1); and second, downstream of mTORC1, both signaling pathways activate signaling components toward mRNA translation mostly in a similar fashion. Although the leucine signaling pathway downstream of mTORC1 is mostly known, the signaling cascades upstream of mTORC1 that facilitate the leucine-induced activation of mTORC1 is still subject to intense investigations.

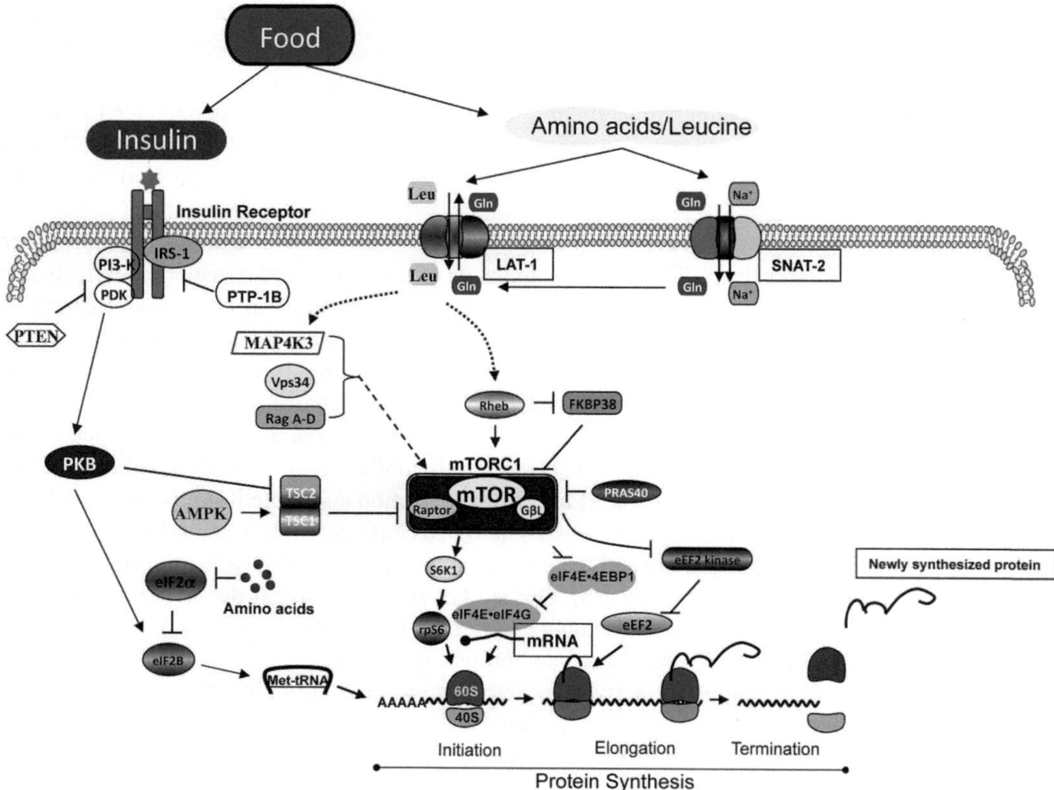

Fig. 16.2 A current model of the insulin and leucine signaling pathways leading to protein synthesis

mTOR is a master kinase that exists in two complexes: mTORC1 and mechanistic target of rapamycin complex 2 (mTORC2) [15]. mTORC1 consists of mTOR, regulatory-associated protein of mTOR (raptor), G protein beta subunit-like (GβL) and proline-rich Akt substrate of 40 kDa (PRAS40) while mTORC2 is composed of mTOR, rapamycin-insensitive companion of mTOR (rictor), and GβL. Whereas mTORC2 has very limited action, mTORC1 has many important physiological functions such as protein synthesis, autophagy, energy metabolism, lipid biosynthesis, and lysosome biogenesis [16].

Insulin and other growth factors initiate their signal by binding to their respective receptors. For the leucine signaling pathway, however, it is still highly debatable as to whether "receptor-like" amino acid or leucine sensing in the cell plasma membrane exists or not. This postulation was based on studies in yeast which showed that there are amino acid sensors, called Ssy1p and Ptr3p, that detect amino acid availability for metabolism and growth [17]. Similar amino acid sensors have not been found in mammals. Early studies in isolated rat hepatocytes showed that treatment with the nontransportable peptide, Leu8-MAP, inhibited autophagic protein degradation as effectively as leucine, suggesting the existence of a leucine sensor in the plasma membrane of hepatocytes. Unfortunately, this finding was not followed by a genetic study to demonstrate the existence of a leucine sensor in vivo. Another setback of this finding is that another study could not repeat Leu8-MAP's effect in isolated adipocytes [18]. A recent study demonstrated that intracellular leucine acts as a regulator of protein synthesis, suggesting the involvement of amino acid transporters that facilitate leucine transport into the cells [19]. Taken together, there seems to be consensus with the notion that leucine acts intercellularly.

An elegant study conducted by Nicklin et al. [19] revealed that in order for leucine to initiate its action on protein synthesis, leucine has to be transported into the cell by coordination of two amino acid transporters, a leucine transporter, the L-type amino acid transporter 1 (LAT1), and a glutamine transporter, such as the sodium-coupled neutral amino acid transporter 2 (SNAT2). A two-step leucine transport mechanism leading to the activation of mTORC1 was proposed (Fig. 16.2). First, SNAT2 facilitates the increase in intracellular glutamine concentrations. Second, LAT1 utilizes intracellular glutamine as an efflux substrate needed for the regulation of extracellular leucine uptake. Thus, LAT1 and SNAT2 work in concert to control the intake flow of leucine into cells resulting in mTORC1 activation.

As important components for leucine signaling, the LAT1 and SNAT2 amino acid transporters have been investigated extensively. The protein abundance of the amino acid transporters reflects their transporter activity. Several studies have shown that LAT1 abundance in various types of cancers is uncharacteristically high, consistent with abnormally high mTORC1 activation that supports cancer growth [20]. For this reason, LAT1 is considered to be one of the important targets for cancer drug discovery. Under amino acid starvation conditions, SNAT2 abundance is increased, suggesting a crucial role of this transporter under nutritional stress conditions to maintain normal cell metabolism [21]. Although there are several glutamine transporters, only inhibition of SNAT2 caused depletion of intercellular leucine that led to impairment of mTORC1 activation and protein synthesis in studies in muscle cells [22]. In our own studies, we have sought to determine possible associations between the protein abundance of these transporters and the enhanced activation of mTORC1 and protein synthesis in skeletal muscle of the neonatal pig [23]. Our studies have shown that the abundance of LAT1 and SNAT2 in skeletal muscle is markedly higher in 6- than in 26-day-old pigs, which parallels the developmental changes in mTORC1 activation and protein synthesis.

Once inside the cell, leucine can activate several amino acid signaling components such as mitogen-activated protein kinase kinase kinase kinase-3 (MAP4K3) [24], vacuolar protein sorting 34 (Vps34) [25] and Ras-related GTPase (Rag GTPases) [26]. As a result, activation of mTORC1 and downstream pathways are initiated. Unlike insulin, which has a clear-cut mechanism underlying its action via the insulin receptor and downstream cascades, the exact mechanism of leucine's action on these "signaling sensors" is unknown. Recent studies show that the leucine-induced activation of mTOR takes place at the lysosome membranes [16]. In this model, leucine enters and accumulates in the lysosomal lumen where, by an unknown mechanism, leucine activates v-ATPase followed by its binding with the Rag GTPase-Rag regulator (Ragulator) complex. In the next step, the active RAG complex binds to mTORC1 and recruits it to the lysosome for mTORC1 activation.

The leucine-induced activation of mTORC1 is one of the crucial steps in the regulation of protein synthesis. Protein synthesis or mRNA translation is a multistep, complex process that starts with translation initiation (reviewed in Kimball and Jefferson [27]). These pathways which occur downstream of mTORC1 are well characterized and both insulin and amino acids/leucine seem to utilize similar signaling cascades. The translation initiation process is regulated by two major steps: (1) the binding of initiator methionyl-tRNA (met-$tRNA_i$) to the 40S ribosomal subunit to form the 43S preinitiation complex, and (2) binding of mRNA to the preinitiation complex (Fig. 16.3). The first step of the translation initiation process is controlled by eukaryotic translation initiation factor (eIF) 2. Under amino acid deprivation conditions, eIF2α is activated by phosphorylation, resulting in the inhibition of eIF2B activation and ultimately, depressed protein synthesis (Fig. 16.2). By contrast, an abundant supply of amino acids suppresses eIF2α activation which allows translation initiation to proceed. We [28] and others [29] have studied the effect of refeeding after an overnight fast on the activity of eIF2B. To our surprise, the activity of eIF2B was not altered by refeeding, indicating that under overnight fasting conditions, intracellular amino acids are not depleted sufficiently to alter this step of translation initiation. In terms of eIF2α activation, a rodent study showed that oral leucine administration does not alter skeletal muscle eIF2α phosphorylation [29]. In our hands, we were able to show that 24 h leucine infusion reduced eIF2α phosphorylation, which was positively associated with enhanced muscle protein synthesis in neonatal pigs [30].

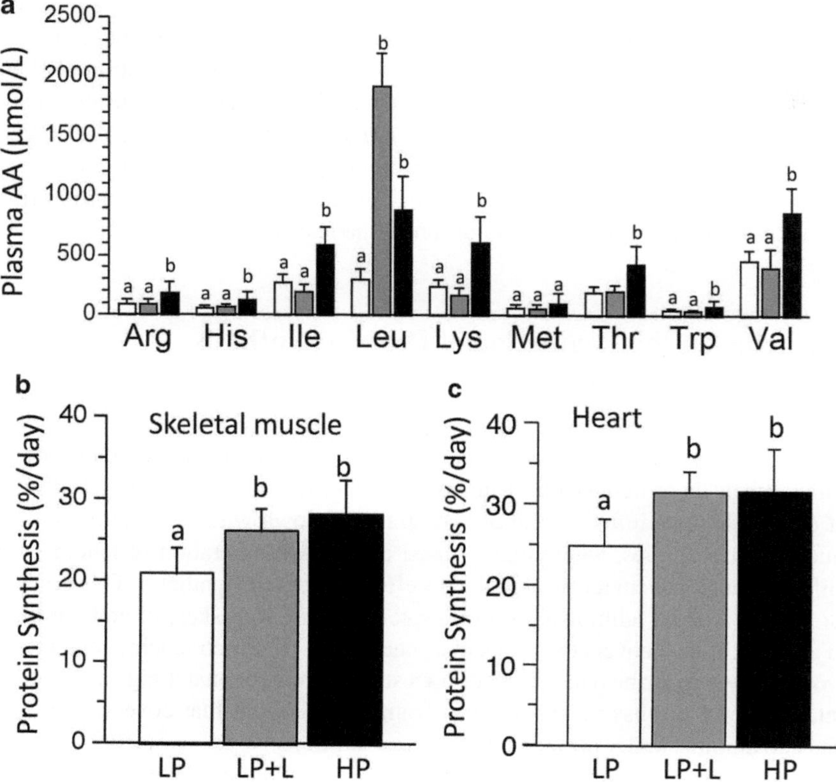

Fig. 16.3 Plasma concentrations of essential amino acids (**a**) and rates of protein synthesis in the skeletal muscle (**b**) and heart (**c**) of piglets fed a LP, LP+L, or HP meal. Values are means ± SEM, $n = 7$. Means without a common letter differ, $P < 0.05$. (Adapted from Murgas et al. [37])

The second step of translation initiation is mediated by the eIF4F complex [27]. The eIF4F complex consists of a RNA helicase (eIF4A), the protein that binds to the m^{7}GTP cap structure at the 5′-end of the mRNA (eIF4E), and a scaffolding protein that binds to eIF4A and eIF4E and also binds to the 43S preinitiation complex (eIF4G). The process of association of eIF4E with eIF4G is obligatory for the binding of the 43S preinitiation complex with mRNA. This step is regulated by eukaryotic initiation factor 4E binding protein 1 (4EBP1), a well-known substrate of mTORC1. Under unphosphorylated conditions, 4EBP1 tightly binds to eIF4E, forming the inactive eIF4E·4EBP1 complex (Fig. 16.2). During anabolic conditions, mTORC1 induces the phosphorylation of 4EBP1, resulting in the dissociation of eIF4E from the inactive complex and allowing eIF4E to form an active complex with eIF4G (Fig. 16.2). Data from the literature and from our own work show that leucine induces the phosphorylation of 4EBP1 leading to stimulation of muscle protein synthesis [12].

Another mTORC1 substrate that participates in the regulation of mRNA translation is p70 S6 kinase (S6K1). Early observations indicated that this kinase plays a crucial role in the regulation of specific mRNA, called terminal oligopyrimidine (TOP) mRNA, which are responsible for the translation of proteins involved in the protein synthetic apparatus, such as ribosomal proteins [31]. However, genetic studies in mice disputed this finding [32]. Nonetheless, deletion of this gene in the mouse reduced overall growth including that of skeletal muscle [33]. Although the physiological function of S6K1 is still controversial, determining the phosphorylation of S6K1 along with 4EBP1 phosphorylation is still a gold standard as a readout for mTORC1 activity leading to protein synthesis. Our studies [12] and others [29] show that leucine alone can stimulate S6K1 phosphorylation.

The elongation step is a high-energy process that is regulated partly by elongation factor 2 (eEF2) (reviewed in Browne and Proud [34]) (Fig. 16.2). Its function is to catalyze the translocation of two tRNA and the mRNA after peptidyl transfer on the 80S ribosome. The regulation of eEF2 activity is carried out by eEF2 kinase. In the phosphorylated state, eEF2 is inactive, resulting in stalling of the elongation process. Anabolic agents, such as insulin and amino acids, activate eEF2 by inhibiting eEF2 kinase activation. Interestingly, data from our studies [35] show no effect of leucine administration in the neonate on the phosphorylation of eEF2, indicating that the elongation process is not a limiting step under normal physiological conditions in neonates.

Effect of Leucine on the Regulation of Protein Synthesis in Skeletal and Cardiac Muscles

Inspired by early in vitro work [11], investigators conducted experiments to determine whether leucine has similar effects on protein synthesis in vivo. Early observations in human studies (1978–2002) showed that chronic leucine infusion reduces protein breakdown without affecting protein synthesis [8]. Subsequent human studies, which mainly used enteral administration of leucine, have reported somewhat different outcomes in terms of leucine's effect on protein synthesis. This section focuses on the effect of enteral leucine administration on protein synthesis in skeletal and cardiac muscles in humans and animals. It also covers the molecular mechanism by which leucine regulates protein synthesis in vivo. First, we describe data from our own studies using neonatal pigs as a model for human development. Then, we discuss recent findings from the literature that covers studies in adult and older humans and animals under normal and disease conditions.

As a proof of principle for the effectiveness of leucine in stimulating skeletal muscle protein synthesis in neonates, we have conducted several experiments using parenteral leucine infusion. Our results show that the physiological rise in leucine enhances protein synthesis in skeletal muscle and cardiac muscle through the activation of mTORC1-dependent translation initiation factors [12, 36]. Importantly, we established that among the members of the BCAA, only leucine has an anabolic effect on protein synthesis in vivo [12]. In subsequent experiments, we examined the effect of enteral leucine supplementation of the diet. To determine whether leucine supplementation of a meal acutely stimulates protein synthesis in the neonate [37], 5-day-old pigs were fed either a low-protein (LP) meal, a low-protein meal supplemented with leucine (LP+L), or a high-protein meal (HP). The diets were isocaloric and the amount of leucine in the LP+L and HP meal was similar. As presented in Fig. 16.3a, we found that acute leucine administration had no effect on the plasma concentrations of the other essential amino acids (EAA), including isoleucine and valine. Importantly, leucine supplementation of the LP meal enhanced skeletal muscle protein synthesis to the rate achieved by feeding a HP meal (Fig. 16.3b). The increase in skeletal muscle protein synthesis was at least in part due to the leucine-induced activation of mTORC1 and downstream signaling proteins leading to translation initiation.

To determine whether the effects of leucine supplementation could be sustained by more prolonged enteral leucine supplementation, neonatal pigs were fed the LP, LP+L, or HP diets every 4 h for 24 h [38]. We found that leucine supplementation of a LP diet for 24 h increased protein synthesis in skeletal muscle, but not to the level achieved with feeding a HP diet (Fig. 16.4). We postulated that differences in the response between the acute and the 24 h leucine supplementation studies can likely be explained by the "leucine paradox" whereby BCAA oxidation is stimulated due to a higher leucine concentration and, consequently, all members of the BCAA are utilized by BCKD for oxidation [39]. Indeed, unlike in the acute meal feeding study, the circulating concentrations of valine and isoleucine were reduced following 24 h of leucine supplementation (Fig. 16.5). Thus, it seems likely that the reduction in circulating levels of valine and isoleucine induces an EAA imbalance that blunts the leucine-induced stimulation of protein synthesis due to substrate limitations. Nonetheless, the results

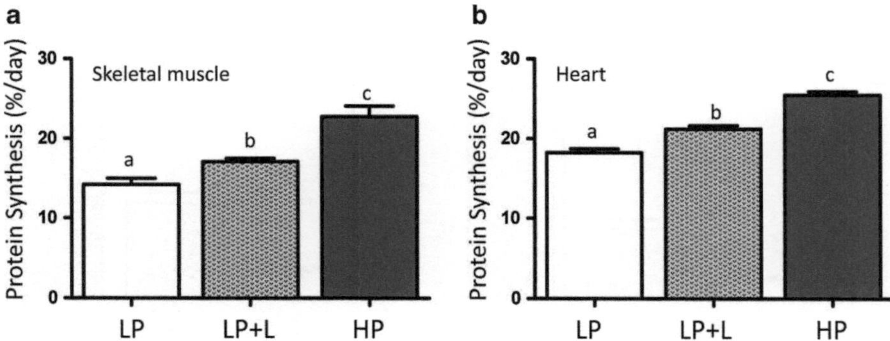

Fig. 16.4 Protein synthesis in skeletal (**a**) and heart (**b**) muscles of piglets fed LP, LP+L, or HP diets for 24 h. Values are means ± SEM, $n=7$–10 per treatment. Means without a common symbol differed, $P<0.05$. (Adapted from Suryawan et al. [38])

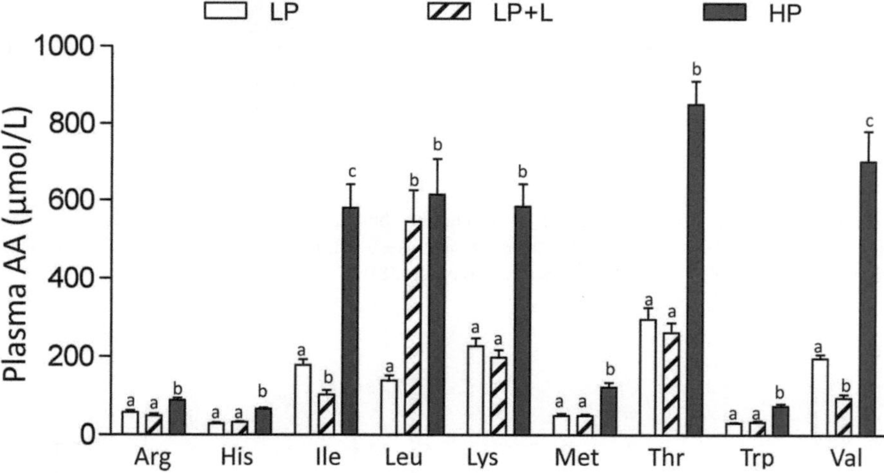

Fig. 16.5 Plasma concentrations of essential AA in piglets fed LP, LP+L, or HP diets for 24 h. Values are means ± SEM, $n=7$–10 per treatment. Means without a common symbol differed, $P<0.05$. (Adapted from Suryawan et al. [38])

of our studies reemphasized leucine's ability to stimulate protein synthesis in neonates and that leucine's effect on protein synthesis is at least in part due to activation of mTORC1 substrates, S6K1 and 4EBP1, and increased abundance of eIF4E·eIF4G (Fig. 16.6). Furthermore, the phosphorylation of eIF2α and eEF2 were not affected by leucine (Fig. 16.7) suggesting that the translation initiation step by which met-tRNA$_i$ binds to the 40S ribosomal subunit to form the 43S preinitiation complex, as well as the elongation process, are not limiting factors for the leucine-induced stimulation of protein synthesis in skeletal muscle of the neonate. So an obvious question is then: will supplementation of the diet with leucine enhance lean growth? We are currently pursuing studies to evaluate the potential benefit of long-term leucine supplementation in neonatal pigs.

In mature animals and adult humans, studies on the effect of leucine have mainly been conducted in exercise conditions or in disease states such as cancer. There are few studies in normal sedentary animals and humans. Anthony et al. [29] reported that enteral leucine administration stimulates muscle protein synthesis in mature rats through an mTORC1-dependent pathway. However, supplementation with leucine via the drinking water for 12 days increased muscle protein synthesis in mature rats without affecting the abundance of mTORC1 signaling components or the abundance of enzymes that catabolize BCAA [18]. Unfortunately, muscle mass or body composition was not measured in that

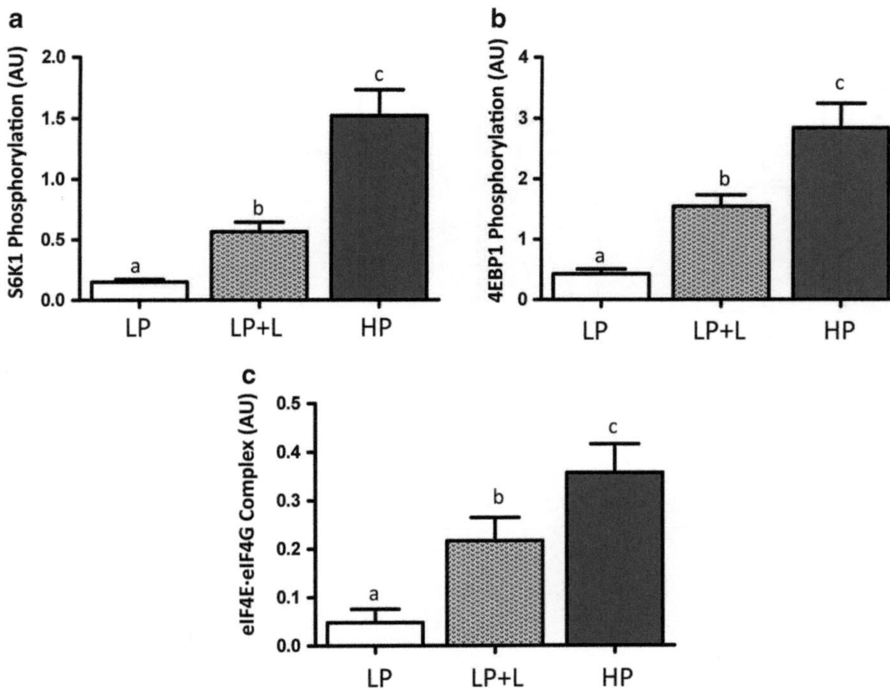

Fig. 16.6 Phosphorylation of S6K1 (**a**) and 4EBP1 (**b**) and the abundance of eIF4E·eIF4G complex in skeletal muscle of piglets fed LP, LP+L, or HP diets for 24 h. Values are means ± SEM, $n = 7$–10 per treatment. Means without a common symbol differed, $P < 0.05$. (Adapted from Suryawan et al. [38])

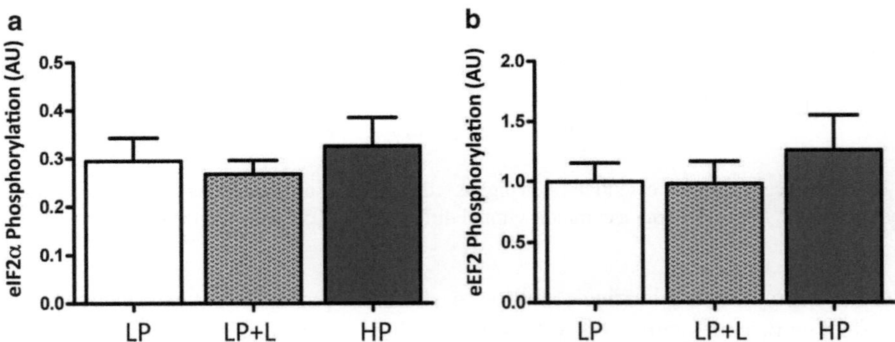

Fig. 16.7 Phosphorylation of eIF2α (**a**) and eEF2 (**b**) in skeletal muscle of piglets fed LP, LP+L, or HP diets for 24 h. Values are means ± SEM, $n = 7$–10 per treatment. (Adapted from Suryawan et al. [38])

study. Recently, the effects of acute enteral leucine administration on muscle protein synthesis were examined in sedentary healthy young men and the results showed that leucine stimulates skeletal muscle protein synthesis [40]. Moreover, this effect was due to the leucine-induced phosphorylation of S6K1 and 4EBP1 without any effect on the phosphorylation of eIF2α or eEF2. To date, there have been no long-term studies on the effect of leucine on muscle mass in adult humans.

Recognition of leucine as an amino acid that possesses anabolic properties towards protein metabolism in skeletal muscle makes leucine supplementation an attractive choice for active people who perform various types of exercises. During intense exercise, the metabolism of BCAA increases substantially. Shimomura et al. [41] reported that exercise promotes BCAA catabolism by inducing the activity of the

BCAA catabolic enzyme, BCKD. As a result, circulating levels of leucine decrease by 30 % following resistance exercise [42]. Furthermore, several studies suggest that BCAA supplementation before and after exercise may decrease exercise-induced muscle damage and increase muscle protein synthesis [41].

Under pathological conditions, such as cancer and sepsis, protein metabolism can be negatively compromised. In extreme disease states, involuntary weight loss and muscle atrophy can occur. BCAA or leucine supplementation has been used in cancer patients to alleviate the reduction of muscle mass, although the effectiveness is still questionable [43]. An important drawback of using BCAA/leucine supplementation in cancer patients is that leucine stimulates cancer growth. This finding is not surprising, since the abundance of several amino acid transporters, including the one for leucine (LAT1), is upregulated in several cancer cell lines [22]. Further studies are necessary to establish the effectiveness of BCAA/leucine in the nutritional management of cancer patients.

Another group for which BCAA/leucine supplementation may have an advantage is the elderly population (reviewed in Fujita and Volpi [44]). Sarcopenia, i.e., the progressive loss of muscle mass due to aging, can significantly affect the overall well-being of the elderly, and thus, maintaining a normal muscle mass through adulthood and old age is critical. Although numerous studies have shown that feeding increases protein synthesis in growing animals [45], Dardevet and coworkers [46] showed no effect of meal consumption on muscle protein synthesis in 23-month-old rats unless the meal was supplemented with leucine. However, in a follow-up study, this group found no effect of 6 months of leucine supplementation in old rats on muscle mass [47]. Studies in elderly men suggest that, although protein metabolism in response to a meal is lower in the elderly compared to their younger counterparts, they still are able to mount an increase in protein synthesis in muscle is response to leucine administration [44]. Taken together, the results suggest that BCAA or leucine supplementation may be an effective intervention to prevent or alleviate the negative effects of sarcopenia.

Optimal protein metabolism in cardiac muscle is crucial for proper heart function. There are limited studies evaluating the postprandial regulation of protein synthesis in cardiac muscle. Our own studies show that feeding modestly stimulates cardiac muscle protein synthesis and this effect decreases with age [45]. Moreover, we showed that enteral leucine administration for 1 h or 24 h also stimulates cardiac muscle protein synthesis in neonatal pigs [37, 38] (Figs. 16.3 and 16.4). These findings are not surprising since all organs undergo rapid growth during early postnatal development, albeit the most rapid growth occurs in skeletal muscle. Interestingly, long-term enteral leucine administration did not affect cardiac muscle protein synthesis in adult rats [40].

Studies show that in pathological conditions such as sepsis, protein synthesis in cardiac muscle is lower than normal, due to the reduction in the activation of signaling pathways leading to translation initiation [48]. Lang and coworkers [48] found that acute enteral leucine administration reversed the sepsis-induced alteration in eIE4E distribution, a rate limiting factor for cardiac muscle protein synthesis in adult rats, but failed to stimulate S6K1 phosphorylation. Another pathological condition that negatively affects cardiac muscle protein synthesis is chronic alcohol consumption [49]. Studies showed that leucine stimulates cardiac muscle protein synthesis through enhancement of eIF4G phosphorylation and increased eIF4G·eIF4E abundance without affecting S6K1 phosphorylation in adult rats after consuming alcohol. Clearly, the role of leucine in the regulation of cardiac muscle protein synthesis in these pathological conditions deserves further study.

Conclusions

The efficacy of enteral leucine administration in stimulating protein synthesis has been known for several years. However, despite recent work, the cellular and molecular mechanisms by which leucine activates mTORC1 leading to protein synthesis remains obscure. Therefore, to better understand leucine's effect on protein synthesis, more studies investigating these signaling cascades will be advantageous. Although with limited success, the use of leucine as an anabolic agent has been shown to be beneficial for various applications such as exercise and sarcopenia. For other applications, available references indicate that further study is needed, particularly the determination of the effectiveness of long-term enteral leucine supplementation. Overall, our understanding of the role of leucine in the regulation of protein synthesis and the benefit of using it as an anabolic agent is not complete and tremendous work needs to be done to ensure that leucine can be used safely and effectively.

Most of experimental investigations on signaling cascades associated with leucine's effect on protein synthesis have been carried out by investigators interested in how anabolic agents, such as hormones, growth factors or amino acids, regulate mTORC1-induced protein synthesis in normal or abnormal cell growth and development. Given the main focus of those investigations is cancer biology, more comprehensive studies on the cellular and molecular mechanisms concerning the action of leucine on the postprandial stimulation of skeletal or cardiac muscle protein synthesis needs to be done. For instance, an important finding indicates that in older rats, leucine does not stimulate the synthesis of myosin heavy chains while the leucine-induced stimulation of mitochondrial and sarcoplasmic protein synthesis is comparable to that of younger rats [50]. The mechanism by which aging alters leucine-induced protein synthesis is currently unknown. Another important aspect regarding signaling pathways is to develop an intervention that can further enhance the high rate of postprandial protein synthesis achieved by the collaboration between the insulin and leucine signaling pathways. Undoubtedly, future studies that address the above concerns are needed.

In term of leucine's physiological application, it is recognizable that long-term study is required. Our studies and those of others indicate that one of the crucial features in coping with long-term leucine administration is a potential leucine antagonism affect. The conditions need to be defined that can ensure the physiological balance of all amino acids is achieved and that substrate availability is fulfilled to achieve maximum rates of protein synthesis. In conclusion, long-term leucine administration could someday be used as a method of choice to enhance and preserve skeletal and cardiac muscle in normal and disease states.

Acknowledgments and Disclosure Dr. Davis is a recipient of research support from the National Institutes of Health (NIH, R01 AR44474 and HD072891), US Department of Agriculture National Institute of Food and Agriculture (USDA NIFA 2013-67015-20438), USDA/Agricultural Research Service (ARS, 6250-51000-055), and Abbott Nutrition. This work is a publication of the USDA/ARS Children's Nutrition Research Center, Department of Pediatrics, Baylor College of Medicine, Houston, Texas. The contents of this publication do not necessarily reflect the views or policies of the USDA.

References

1. Rasmussen BB, Phillips SM. Contractile and nutritional regulation of human muscle growth. Exerc Sport Sci Rev. 2003;31:127–31.
2. Frey N, Katus HA, Olson EN, Hill JA. Hypertrophy of the heart: a new therapeutic target? Circulation. 2004;109(13):1580–9.
3. Preedy VR, Smith DM, Kearney NF, Sugden PH. Rates of protein turnover in vivo and in vitro in ventricular muscle of hearts from fed and starved rats. Biochem J. 1984;222:395–400.
4. Davis TA, Suryawan A, Orellana RA, Fiorotto ML, Burrin DG. Amino acids and insulin are regulators of muscle protein synthesis in neonatal pigs. Animal. 2010;4:1790–6.
5. Brosnan JT, Brosnan ME. Branched-chain amino acids: enzyme and substrate regulation. J Nutr. 2006;136:207S–11.
6. May ME, Buse MG. Effects of branched-chain amino acids on protein turnover. Diabetes Metab Rev. 1989;5(3):227–45.
7. Harper AE, Miller RH, Block KP. Branched-chain amino acid metabolism. Annu Rev Nutr. 1984;4:409–54.
8. Matthews DE. Observations of branched-chain amino acid administration in humans. J Nutr. 2005;135:1580S–4.

9. Suryawan A, Hawes JW, Harris RA, Shimomura Y, Jenkins AE, Hutson SM. A molecular model of human branched-chain amino acid metabolism. Am J Clin Nutr. 1998;68:72–81.
10. Hutson SM, Sweatt AJ, Lanoue KF. Branched-chain amino acid metabolism: implications for establishing safe intakes. J Nutr. 2005;135:1557S–64.
11. Buse MG, Reid SS. Leucine. A possible regulator of protein turnover in muscle. J Clin Invest. 1975;56:1250–61.
12. Escobar J, Frank JW, Suryawan A, Nguyen HV, Kimball SR, Jefferson LS, Davis TA. Regulation of cardiac and skeletal muscle protein synthesis by individual branched-chain amino acids in neonatal pigs. Am J Physiol Endocrinol Metab. 2006;290:E612–21.
13. Proud CG. Regulation of protein synthesis by insulin. Biochem Soc Trans. 2006;34:213–6.
14. O'Connor PM, Bush JA, Suryawan A, Nguyen HV, Davis TA. Insulin and amino acids independently stimulate skeletal muscle protein synthesis in neonatal pigs. Am J Physiol Endocrinol Metab. 2003;284:E110–9.
15. Sabatini DM. mTOR and cancer: insights into a complex relationship. Nat Rev Cancer. 2006;6:729–34.
16. Bar-Peled L, Sabatini DM. SnapShot: mTORC1 signaling at the lysosomal surface. Cell. 2012;151(6):1390–1390.e1.
17. Klasson H, Fink GR, Ljungdahl PO. Ssy1p and Ptr3p are plasma membrane components of a yeast system that senses extracellular amino acids. Mol Cell Biol. 1999;19:5405–16.
18. Lynch CJ, Hutson SM, Patson BJ, Vaval A, Vary TC. Tissue-specific effects of chronic dietary leucine and norleucine supplementation on protein synthesis in rats. Am J Physiol Endocrinol Metab. 2002;283:E824–35.
19. Nicklin P, Bergman P, Zhang B, Triantafellow E, Wang H, Nyfeler B, Yang H, Hild M, Kung C, Wilson C, Myer VE, MacKeigan JP, Porter JA, Wang YK, Cantley LC, Finan PM, Murphy LO. Bidirectional transport of amino acids regulates mTOR and autophagy. Cell. 2009;136:521–34.
20. Fuchs BC, Bode BP. Amino acid transporters ASCT2 and LAT1 in cancer: partners in crime? Semin Cancer Biol. 2005;15:254–66.
21. Gaccioli F, Huang CC, Wang C, Bevilacqua E, Franchi-Gazzola R, Gazzola GC, Bussolati O, Snider MD, Hatzoglou M. Amino acid starvation induces the SNAT2 neutral amino acid transporter by a mechanism that involves eukaryotic initiation factor 2alpha phosphorylation and cap-independent translation. J Biol Chem. 2006;281:17929–40.
22. Evans K, Nasim Z, Brown J, Butler H, Kauser S, Varoqui H, Erickson JD, Herbert TP, Bevington A. Acidosis-sensing glutamine pump SNAT2 determines amino acid levels and mammalian target of rapamycin signalling to protein synthesis in L6 muscle cells. J Am Soc Nephrol. 2007;18:1426–36.
23. Suryawan A, Nguyen HV, Almonaci RD, Davis TA. Abundance of amino acid transporters involved in mTORC1 activation in skeletal muscle of neonatal pigs is developmentally regulated. Amino Acids. 2013;45(3):523–30.
24. Findlay GM, Yan L, Procter J, Mieulet V, Lamb RF. A MAP4 kinase related to Ste20 is a nutrient-sensitive regulator of mTOR signalling. Biochem J. 2007;403:13–20.
25. Backer JM. The regulation and function of Class III PI3Ks: novel roles for Vps34. Biochem J. 2008;410:1–17.
26. Sancak Y, Sabatini DM. Rag proteins regulate amino-acid-induced mTORC1 signalling. Biochem Soc Trans. 2009;37:289–90.
27. Kimball SR, Jefferson LS. Signaling pathways and molecular mechanisms through which branched-chain amino acids mediate translational control of protein synthesis. J Nutr. 2006;136:227S–31.
28. Davis TA, Nguyen HV, Suryawan A, Bush JA, Jefferson LS, Kimball SR. Developmental changes in the feeding-induced stimulation of translation initiation in muscles of neonatal pigs. Am J Physiol Endocrinol Metab. 2000;279(6):E1226–34.
29. Anthony JC, Anthony TG, Kimball SR, Vary TC, Jefferson LS. Orally administered leucine stimulates protein synthesis in skeletal muscle of postabsorptive rats in association with increased eIF4F formation. J Nutr. 2000;130:139–45.
30. Wilson FA, Suryawan A, Gazzaneo MC, Orellana RA, Nguyen HV, Davis TA. Stimulation of muscle protein synthesis by prolonged parenteral infusion of leucine is dependent on amino acid availability in neonatal pigs. J Nutr. 2010;140:264–70.
31. Loreni F, Thomas G, Amaldi F. Transcription inhibitors stimulate translation of 5′ TOP mRNAs through activation of S6 kinase and the mTOR/FRAP signalling pathway. Eur J Biochem. 2000;267(22):6594–601.
32. Ruvinsky I, Meyuhas O. Ribosomal protein S6 phosphorylation: from protein synthesis to cell size. Trends Biochem Sci. 2006;31:342–8.
33. Pende M, Um SH, Mieulet V, Sticker M, Goss VL, Mestan J, Mueller M, Fumagalli S, Kozma SC, Thomas G. S6K1(−/−)/S6K2(−/−) mice exhibit perinatal lethality and rapamycin-sensitive 5′-terminal oligopyrimidine mRNA translation and reveal a mitogen-activated protein kinase-dependent S6 kinase pathway. Mol Cell Biol. 2004;24:3112–24.
34. Browne GJ, Proud CG. Regulation of peptide-chain elongation in mammalian cells. Eur J Biochem. 2002;269(22):5360–8.
35. Suryawan A, Jeyapalan AS, Orellana RA, Wilson FA, Nguyen HV, Davis TA. Leucine stimulates protein synthesis in skeletal muscle of neonatal pigs by enhancing mTORC1 activation. Am J Physiol Endocrinol Metab. 2008;295(4):E868–75.

36. Escobar J, Frank JW, Suryawan A, Nguyen HV, Kimball SR, Jefferson LS, Davis TA. Physiological rise in plasma leucine stimulates muscle protein synthesis in neonatal pigs by enhancing translation initiation factor activation. Am J Physiol Endocrinol Metab. 2005;288(5):E914–21.
37. Murgas Torrazza R, Suryawan A, Gazzaneo MC, Orellana RA, Frank JW, Nguyen HV, Fiorotto ML, El-Kadi S, Davis TA. Leucine supplementation of a low-protein meal increases skeletal muscle and visceral tissue protein synthesis in neonatal pigs by stimulating mTOR-dependent translation initiation. J Nutr. 2010;140(12):2145–52.
38. Suryawan A, Torrazza RM, Gazzaneo MC, Orellana RA, Fiorotto ML, El-Kadi SW, Srivastava N, Nguyen HV, Davis TA. Enteral leucine supplementation increases protein synthesis in skeletal and cardiac muscles and visceral tissues of neonatal pigs through mTORC1-dependent pathways. Pediatr Res. 2012;71:324–31.
39. Shimomura Y, Harris RA. Metabolism and physiological function of branched-chain amino acids: discussion of session 1. J Nutr. 2006;136:232S–3.
40. Wilkinson DJ, Hossain T, Hill DS, Phillips BE, Crossland H, Williams J, Loughna P, Churchward-Venne TA, Breen L, Phillips SM, Etheridge T, Rathmacher JA, Smith K, Szewczyk NJ, Atherton PJ. Effects of leucine and its metabolite, β-hydroxy-β-methylbutyrate (HMB) on human skeletal muscle protein metabolism. J Physiol. 2013;591(Pt 11):2911–23.
41. Shimomura Y, Murakami T, Nakai N, Nagasaki M, Harris RA. Exercise promotes BCAA catabolism: effects of BCAA supplementation on skeletal muscle during exercise. J Nutr. 2004;134:1583S–7.
42. Mero A. Leucine supplementation and intensive training. Sports Med. 1999;27:347–58.
43. Baracos VE, Mackenzie ML. Investigations of branched-chain amino acids and their metabolites in animal models of cancer. J Nutr. 2006;136:237S–42.
44. Fujita S, Volpi E. Amino acids and muscle loss with aging. J Nutr. 2006;136:277S–80.
45. Davis TA, Burrin DG, Fiorotto ML, Nguyen HV. Protein synthesis in skeletal muscle and jejunum is more responsive to feeding in 7-than in 26-day-old pigs. Am J Physiol. 1996;270:E802–9.
46. Dardevet D, Sornet C, Bayle G, Prugnaud J, Pouyet C, Grizard J. Postprandial stimulation of muscle protein synthesis in old rats can be restored by a leucine-supplemented meal. J Nutr. 2002;132:95–100.
47. Zeanandin G, Balage M, Schneider SM, Dupont J, Hébuterne X, Mothe-Satney I, Dardevet D. Differential effect of long-term leucine supplementation on skeletal muscle and adipose tissue in old rats: an insulin signaling pathway approach. Age. 2012;34:371–87.
48. Lang CH, Pruznak AM, Frost RA. TNFalpha mediates sepsis-induced impairment of basal and leucine-stimulated signaling via S6K1 and eIF4E in cardiac muscle. J Cell Biochem. 2005;94:419–31.
49. Vary T. Oral leucine enhances myocardial protein synthesis in rats acutely administered ethanol. J Nutr. 2009;139:1439–44.
50. Guillet C, Zangarelli A, Mishellany A, Rousset P, Sornet C, Dardevet D, Boirie Y. Mitochondrial and sarcoplasmic proteins, but not myosin heavy chain, are sensitive to leucine supplementation in old rat skeletal muscle. Exp Gerontol. 2004;39:745–51.

Chapter 17
Use of Branched Chain Amino Acids Granules in Experimental Models of Diet-Induced Obesity

Takayuki Masaki

Key Points

- Obesity is the effect of imbalance between energy intake and expenditure.
- Diet-induced obesity is associated with several metabolic disorders, including insulin resistance, hyperlipidemia, and fatty liver.
- Branched chain amino acid (BCAA) treatment prevents fat accumulation in muscle, and liver in diet-induced obesity.
- BCAA treatment ameliorates diet-induced metabolic disorders through coordinate activation of uncoupling protein in diet-induced obesity.
- BCAA supplementation might be useful in the treatment of metabolic syndrome.

Keywords Branched chain amino acid • Metabolic syndrome • Diet • Diabetes • Uncoupling protein

Abbreviations

BCAA	Branched chain amino acid
PPAR	Peroxisome proliferator-activated receptors
UCP	Uncoupling protein

Introduction

Obesity is the effect of imbalance between food intake and energy expenditure. A number of substances and factors have implicated in the regulation of energy and lipid metabolism. Branched chain amino acids (BCAAs) have been shown to involve in diet-induced obesity. In addition, BCAAs increase free fatty acid oxidation in peripheral tissue.

T. Masaki, M.D., Ph.D. (✉)
Department of Endocrinology and Metabolism, Faculty of Medicine, Oita University,
Idaigaoka 1-1, Hasama, Oita 879-5593, Japan
e-mail: masaki@oita-u.ac.jp

R. Rajendram et al. (eds.), *Branched Chain Amino Acids in Clinical Nutrition: Volume 1*, Nutrition and Health, DOI 10.1007/978-1-4939-1923-9_17,

The Effects of BCAA on Peripheral Tissue

Branched chain amino acids (BCAAs) have aliphatic side chains with a branch point, and comprise valine, leucine, and isoleucine. BCAAs play important roles, particularly during exercise and growth of skeletal muscle. BCAA supplementation improves disorders of albumin metabolism, quality of life, subjective symptoms, and prognosis in patients with chronic hepatitis [1].

BCAAs induce albumin synthesis and activate the mammalian target of rapamycin (mTOR) and subsequently upregulate the downstream eukaryotic initiation factor 4E-binding protein-1 and ribosomal protein S6 (Fig. 17.1).

In addition, BCAAs have been reported to modulate glucose metabolism [2]. They promote glucose uptake in skeletal muscle in a rat model of liver cirrhosis and reduce hepatic glucose production in insulin resistant rats. In skeletal muscle isolated from nondiabetic rats, BCAAs promote glucose uptake by enhancing translocation of glucose transporter 4 to the plasma membrane via phosphatidylinositol 3-kinase and protein kinase C pathways. BCAAs upregulate mTOR in hepatocytes and the liver [3]. mTOR negatively regulates insulin-mediated PI3K activity by degrading insulin receptor substrates in skeletal muscle cells during long-term insulin treatment.

Food Intake, Energy Expenditure and Obesity

The prevalence of obesity is increasing worldwide and associated with increased morbidity and mortality. The increase in the prevalence of the disorder is mostly likely due to changes in diet and exercise. Studies have revealed that the hypothalamic systems that regulate food intake and energy expenditure have a central role in the development of obesity [4] (Fig. 17.2). Several orexigenic and anorexigenic

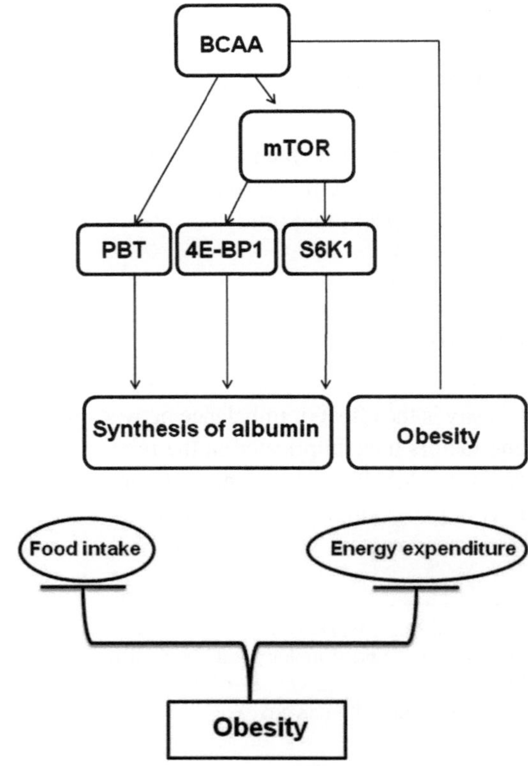

Fig. 17.1 BCAA-induced albumin synthesis. BCAA activates the mTOR and subsequently upregulates eukaryotic initiation factor 4E-binding protein-1 (4EBP1) and S6 kinase (S6K1). BCAAs also stimulate the polypyrimidine-tract-binding protein (PBT), which increases albumin translation

Fig. 17.2 Obesity is regulated by food intake and energy expenditure. Several orexigenic and anorexigenic neuropeptides in the hypothalamus are involved in food intake and expenditure

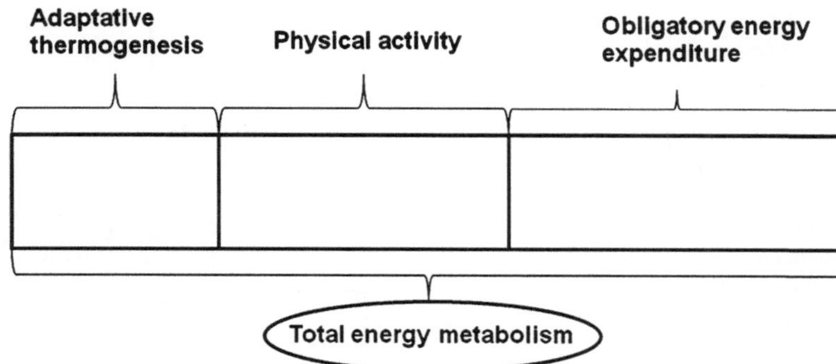

Fig. 17.3 A thermodynamic perspective of energy expenditure. Total energy expenditure can be divided into adaptive thermogenesis, physical activity, and obligatory energy expenditure

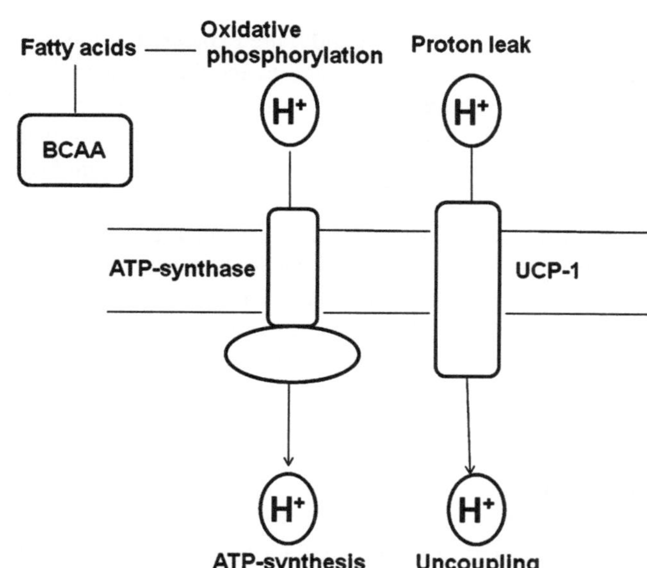

Fig. 17.4 UCP-1 location and function in the mitochondrial respiratory chain. The proton gradient drives the synthesis of ATP. UCP1 catalyzes a regulated reentry of protons into matrix and generating heat

neuropeptides in the hypothalamus are involved in food intake and obesity, although the contribution of each peptide to the development of obesity is different. Since the discovery of *ob* gene product leptin and its receptors, research into obesity has progressed throughout the world. There is mounting evidence that the effects of leptin are governed by several hypothalamic mediators, including orexigenic substances and anorexigenic substances. Energy expenditure can be subdivided into: (1) basal energy expenditure or resting metabolic rate, measured under resting conditions; (2) energy expenditure resulting from physical activity; and (3) energy expenditure attributed to adaptive thermogenesis (Fig. 17.3). A number of studies have been focused on the genes related to energy expenditure, such as those encoding mitochondrial uncoupling proteins (UCPs) and peroxisome proliferator-activated receptors (PPARs) [5, 6].

UCPs in Peripheral Tissue

UCPs are members of an anion-carrier protein family, and are located in the inner mitochondrial membrane [5]. The original UCP, UCP1, is mainly expressed in brown adipose tissue (BAT). The proton gradient drives the synthesis of ATP. UCP1 catalyzes a regulated reentry of protons into matrix and generating heat (Fig. 17.4). UCP2 is widely distributed, whereas UCP3 is mainly restricted to the

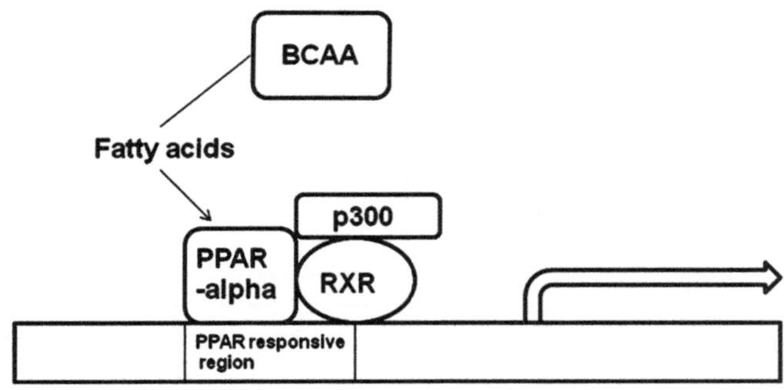

Fig. 17.5 The proximal region responsive to PPAR-alpha activation via PPAR/RXR heterodimers. The proximal region responsive to PPAR-alpha activation via PPAR/RXR heterodimers is shown. P300, the main coactivator linking ligand-dependent activation of PPARs with transcriptional activation is shown

skeletal muscle, and UCP4 and 5 are expressed in the brain. UCP1 plays important roles in energy balance and regulation, cold-induced and diet-induced thermogenesis, which are mechanisms associated with the pathogenesis of obesity. BAT is a metabolically active tissue, which consists of adipocytes rich in mitochondria and numerous small lipid droplets, and is innervated by sympathetic nerves. Especially, thermogenesis in BAT UCP1 has important roles in thermal and energetic balance and, when deficient, may lead to obesity.

At birth, human newborns have considerable amounts of BAT, corresponding to 1–5 % of total body weight. This amount takes care of heat generation for the body when the skeletal muscles are yet not able to make any controlled movements and thus produce heat. In addition, BAT in adults is highly active both functionally and metabolically, especially after exposure to cold.

Like UCP1, UCP2 and UCP3 are located in the mitochondrial membrane, but it is unclear whether dissipation of metabolic energy as heat is their primary biological function. However, their capacity to protect against obesity has been demonstrated, at least for UCP3, in experimental settings based on transgenic mice overexpressing the protein in muscle. The specific involvement of UCP2 and UCP3 in fatty acid oxidation has been proposed [7, 8]. The transcriptional control of gene expression of UCPs determines the levels of the corresponding proteins in tissues and cells. As well as providing a basis for insight into the regulation of transcription of UCP genes in response to PPARs [6], an understanding of the mechanisms and the PPAR subtypes involved in this regulation would provide the possibility of the development of pharmacological approaches to modulate the levels of UCPs, given the availability of drugs acting selectively on PPAR subtypes, such as fibrates and thiazolidinediones. PPAR-alpha is mainly expressed in muscle and liver, where it regulates various target genes, including those involved in fatty acid oxidation and lipid metabolism. Fatty acids act the proximal region responsive to PPAR-alpha activation via PPAR/RXR heterodimers (Fig. 17.5). Activation of PPAR-alpha accelerates fatty acid oxidation in skeletal muscle and liver tissues. PPAR-alpha regulates the UCPs, which are mitochondrial membrane transporters involved in the control of energy conversion.

BCAA Metabolism in Obesity and Insulin Resistance

It has been recognized that obesity is associated with resistance to insulin action on skeletal muscle and impaired glucose disposal. The majority of the studies demonstrate the existence of an insulin-resistant state pertaining to protein metabolism. Disturbances of BCAA metabolism have been associated with obesity, diabetes mellitus, and insulin-resistant states [9–11]. The plasma concentration of BCAAs in obese model is higher than compared with normal controls. In obese patients, a much greater increase in insulin secretion than in controls is required to produce a comparable fall in BCAA concentration after glucose infusion. During euglycemic insulin infusion, insulin reduces plasma BCAA levels less efficiently in obese subjects compared with controls. In diabetes mellitus, there is an increase in the plasma concentration of BCAAs compared with healthy individuals [9]. The urinary excretion of BCAAs is also elevated in diabetic patients before insulin treatment and normalizes after initiation of insulin. Accordingly with the occurrence of an insulin-resistant state, post absorptive serum insulin levels in obesity have been consistently found elevated and insulin has been observed to be less effective in blocking the post absorptive outflow of BCAAs from muscle [10]. Consequently, the post absorptive plasma concentration of BCAAs has been detected elevated in obese patients compared with nonobese individuals [11]. The elevated concentration of BCAAs in overweight and obese patients has been independently associated with insulin resistance and linearly related to the homeostasis model assessment index of insulin resistance. Additionally, weight loss results in a fall in serum insulin and a concomitant reduction in the concentration of BCAAs.

The decline in BCAAs plasma level is blunted in insulin-resistant subjects.

Further, in normal weight subjects, a definite association between increased blood levels of BCAAs and insulin resistance has been detected, suggesting that the increased concentration of BCAAs found in obese subjects is related to insulin resistance itself rather than obesity.

Rats were fed a high-fat diet enriched in BCAA that led to increase mTOR phosphorylation in muscle. The insulin resistance and impaired glucose tolerance of non obese BCAA-HF rats were equal to those of obese high-fat diet fed rats. No effect of high dietary BCAA on glucose homeostasis or muscle mTOR activation was apparent when rats were fed low-fat diets, indicating an interaction between BCAA effects and dietary lipid. These results support the idea that high BCAA tissue exposure contributes to the insulin resistance phenotype in diet-induced obese rats through activation of mTOR but that this effect is manifested only in the context of high-fat diet feeding.

Despite such observations, the evidence that BCAA promote obesity-associated insulin resistance and the T2DM metabolic phenotype is not clear-cut.

For instance, the post surgery drop in blood BCAA concentrations in bariatric surgery patients [12] is concurrent with improved insulin sensitivity, but this association could also result if inhibition of BCAA catabolism is a consequence and not a cause of insulin resistance.

In summary, evidence that elevated blood BCAA concentrations contribute to the insulin resistance often associated with obesity remains equivocal and the weight of evidence points to protein- and BCAA-rich diets as actually beneficial to whole-body metabolic health.

Effects of BCAAs on Triglyceride (TG) Content in Skeletal Muscle

Knockout the gene of mitochondrial BCAA aminotransferase (BCATm), which catalyzes the first step of BCAA catabolism lead to an elevation in the serum BCAA level [13]. In BCATm knockout mice, fasting blood glucose and fasting serum insulin levels were decreased, respectively, and the Homeostasis Model Assessment for Insulin Resistance index was significantly lower than that of wild-type mice. In skeletal muscle, BCAAs promote glucose uptake through activation of PI3K and protein kinase C and subsequent translocation of GLUT4 to the plasma membrane. In the liver, BCAAs upregulate the liver X receptor a/sterol regulatory element binding protein-1c pathway. In addition, BCAA suppresses hepatic expression of glucose-6-phosphatase, which catalyzes the final steps of gluconeogenesis [14].

BCAA supplementation has been also reported to improve insulin resistance by increasing oxidation of free fatty acids [15]. BCAAs increase PPAR-alpha expression and subsequent expression of uncoupling proteins 2 in liver and uncoupling proteins 3 in muscle [15] (Fig. 17.6).

The TG content in the liver and skeletal muscle was also less in BCAA-treated group, and the lowering of fat deposition in these two tissues. The adiposity of WAT was less in BCAA-treated

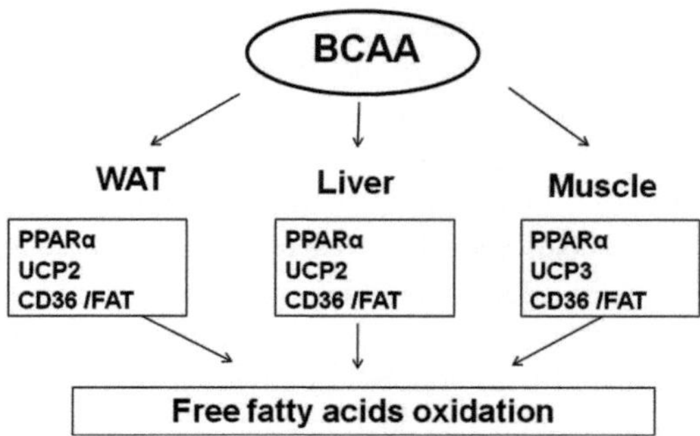

Fig. 17.6 Distinctive molecular pathway for BCAA-induced fatty acids oxidation. BCAAs increase PPAR-alpha expression and subsequent UCP2 in liver and UCP3 in muscle. Upregulation of UCP2 and UCP3 expression increases oxidation of free fatty acids

group, as assessed by changes in tissue weight and the adipocyte morphology. Hence, BCAA may play a protective role against fat accumulation in tissues [15]. An increased tissue TG content has been reported to interfere with insulin-stimulated PI3K activation and/or subsequent GLUT4 translocation and glucose uptake, leading to insulin resistance. The treatment of PPAR-alpha inhibitor partially attenuated the effects of BCAA-induced the reduction of TG accumulation of those tissues in obese mice [15]. Thus, the results indicated that the activation of PPAR-alpha may have contributed to the reduction of TG accumulation in those tissues.

PPAR-alpha expression in WAT in the BCAA-treated group was higher compared with the controls. PPAR-alpha regulates the UCP3 gene in muscle and controls the UCP2 gene in liver, and both of these genes in other tissues. BCAA treatment prevents fat accumulation in muscle, liver, and WAT, accompanied by improved insulin resistance, in DIO mice (Fig. 17.6).

Conclusions

The BCAA is sensitive to the effects of insulin and therefore their metabolism is affected in clinical conditions featuring insufficient insulin action related. Further studies are needed to clarify the clinical role of BCAA in diabetes and obesity.

References

1. Marchesini G, Bianchi G, Merli M, et al. Nutritional supplementation with branched-chain amino acids in advanced cirrhosis: a double-blind, randomized trial. Gastroenterology. 2003;124:1792–801.
2. Nishitani S, Takehana K, Fujitani S, et al. Branched chain amino acids improve glucose metabolism in rats with liver cirrhosis. Am J Physiol Gastrointest Liver Physiol. 2005;288:G1292–300.
3. Honda M, Takehana K, Sakai A, et al. Malnutrition impairs interferon signaling through mTOR and FoxO pathways in patients with chronic hepatitis C. Gastroenterology. 2011;141:128–40.
4. Velloso LA, Schwartz MW. Altered hypothalamic function in diet-induced obesity. Int J Obes (Lond). 2011;35(12):1455–65.
5. Azzu V, Jastroch M, Divakaruni AS, Brand MD. The regulation and turnover of mitochondrial uncoupling proteins. Biochim Biophys Acta. 2010;1797:785–91.
6. Villarroya F, Iglesias R, Giralt M. PPARs in the control of uncoupling proteins gene expression. PPAR Res. 2007;2007:74364.
7. Thompson MP, Kim D. Links between fatty acids and expression of UCP2 and UCP3 mRNAs. FEBS Lett. 2004;568:4–9.
8. Chevalier S, Burgess SC, Malloy CR, et al. The greater contribution of gluconeogenesis to glucose production in obesity is related to increased whole-body protein catabolism. Diabetes. 2006;55:675–81.

9. Manders RJ, Koopman R, Sluijsmans WE, et al. Co-ingestion of a protein hydrolysate with or without additional leucine effectively reduces postprandial blood glucose excursions in type 2 diabetic men. J Nutr. 2006; 136:1294–9.
10. She P, Van Horn C, Reid T, et al. Obesity-related elevations in plasma leucine are associated with alterations in enzymes involved in branched-chain amino acid metabolism. Am J Physiol Endocrinol Metab. 2007;293:E1552–63.
11. Newgard CB, An J, Bain JR, et al. A branched-chain amino acid-related metabolic signature that differentiates obese and lean humans and contributes to insulin resistance. Cell Metab. 2009;9:311–26.
12. Laferrère B, Reilly D, Arias S. Differential metabolic impact of gastric bypass surgery versus dietary intervention in obese diabetic subjects despite identical weight loss. Sci Transl Med. 2011;3:80re2.
13. She P, Reid TM, Bronson SK, et al. Disruption of BCATm in mice leads to increased energy expenditure associated with the activation of a futile protein turnover cycle. Cell Metab. 2007;6:181–94.
14. Higuchi N, Kato M, Miyazaki M, et al. Potential role of branched-chain amino acids in glucose metabolism through the accelerated induction of the glucose-sensing apparatus in the liver. J Cell Biochem. 2011;112:30–8.
15. Arakawa M, Masaki T, Nishimura J, et al. The effects of branched-chain amino acid granules on the accumulation of tissue triglycerides and uncoupling proteins in diet-induced obese mice. Endocr J. 2011;58:161–70.

Chapter 18
Experimental Models of High Fat Obesity and Leucine Supplementation

Yuran Xie and Zhonglin Xie

Key Points

- Excess consumption of a high fat diet is a major cause of obesity, which leads to lipid accumulation in fat tissue, skeletal muscle, and liver.
- High fat diet-induced obesity is associated with chronic metabolic disorders, including mitochondrial dysfunction, insulin resistance, and type 2 diabetes.
- Dietary leucine prevents high fat diet-induced obesity, mitochondrial dysfunction, and insulin resistance.
- Leucine supplementation ameliorates high fat diet-induced metabolic disorders through coordinate activation of mammalian target of rapamycin, AMP-activated protein kinase, and sirtuin1 signaling pathways.
- Dietary supplementation of leucine may provide an adjunct in the prevention and treatment of high fat diet-induced chronic metabolic disorders.

Keywords Leucine supplementation • Obesity • Insulin resistance • AMPK • SIRT1

Abbreviations

AMPK	AMP-activated protein kinase
BACC	Branched chain amino acids
4EBP-1	4E binding protein 1
ER	Endoplasmic reticulum
FAO	Fatty acid oxidation
FoxO1	Forkhead box O1
HFD	High fat diet
IRS-1	Insulin receptor substrate 1
mTOR	Mammalian target of rapamycin
NAMPT	Nicotinamide phosphoribosyltransferase

Y. Xie, B.S. • Z. Xie, M.D., Ph.D. (✉)
Section of Molecular Medicine, Department of Medicine, University of Oklahoma Health Sciences Center, 941 Stanton L Young Blvd. Basic Science Education Building 302B, Oklahoma City, OK 73104, USA
e-mail: Yuran-xie@ouhsc.edu; zxie@ouhsc.edu

R. Rajendram et al. (eds.), *Branched Chain Amino Acids in Clinical Nutrition: Volume 1*, Nutrition and Health, DOI 10.1007/978-1-4939-1923-9_18,

PGC1α	Peroxisome proliferator-activated receptor γ coactivator-1α
ROS	Reactive oxygen species
Sirt1	Sirtuin1
TRB3	Tribbles homolog 3

Introduction

Obesity, the most common nutritional disorders, affects the majority of adults in Western society. It has become a leading health concern due to its link to insulin resistance, diabetes, and cardiovascular disease [1]. Traditionally, prevention and treatment of obesity mainly depend on caloric restriction and increasing physical activity. Although short-term weight loss can be achieved by various dietary approaches, sustainability of weight loss seems to be difficult [2]. Recently, several studies have shown that dietary manipulation of essential amino acids, including leucine, arginine, and glutamine, improves lipid and glucose metabolism [3]. Specifically, dietary supplementation of leucine prevents HFD-induced obesity, mitochondrial dysfunction, and insulin resistance, suggesting the potential importance of dietary supplementation of leucine in the prevention of HFD-induced metabolic disorders [4]. In this article, we review the metabolic roles of leucine and explore the underlying mechanisms by which leucine supplementation ameliorates HFD-induced metabolic disorders.

High Fat Diet Induces Obesity

In industrial countries, excess consumption of a HFD plays an important role in the development of obesity because a HFD stimulates voluntary energy intake due to its high energy intensity and high palatability [5]. In addition, the excess consumption of diets rich in fatty acids does not stimulate FAO even when dietary fat is given in excess of energy expenditure. Thus, excess consumption of a HFD results in lipid accumulation in adipose tissue, muscle, and liver. Several studies have demonstrated that a HFD reliably produces obesity in rats, mice, dogs, and primates [6], while a low fat diet rarely induces obesity in animals, even when animals are maintained in small cages to limit physical activity. Moreover, switching rodents from a HFD to a low fat diet can ameliorate the HFD-induced obesity [7], suggesting a close correlation between the excess consumption of a HFD and the development of obesity. Observations in humans also suggest that a HFD promotes the development of obesity through increasing energy intake and reducing energy expenditure. In both lean and obese subjects, a HFD enhances fat intake much more than FAO, resulting in a positive fat balance. Furthermore, a HFD may disrupt the balance between the intake and oxidation of carbohydrate, making the obese individuals particularly prone to increase in body weight and fat mass. Thus, dietary fat plays an important role in the development of obesity.

The endoplasmic reticulum (ER) is a major organelle in maintaining lipid metabolic homeostasis. Excessive intake of nutrients may stimulate the ER stress response or the unfolded protein response through three ER membrane proteins; inositol-requiring enzyme-1, activating transcription factor, and protein kinase-like ER kinase. Deregulation of the ER stress response has been implicated in obesity [8]. In pancreatic β cells, the activation of protein kinase-like ER kinase and eukaryotic translation initiation factor 2 subunit alpha upregulates the expression of sterol-regulatory binding proteins, which are major regulators of cholesterol and fatty acid synthesis [9]. In mammary epithelial cells, the loss of protein kinase-like ER kinase reduces the sterol-regulatory binding protein activity and lipogenesis [10]. Moreover, in cultured hepatocellular carcinoma HepG2 cells, elevated palmitate levels, which mimics fatty acid overload conditions, increases ER stress markers, upregulates the proteins related to fatty acid synthesis, and induces lipid accumulation. Similarly, feeding a HFD to mice stimulates the ER stress response, induces lipid accumulation, and reduces insulin sensitivity in mice. Importantly, inhibition of the unfolded protein response by pharmacological and genetic means prevents HFD-induced ER stress response and lipid accumulation. Thus, excess consumption of a HFD may induce obesity through stimulating the ER stress response.

HFD-Induced Obesity Is Associated with Mitochondrial Dysfunction and Insulin Resistance

The primary function of the mitochondria is to produce energy from carbohydrate, fat, and protein. As a major site of FAO, the mitochondria contain the enzymes essential for lipid metabolism. Excessive intake of dietary fat leads to mitochondrial dysfunction with consequential impaired lipid and glucose metabolism. Exposure of C2C12 myotubes to saturated free fatty acid impairs mitochondrial function as evidenced by decreases in both mitochondrial hyperpolarization and ATP generation. Meanwhile saturated free fatty acids reduce insulin-enhanced glycogen synthesis, glucose oxidation, and lactate production. The inhibition of mitochondrial respiration reduces FAO and increases triglyceride accumulation in 3T3L1 preadipocytes. These data suggest a close correlation among the consumption of a HFD, lipid accumulation, and mitochondrial dysfunction.

The excessive intake of dietary fat can also impair mitochondrial function through the inhibition of mitochondrial biogenesis. The genes encoding proteins involved in oxidative phosphorylation and mitochondrial biogenesis are downregulated in young men who consumed a HFD [11]. Likewise, reductions in the genes related to oxidative phosphorylation and mitochondrial biogenesis are also found in mouse models of HFD-induced obesity and insulin resistance [11]. Furthermore, exposure of rats to a HFD significantly impaired mitochondrial function by rapidly reducing ATP synthesis. The defects in mitochondrial oxidative capacity may further promote the development of obesity because a deficiency in β-oxidation and a lower oxidative metabolism led to lipid accumulation in non-adipose tissues.

The accumulation of fatty acids and their metabolites, including acyl-CoAs, ceramides, and diacylglycerol in non-adipose tissues, particularly in the muscle and liver, is closely associated with insulin resistance, because these "toxic" lipids can serve as signaling molecules that activate protein kinases such as protein kinase C, c-Jun N-terminal kinases, and the inhibitor of nuclear factor-κB kinase-β. These kinases, in turn, can phosphorylate IRS-1, leading to degradation of IRS-1 and impaired PI3K signaling. In addition, a HFD may upregulate TRB3 protein and promote the association between Akt and TRB3, which inhibits insulin-stimulated Akt phosphorylation (Ser473), leading to high blood glucose, impaired insulin tolerance, and low glucose infusion rate during a clamp [4].

Mitochondrial dysfunction is considered a major factor contributing to the development of insulin resistance and type 2 diabetes. Compared to insulin-sensitive control subjects, type 2 diabetic patients and insulin-resistant individuals with impaired glucose tolerance have fewer mitochondria, lower levels of mitochondrial oxidative enzymes, and lower ATP synthesis in their muscles [12]. The expression of genes related to PGC1-α and various mitochondrial constituents [13] is also lower in muscles from these patients. In addition, the number of mitochondria and the expression of the genes that regulate mitochondrial biogenesis are significantly lower in adipocytes from type 2 diabetic patients and obese individuals. However, other groups reported that consumption of a HFD increases mitochondrial biogenesis and fatty acid oxidative capacity in skeletal muscle [14]. Recently, Hancock et al. [15] showed that insulin resistance can develop in animals maintained on a HFD, despite a significant increase in the mitochondrial contents. Thus, the role of mitochondrial metabolism in the etiology of insulin resistance warrants further investigation.

Leucine Supplementation Prevents HFD-Induced Metabolic Disorders

Leucine is an essential BCAA. In humans it cannot be produced by the body and has to be obtained from dietary source. In contrast to other essential amino acids that are mainly metabolized in the liver, leucine is primarily metabolized in peripheral tissues, such as muscle. Leucine serves not only as a building block for protein synthesis but also as a signal to activate mTOR kinase and its downstream targets [16] to regulate many cellular processes, including protein synthesis, cell growth, and metabolism. Like dietary protein, leucine supplementation has been implicated in the regulation of satiety, because leucine can directly stimulate mTOR signaling in the hypothalamus and induce the secretion of leptin [17], an important adipokine in regulating hunger and food consumption, leading to decreased food intake. Furthermore, increasing dietary intake of leucine has been shown to reduce body weight, plasma levels of cholesterol, and lipid accumulation in the mice subjected to a HFD [4, 18]. The beneficial effects of leucine supplementation are associated with the upregulation of uncoupling protein 3 in brown and white adipose tissues and skeletal muscle, which can increase resting energy expenditure.

Dietary supplementation of leucine also increases mitochondrial biogenesis and improves mitochondrial function in the liver, brown adipose tissue, and muscle. Compared with normal chow diet-fed mice, HFD plus leucine-fed mice had greater expression of the genes related to mitochondrial biogenesis, such as PGC1-α, nuclear respiratory factor-1, mitochondrial DNA transcription factor A, and NADH dehydrogenase [ubiquinone] iron-sulfur protein 8. The high expression of mitochondrial biogenesis genes was associated with an increase in mitochondrial mass and an improvement of mitochondrial function, as determined by more citrate synthase activity and greater ATP content [4]. Moreover, addition of leucine to the HFD resulted in a significantly higher mRNA level of PPARα, an important enzyme controlling fatty acid metabolism. The functional relevance of this induction was validated by detecting more expression of two important genes regulating FAO, including carnitine palmitoyltransferase-1b (CPT-1b) and medium-chain acyl CoA dehydrogenase, which may stimulate FAO.

The mitochondria are essential organelles responsible for processing oxygen and converting substances from the foods into energy for essential cellular functions. Meanwhile, they are also a major source of ROS, which are associated with a wide variety of inflammatory and metabolic diseases [19]. The excessive intake of dietary fat may enhance the electron flow in the mitochondrial respiratory chain, which increases ROS generation, thereby inducing oxidative stress [20]. This may cause structural and functional damage to the liver, kidney, and heart. Thus, the improvement of mitochondrial biogenesis and function by leucine supplementation may also attenuate HFD-induced oxidative stress. As assessed by immunohistochemistry, the consumption of a HFD increased the formation of 3-nitrotyrosine, a footprint of oxidative stress. The increase was diminished by addition of leucine to the HFD [4].

Another beneficial effect of leucine supplementation is improvement of insulin sensitivity and glucose metabolism. In normal diet-fed mice, insulin stimulated Akt phosphorylation at Ser473, but in HFD-fed mice insulin failed to stimulate the Akt phosphorylation. The consumption of additional leucine with a HFD recovered the effect of insulin-stimulated Akt phosphorylation. Consistent with the improvement of insulin sensitivity, leucine supplementation normalized blood glucose levels, improved glucose tolerance, and enhanced glucose infusion rate during a hyperinsulinemic–euglycemic clamp study in HFD-fed mice.

In addition to improvement of glucose and lipid metabolism in HFD-induced obese mouse model, chronic leucine supplementation improves glucose-insulin homeostasis in other mouse models of obesity and diabetes with different etiologies. In RCS10 mice [21], a polygenic model predisposed to beta cell failure and type 2 diabetes, chronic supplementation of leucine prevents the development of overt diabetes through increasing insulin secretion. In yellow agouti mice, which carry a mutation of the agouti gene in chromosome 2 and exhibit phenotypes of mild hyperphagia, hyper-metabolism and insulin resistance, dietary leucine improves insulin sensitivity as estimated by lower insulin levels together with lower HbA1c and plasma glucose levels in leucine-treated yellow agouti mice [21]. Meanwhile, leucine also attenuates adipose tissue inflammation, increases resting metabolic rate, and upregulates the genes related to mitochondrial function and energy metabolism in these mice.

The role of dietary leucine in the regulation of energy metabolism remains controversial in the literature because some groups reported that dietary supplementation of leucine had no effect on lipid metabolism and did not alter susceptibility to diet-induced obesity in mice. In the in vitro experiments, incubation of extensor digitorum longus muscle with leucine reduces insulin-stimulated phosphorylation of Akt at Ser473 [22]. Moreover, supplemental BCAA worsens insulin resistance in HFD-fed rats even though the supplementation of BCAA reduces food intake and decreases body weight. Thus, further investigations are necessary for establishing the role of dietary leucine in energy metabolism.

Leucine and mTOR Signaling

The mammalian target of rapamycin, a coordinator between nutritional stress and cellular growth machinery, functions in an intracellular signaling pathway that senses the availability of amino acids. mTOR exists as two distinct protein complexes, mTOR complex1 and mTOR complex2 [23]. Activation of mTOR complex1 increased protein synthesis and ribosomal biogenesis, thereby playing a key role in coupling nutrients to protein synthesis [24].

In skeletal muscle, an increase in the leucine concentration stimulates the mTOR signaling pathway, which phosphorylates the inhibitory binding protein 4EBP-1, causing the binding protein to dissociate from the translational initiation factor, eukaryotic initiation factor-4E [25]. In addition, leucine-activated mTOR phosphorylates P70S6 kinase, leading to the phosphorylation of the S6 ribosomal protein [26]. P70S6 kinase and 4EBP-1 are two proteins essential in the regulation of protein synthesis. In adipose tissue, leucine-stimulated mTOR signaling regulates preadipocytes differentiation, adipose tissue morphogenesis, and leptin secretion [27]. For example, a high concentration of leucine in primary cultures of rat adipocytes activates mTOR signaling, leading to preadipocyte differentiation and adipogenesis [28].

Activation of the mTOR/S6K1 pathway increases IRS-1 phosphorylation at Ser1101 while it inhibits IRS-1 phosphorylation at tyr612, resulting in the degradation of IRS-1 and the impairment of PI3K signaling [29], critical events in the development of insulin resistance [30]. In one study, long-term supplementation of BCAA including leucine, isoleucine, and valine, to HFD-fed mice induced insulin resistance that was accompanied by higher phosphorylation of mTOR and IRS-1 on Ser307, all of which were attenuated by administration of rapamycin, an inhibitor of mTOR. In contrast, administration of leucine diminished the insulin resistance caused by consuming a HFD even though leucine increased insulin-stimulated phosphorylation of P70S6K in the muscle, liver, and brown adipose tissue [31]. Thus the role of leucine in insulin resistance needed to be further investigated.

Leucine and AMPK Signaling Pathway

AMPK is a heterotrimer comprising of α, β, and γ subunits [32]. The α subunit contains the catalytic domain. Increases in the ratio of AMP/ATP can activate AMPK through an allosteric effect and the inhibition of the dephosphorylation of Thr172 in the activation loop in the kinase domain [33]. To date, several upstream kinases have been identified to phosphorylate AMPK; these include the tumor suppressor kinase LKB1 [34, 35] and two calmodulin-dependent protein kinase kinases, CaMKK-α and CaMKK-β. AMPK is a fuel gauge that senses the intracellular energy status [36] and plays an important role in the regulation of glucose and lipid metabolism [37]. Activation of AMPK phosphorylates several target molecules, leading to the reduction of energy demands and an increase in energy supply [37]. In addition, activation of AMPK provides an important cellular protective response in various stress conditions, including hypoxia, oxidative stress, exercise, and starvation [38, 39].

Recent evidence suggests that the inhibition of AMPK may also actively participate in the regulation of many cellular processes. In OVE26 mice, a transgenic model of severe early-onset type 1 diabetes, and streptozotocin-induced diabetic mice, both the AMPK activity and the phosphorylation at Thr172 are significantly reduced in the diabetic hearts, which is accompanied by the suppression of autophagy and by impairment of cardiac structure and function [40, 41]. Feeding mice a HFD results in the dysregulation of AMPK, detected by both a reduction in AMPK protein expression and an inhibition of AMPK phosphorylation in skeletal muscle, heart, liver, aortic endothelium, and hypothalamus [42]. Incubation of endothelial cells in a medium containing palmitate to mimic fatty acid overload conditions results in an increase in ceramide production and the inhibition of the phosphorylation of AMPK and its downstream molecule acetyl-CoA carboxylase through activation of protein phosphatase 2 [43]. Similarly, feeding mice a HFD rich in palmitate also inhibits AMPK activity [43], suggesting that inhibition of AMPK activity may be an important mechanism underlying HFD-induced metabolic disorders. The consumption of additional leucine with a HFD restored AMKP phosphorylation, which was associated with a decrease in body weight and fat mass, as well as an improvement of insulin sensitivity and glucose metabolism. The activation of AMPK also stimulated the SIRT1 signaling pathway through increasing the expression of SIRT1 and NAMPT, a rate-limiting enzyme responsible for NAD^+ biosynthesis [4]. These results suggest that dietary leucine can prevent HFD-induced metabolic disorders through coordinately regulating the AMPK and SIRT1 signaling pathways.

Leucine Activates SIRT1 Signaling

SIRT1, an NAD^+-dependent deacetylase, can enhance glucose utilization, increase mitochondrial FAO, and improve insulin sensitivity [44]. In the mouse liver, SIRT1 is required for the activation of PGC1α and induction of gluconeogenic genes in response to fasting signals [45]. Under starvation conditions, knockdown of SIRT1 by transfection of siRNA results in mild hypoglycemia, increased glucose tolerance, improved insulin sensitivity, and decreased hepatic glucose production [46]. In addition, knockdown of SIRT1 positively regulates the nuclear receptor LXR (liver X receptor) proteins, leading to the accumulation of free fatty acids and cholesterol in the liver, which can be reversed by SIRT1 overexpression [46].

The association between the low expression of SIRT1 protein and insulin resistance in skeletal muscle has been reported for obese and aged individuals and for type 2 diabetic patients [47]. Glucose restriction activates AMPK by transcriptional repression of the protein tyrosine phosphatase 1B [47], resulting in high expression of the NAMPT gene and activation of the SIRT1 signaling pathway, which in turn improves insulin sensitivity and glucose homeostasis. In addition, the activation of SIRT1 decreases acetylation of PGC1α and FoxO1, stimulating mitochondrial biogenesis, FAO [48], and gluconeogenesis [49]. Thus, activation of SIRT1 signaling is essential for the prevention of metabolic disorders.

Leucine supplementation in HFD-fed mice activated SIRT1 signaling through AMPK-mediated upregulation of SIRT1 and NAMPT. The activation of SIRT1 reduced acetylation of PGC1α and FoxO1 [50], which increased the expression of genes related to mitochondria biogenesis, leading to enhanced mitochondrial contents and improved mitochondrial function. In addition, the reduction in acetylation of PGC1α was associated with upregulation of the genes related to FAO and activate the signaling pathway that controls FAO, thereby normalizing plasma lipid profile, reducing subcutaneous and visceral fat mass, and preventing lipid accumulation [4].

FoxO1 was originally identified as a negative regulator of insulin signaling [51], but recent evidence suggests that it can improve hepatic insulin sensitivity through upregulation of TRB3. TRB3 is an endogenous inhibitor of Akt, which is a critical regulator in insulin signaling [52]. The expression of TRB3 has been implicated in the regulation of insulin signaling and glucose metabolism. For instance, the expression of TRB3 at the mRNA and protein levels is higher in livers from mice with diabetes than in mice without diabetes [53]. In cultured hepatocytes isolated from HFD-fed mice, overexpression of TRB3 disrupts insulin signaling by directly binding to Akt and prevents Akt phosphorylation, resulting in hyperglycemia and glucose intolerance in the mice [52]. Moreover, in the individuals susceptible to type 2 diabetes, TRB3 contributes to the development of insulin resistance by interfering with Akt activation [54]. These data suggested that upregulation of TRB3 suppresses insulin sensitivity via Akt inhibition. In support of this hypothesis, a recent study showed that the consumption of leucine to a HFD reduced the association of TRB3 and Akt and enhanced insulin-stimulated Akt phosphorylation, which was accompanied by lower blood glucose and increased glucose infusion rate during a clamp in HFD-fed mice. These studies suggest that leucine supplementation restores insulin sensitivity in HFD-fed mice through suppressing TRB3 expression and interfering with the interaction between TRB3 and Akt [4].

Conclusions

Leucine functions as a nutrient signal to coordinately regulate mTOR, AMPK, and SIRT1 signaling pathways in the liver, skeletal muscle, and adipose tissue. Dietary supplementation of leucine significantly ameliorates the deleterious effects of consumption of a HFD, including obesity, hepatic lipid accumulation, mitochondrial dysfunction, and insulin resistance. Therefore, leucine supplementation may be beneficial to obese individuals and type 2 diabetic patients. The metabolic benefits of leucine supplementation are associated with the upregulation of genes related to mitochondrial biogenesis and FAO, increases in metabolic rates, and suppression of inflammation in adipose tissue. Understanding the molecular mechanism by which dietary supplementation of leucine improves glucose and lipid metabolism may help to define novel nutritional and pharmacological approaches for the treatment of obesity, insulin resistance, and type 2 diabetes. Thus, further investigations are needed to clearly define the beneficial effects of dietary leucine on energy metabolism in obese individuals and diabetic patients.

References

1. Poirier P, Giles TD, Bray GA, Hong Y, Stern JS, Pi-Sunyer FX, Eckel RH. Obesity and cardiovascular disease: pathophysiology, evaluation, and effect of weight loss. Arterioscler Thromb Vasc Biol. 2006;26:968–76.
2. Katz DL. Competing dietary claims for weight loss: finding the forest through truculent trees. Annu Rev Public Health. 2005;26:61–88.
3. Fu WJ, Haynes TE, Kohli R, Hu J, Shi W, Spencer TE, Carroll RJ, Meininger CJ, Wu G. Dietary L-arginine supplementation reduces fat mass in Zucker diabetic fatty rats. J Nutr. 2005;135:714–21.
4. Li H, Xu M, Lee J, He C, Xie Z. Leucine supplementation increases SIRT1 expression and prevents mitochondrial dysfunction and metabolic disorders in high-fat diet-induced obese mice. Am J Physiol Endocrinol Metab. 2012;303:E1234–44.
5. Miller WC, Lindeman AK, Wallace J, Niederpruem M. Diet composition, energy intake, and exercise in relation to body fat in men and women. Am J Clin Nutr. 1990;52:426–30.
6. Atkins CE, LeCompte PM, Chin HP, Hill JR, Ownby CL, Brownfield MS. Morphologic and immunocytochemical study of young dogs with diabetes mellitus associated with pancreatic islet hypoplasia. Am J Vet Res. 1988; 49:1577–81.

7. Hill JO, Dorton J, Sykes MN, Digirolamo M. Reversal of dietary obesity is influenced by its duration and severity. Int J Obes. 1989;13:711–22.
8. Dong Y, Zhang M, Liang B, Xie Z, Zhao Z, Asfa S, Choi HC, Zou MH. Reduction of AMP-activated protein kinase alpha2 increases endoplasmic reticulum stress and atherosclerosis in vivo. Circulation. 2010;121:792–803.
9. Colgan SM, Tang D, Werstuck GH, Austin RC. Endoplasmic reticulum stress causes the activation of sterol regulatory element binding protein-2. Int J Biochem Cell Biol. 2007;39:1843–51.
10. Bobrovnikova-Marjon E, Hatzivassiliou G, Grigoriadou C, Romero M, Cavener DR, Thompson CB, Diehl JA. PERK-dependent regulation of lipogenesis during mouse mammary gland development and adipocyte differentiation. Proc Natl Acad Sci U S A. 2008;105:16314–9.
11. Sparks LM, Xie H, Koza RA, Mynatt R, Hulver MW, Bray GA, Smith SR. A high-fat diet coordinately downregulates genes required for mitochondrial oxidative phosphorylation in skeletal muscle. Diabetes. 2005;54:1926–33.
12. Patti ME, Butte AJ, Crunkhorn S, Cusi K, Berria R, Kashyap S, Miyazaki Y, Kohane I, Costello M, Saccone R, Landaker EJ, Goldfine AB, Mun E, DeFronzo R, Finlayson J, Kahn CR, Mandarino LJ. Coordinated reduction of genes of oxidative metabolism in humans with insulin resistance and diabetes: potential role of PGC1 and NRF1. Proc Natl Acad Sci U S A. 2003;100:8466–71.
13. Crunkhorn S, Dearie F, Mantzoros C, Gami H, da Silva WS, Espinoza D, Faucette R, Barry K, Bianco AC, Patti ME. Peroxisome proliferator activator receptor gamma coactivator-1 expression is reduced in obesity: potential pathogenic role of saturated fatty acids and p38 mitogen-activated protein kinase activation. J Biol Chem. 2007;282:15439–50.
14. Turner N, Bruce CR, Beale SM, Hoehn KL, So T, Rolph MS, Cooney GJ. Excess lipid availability increases mitochondrial fatty acid oxidative capacity in muscle: evidence against a role for reduced fatty acid oxidation in lipid-induced insulin resistance in rodents. Diabetes. 2007;56:2085–92.
15. Hancock CR, Han DH, Chen M, Terada S, Yasuda T, Wright DC, Holloszy JO. High-fat diets cause insulin resistance despite an increase in muscle mitochondria. Proc Natl Acad Sci U S A. 2008;105:7815–20.
16. Li F, Yin Y, Tan B, Kong X, Wu G. Leucine nutrition in animals and humans: mTOR signaling and beyond. Amino Acids. 2011;41:1185–93.
17. Lynch CJ, Gern B, Lloyd C, Hutson SM, Eicher R, Vary TC. Leucine in food mediates some of the postprandial rise in plasma leptin concentrations. Am J Physiol Endocrinol Metab. 2006;291:E621–30.
18. Zhang Y, Guo K, LeBlanc RE, Loh D, Schwartz GJ, Yu YH. Increasing dietary leucine intake reduces diet-induced obesity and improves glucose and cholesterol metabolism in mice via multimechanisms. Diabetes. 2007;56:1647–54.
19. Furukawa S, Fujita T, Shimabukuro M, Iwaki M, Yamada Y, Nakajima Y, Nakayama O, Makishima M, Matsuda M, Shimomura I. Increased oxidative stress in obesity and its impact on metabolic syndrome. J Clin Invest. 2004;114:1752–61.
20. Reaven G, Abbasi F, McLaughlin T. Obesity, insulin resistance, and cardiovascular disease. Recent Prog Horm Res. 2004;59:207–23.
21. Guo K, Yu YH, Hou J, Zhang Y. Chronic leucine supplementation improves glycemic control in etiologically distinct mouse models of obesity and diabetes mellitus. Nutr Metab (Lond). 2010;7:57.
22. Saha AK, Xu XJ, Lawson E, Deoliveira R, Brandon AE, Kraegen EW, Ruderman NB. Downregulation of AMPK accompanies leucine- and glucose-induced increases in protein synthesis and insulin resistance in rat skeletal muscle. Diabetes. 2010;59:2426–34.
23. Sarbassov DD, Ali SM, Kim DH, Guertin DA, Latek RR, Erdjument-Bromage H, Tempst P, Sabatini DM. Rictor, a novel binding partner of mTOR, defines a rapamycin-insensitive and raptor-independent pathway that regulates the cytoskeleton. Curr Biol. 2004;14:1296–302.
24. Kapahi P, Chen D, Rogers AN, Katewa SD, Li PW, Thomas EL, Kockel L. With TOR, less is more: a key role for the conserved nutrient-sensing TOR pathway in aging. Cell Metab. 2010;11:453–65.
25. Kimball SR, Jefferson LS. Regulation of protein synthesis by branched-chain amino acids. Curr Opin Clin Nutr Metab Care. 2001;4:39–43.
26. Crozier SJ, Kimball SR, Emmert SW, Anthony JC, Jefferson LS. Oral leucine administration stimulates protein synthesis in rat skeletal muscle. J Nutr. 2005;135:376–82.
27. Lynch CJ. Role of leucine in the regulation of mTOR by amino acids: revelations from structure-activity studies. J Nutr. 2001;131:861S–5.
28. Tsukiyama-Kohara K, Poulin F, Kohara M, DeMaria CT, Cheng A, Wu Z, Gingras AC, Katsume A, Elchebly M, Spiegelman BM, Harper ME, Tremblay ML, Sonenberg N. Adipose tissue reduction in mice lacking the translational inhibitor 4E-BP1. Nat Med. 2001;7:1128–32.
29. Hartley D, Cooper GM. Role of mTOR in the degradation of IRS-1: regulation of PP2A activity. J Cell Biochem. 2002;85:304–14.
30. Tremblay F, Lavigne C, Jacques H, Marette A. Role of dietary proteins and amino acids in the pathogenesis of insulin resistance. Annu Rev Nutr. 2007;27:293–310.
31. Macotela Y, Emanuelli B, Bang AM, Espinoza DO, Boucher J, Beebe K, Gall W, Kahn CR. Dietary leucine—an environmental modifier of insulin resistance acting on multiple levels of metabolism. PLoS One. 2011;6:e21187.

32. Hardie DG, Carling D, Halford N. Roles of the Snf1/Rkin1/AMP-activated protein kinase family in the response to environmental and nutritional stress. Semin Cell Biol. 1994;5:409–16.
33. Stein SC, Woods A, Jones NA, Davison MD, Carling D. The regulation of AMP-activated protein kinase by phosphorylation. Biochem J. 2000;345(Pt 3):437–43.
34. Xie Z, Dong Y, Scholz R, Neumann D, Zou MH. Phosphorylation of LKB1 at serine 428 by protein kinase C-zeta is required for metformin-enhanced activation of the AMP-activated protein kinase in endothelial cells. Circulation. 2008;117:952–62.
35. Xie Z, Dong Y, Zhang J, Scholz R, Neumann D, Zou MH. Identification of the serine 307 of LKB1 as a novel phosphorylation site essential for its nucleocytoplasmic transport and endothelial cell angiogenesis. Mol Cell Biol. 2009;29:3582–96.
36. Hardie DG. Minireview: the AMP-activated protein kinase cascade: the key sensor of cellular energy status. Endocrinology. 2003;144:5179–83.
37. Carling D. The AMP-activated protein kinase cascade—a unifying system for energy control. Trends Biochem Sci. 2004;29:18–24.
38. Xie Z, Dong Y, Zhang M, Cui MZ, Cohen RA, Riek U, Neumann D, Schlattner U, Zou MH. Activation of protein kinase C zeta by peroxynitrite regulates LKB1-dependent AMP-activated protein kinase in cultured endothelial cells. J Biol Chem. 2006;281:6366–75.
39. Taleux N, De Potter I, Deransart C, Lacraz G, Favier R, Leverve XM, Hue L, Guigas B. Lack of starvation-induced activation of AMP-activated protein kinase in the hypothalamus of the Lou/C rats resistant to obesity. Int J Obes (Lond). 2008;32:639–47.
40. Xie Z, Lau K, Eby B, Lozano P, He C, Pennington B, Li H, Rathi S, Dong Y, Tian R, Kem D, Zou MH. Improvement of cardiac functions by chronic metformin treatment is associated with enhanced cardiac autophagy in diabetic OVE26 mice. Diabetes. 2011;60:1770–8.
41. He C, Zhu H, Li H, Zou MH, Xie Z. Dissociation of Bcl-2-Beclin1 complex by activated AMPK enhances cardiac autophagy and protects against cardiomyocyte apoptosis in diabetes. Diabetes. 2013;62:1270–81.
42. Viollet B, Horman S, Leclerc J, Lantier L, Foretz M, Billaud M, Giri S, Andreelli F. AMPK inhibition in health and disease. Crit Rev Biochem Mol Biol. 2010;45:276–95.
43. Wu Y, Song P, Xu J, Zhang M, Zou MH. Activation of protein phosphatase 2A by palmitate inhibits AMP-activated protein kinase. J Biol Chem. 2007;282:9777–88.
44. Jiang WJ. Sirtuins: novel targets for metabolic disease in drug development. Biochem Biophys Res Commun. 2008;373:341–4.
45. Rodgers JT, Lerin C, Haas W, Gygi SP, Spiegelman BM, Puigserver P. Nutrient control of glucose homeostasis through a complex of PGC-1alpha and SIRT1. Nature. 2005;434:113–8.
46. Rodgers JT, Puigserver P. Fasting-dependent glucose and lipid metabolic response through hepatic sirtuin 1. Proc Natl Acad Sci U S A. 2007;104:12861–6.
47. Sun C, Zhang F, Ge X, Yan T, Chen X, Shi X, Zhai Q. SIRT1 improves insulin sensitivity under insulin-resistant conditions by repressing PTP1B. Cell Metab. 2007;6:307–19.
48. Gerhart-Hines Z, Rodgers JT, Bare O, Lerin C, Kim SH, Mostoslavsky R, Alt FW, Wu Z, Puigserver P. Metabolic control of muscle mitochondrial function and fatty acid oxidation through SIRT1/PGC-1alpha. EMBO J. 2007;26:1913–23.
49. Frescas D, Valenti L, Accili D. Nuclear trapping of the forkhead transcription factor FoxO1 via Sirt-dependent deacetylation promotes expression of glucogenetic genes. J Biol Chem. 2005;280:20589–95.
50. Yu J, Auwerx J. Protein deacetylation by SIRT1: an emerging key post-translational modification in metabolic regulation. Pharmacol Res. 2010;62:35–41.
51. Ogg S, Paradis S, Gottlieb S, Patterson GI, Lee L, Tissenbaum HA, Ruvkun G. The Fork head transcription factor DAF-16 transduces insulin-like metabolic and longevity signals in C. elegans. Nature. 1997;389:994–9.
52. Du K, Herzig S, Kulkarni RN, Montminy M. TRB3: a tribbles homolog that inhibits Akt/PKB activation by insulin in liver. Science. 2003;300:1574–7.
53. Matsushima R, Harada N, Webster NJ, Tsutsumi YM, Nakaya Y. Effect of TRB3 on insulin and nutrient-stimulated hepatic p70 S6 kinase activity. J Biol Chem. 2006;281:29719–29.
54. Prudente S, Hribal ML, Flex E, Turchi F, Morini E, De CS, Bacci S, Tassi V, Cardellini M, Lauro R, Sesti G, Dallapiccola B, Trischitta V. The functional Q84R polymorphism of mammalian Tribbles homolog TRB3 is associated with insulin resistance and related cardiovascular risk in Caucasians from Italy. Diabetes. 2005;54:2807–11.

Chapter 19
Branched Chain Amino Acids in Experimental Models of Amyotrophic Lateral Sclerosis

Alessia De Felice, Annamaria Confaloni, Alessio Crestini, Roberta De Simone, Fiorella Malchiodi-Albedi, Alberto Martire, Andrea Matteucci, Luisa Minghetti, Patrizia Popoli, Aldina Venerosi, and Gemma Calamandrei

Key Points

- Dietary integrators containing branched chain amino acids might constitute a risk factor in the etiology of Amyotrophic Lateral Sclerosis, an adult-onset neurodegenerative disease.
- Excessive intake of branched chain amino acids might favor the establishment of neurotoxic conditions, as branched chain amino acids contribute to de novo synthesis of glutamate, the principal excitatory neurotransmitter in central nervous system.
- Excessive stimulation of glutamate receptor is responsible for excitotoxicity processes.
- Branched chain amino acids are neurotoxic in neuronal/astrocytic cultures through a mechanism that involves overstimulation of glutamate *N*-Methyl-D-Aspartate receptor.

A. De Felice, Master Degree in Biology • G. Calamandrei, Master Degree in Biology (✉)
Unit of Neurotoxicology and Neuroendocrinology, Department of Cell Biology and Neurosciences,
Istituto Superiore di Sanità, Viale Regina Elena, 299, Rome 00161, Italy
e-mail: alessiadefelice5@gmail.com; gemma.calamandrei@iss.it

A. Confaloni, Ph.D. • A. Crestini, Ph.D., M.S., M.Sc. Res.
Master of Science, Post-graduate in applied genetics and molecular biology, M.Sc.Res, Master Degree in Biology,
Section of Clinic, Diagnostic and Therapy of Degenerative Diseases of the Central Nervous System,
Department of Cell Biology and Neurosciences, Istituto Superiore di Sanità, Viale Regina Elena, 299,
Rome 00161, Italy
e-mail: annamaria.confaloni@iss.it; alessio.crestini@iss.it

R. De Simone, M.Sc. • L. Minghetti, Ph.D.
Section of Experimental Neurology, Department of Cell Biology and Neurosciences,
Istituto Superiore di Sanità, Viale Regina Elena, 299, Rome 00161, Italy
e-mail: roberta.desimone@iss.it; luisa.minghetti@iss.it

F. Malchiodi-Albedi, M.D. • A. Matteucci, Ph.D., Biology
Section of Molecular Neurobiology, Department of Cell Biology and Neurosciences,
Istituto Superiore di Sanità, Viale Regina Elena, 299, Rome 00161, Italy
e-mail: fiorella.malchiodialbedi@iss.it; andrea.matteucci@iss.it

A. Martire, Ph.D. • P. Popoli, M.D.
Section of Central Nervous System Pharmacology, Department of Therapeutic Research and Medicines
Evaluation, Istituto Superiore di Sanità, Viale Regina Elena, 299, Rome 00161, Italy
e-mail: alberto.martire@iss.it; patrizia.popoli@iss.it

A. Venerosi, Ph.D.
Section of Neurotoxicology and Neuroendocrinology, Department of Cell Biology and Neurosciences,
Istituto Superiore di Sanità, Viale Regina Elena, 299, Rome 00161, Italy
e-mail: venerosi@iss.it

R. Rajendram et al. (eds.), *Branched Chain Amino Acids in Clinical Nutrition: Volume 1*, Nutrition and Health, DOI 10.1007/978-1-4939-1923-9_19,

- High doses of branched chain amino acids influence gene expression and immune properties of microglial cells and may promote low-grade chronic inflammation favoring neurodegeneration.
- Chronic dietary intake of branched chain amino acids can modulate expression profiles of genes linked to oxidative stress or neurodegenerative pathways in murine central nervous system.
- Chronic dietary intake of branched chain amino acids induces hyperactivity and hyperalgesia in wild-type mice and worsens the motor coordination deficit characteristic of the transgenic mouse model of amyotrophic lateral sclerosis, the G93A-SOD1 mice.
- *Ex vivo* electrophysiology results in transgenic mouse model of amyotrophic lateral sclerosis: G93A-SOD1 mice chronically fed with a branched chain amino acids enriched diet suggest a beneficial influence of such supplementation on the perturbation of glutamate neurotransmission normally reported in this strain of mutated mice.
- The experimental evidence here reported provides a possible link between an exogenous factor (high intake of branched chain aminoacids) and the increased risk of neurotoxic/neurodegenerative damage especially in individuals with a genetic predisposition.

Keywords Neurodegenerative diseases • Excitotoxicity • Neuronal cultures • Microglia • Gene expression • Oxidative stress • Behavior • Electrophysiology • G93A mice

Abbreviations

ALS	Amyotrophic lateral sclerosis
SOD1	Superoxide dismutase 1
BCAA	Branched chain amino acid
GDH	Glutamate dehydrogenase
MSUD	Maple syrup urine disease
AMPA	2-Amino-3-(3-hydroxy-5-methyl-isoxazol-4-yl) propanoic acid
MK801	Dizocilpine non competitive antagonist of NMDA receptor
NMDA	*N*-Methyl-D-Aspartate receptor
H-BCAA	High dose branched chain amino acids
LPS	Lipopolysaccharide
IL-1β	Interleukin 1β
TNF-α	Tumor necrosis factor α
iNOS	Inducible nitric oxide synthase
MRC-1	Mannose receptor C1
15-F2t-IsoP	15-F_{2t}-Isoprostane
IGF-1	Insulin-like growth factor-1
FALS	Familiar amyotrophic lateral sclerosis
C57BL6/J	C57 Black 6 Jackson mouse strain
RT-PCR	Real time polymerases chain reaction
Park 7	Parkinson disease 7
ApoE	Apolipoprotein E
FANCC	Fanconi anemia complementary group C
EAAT2	Excitatory amino acid transporter
mGluR5	Metabotropic glutamate receptor 5
WT	Wild type
STD	Standard
G93A	Mutant form of human SOD1
GLT-1	Glial glutamate transporter

Introduction

Amyotrophic lateral sclerosis (ALS) is an adult-onset neurodegenerative disease, also known as Lou Gehrig disease, characterized by degeneration of neurons in the cortex, bulbus, and spinal cord, leading to progressive paralysis of bulbar, respiratory and limb muscles [1]. The etiology of the disease is still obscure. The most common mutations found in familiar ALS (10 % of total cases) involve the gene coding for the enzyme copper–zinc superoxide dismutase 1 (SOD1), which, however, explain only about 20 % of familiar ALS cases and 2 % of the sporadic form of this disease [2, 3]. The observation of glutamate increase in plasma and subarachnoid fluid in patients with ALS has suggested that excitotoxicity can have a role in triggering neurodegeneration in ALS [4]. The vast majority of patients is affected in a sporadic form and this strongly supports the involvement of several genes and the possible role of environmental factors that might trigger the pathogenic mechanisms in vulnerable individuals [5]. Notably, epidemiological studies have shown a much higher risk for ALS among Italian professional soccer players, in comparison to the rest of the population [6, 7] and higher prevalence of the disease has also been reported among football players in the United States [8]. Several hypotheses have been formulated to explain the excess of deaths among these athletes, such as recurring traumas, chronic intake of anti-inflammatory drugs, exposure to pesticides used on playing fields, use of illegal toxic substances or dietary integrators, intense physical activity, but so far no sound scientific evidence has been collected in support of any of them.

Among the potential risk factors implicated in ALS pathogenesis, one line of research has been focusing on the excessive use of dietary integrators. Specifically, the branched chain amino acids (BCAAs) valine, isoleucine and leucine are widely used by both professional and non-professional athletes as integrators to enhance both physical and mental capacity. BCAAs play a vital role in regulating gluconeogenesis as well as maintaining the growth of skeletal muscle in addition to influencing the production of neurotransmitters within the brain [9, 10].

The hypothesis that BCAAs could be implicated in ALS pathogenesis was supported by the failure of clinical trials that explored the potential beneficial effects of BCAA supplementation in ALS progression and motor symptoms. ALS patients have a defect in glutamate dehydrogenase (GDH) and BCAAs act as nitrogen donors restoring GDH levels. However, studies did not yield favorable results with BCAAs and a large clinical trial was interrupted because there was an excess mortality in BCAA-treated ALS patients [11].

In the light of these results, it has been suggested that excessive intake of BCAAs might favor the establishment of neurotoxic conditions and exert negative consequences on brain functions. The hypothesis of potential neurotoxic effects of BCAAs or their catabolic products are supported by the severe neurological symptoms characterizing inherited disorders of BCAA catabolism, such as maple syrup urine disease (MSUD), in which levels of BCAAs, as well as their branched chain keto-acids, increase up to about 30-fold in blood, urine, and cerebrospinal fluid, as compared to control subjects [12, 13]. BCAAs can enter the brain through a transporter, located at the blood–brain barrier, which is competitive and shared by amino acids such as tryptophan, precursor of serotonin, and tyrosine and phenylalanine, precursors of catecholamines. As a consequence, if BCAAs rise in the plasma, uptake of the other amino acids that share the transporter decrease and synthesis and release of serotonin and catecholamines are reduced [14]. In addition BCAAs contribute to de novo synthesis of glutamate, the principal excitatory neurotransmitter in CNS, and it is well known that excessive stimulation of glutamate receptors is responsible for excitotoxicity processes.

Results of in vivo studies on rat brain, as well as investigations on retinal explants and cultured rat astroglia and neurons, have shown that the maintenance of glutamate levels in the brain depends for a substantial proportion on de novo synthesis of glutamate [15], besides on the glutamine/glutamate cycle. It is tempting to speculate that a high intake of BCAAs could alter glutamate metabolism, creating a higher susceptibility to excitotoxicity and facilitating the onset of ALS.

This hypothesis is of particular concern since there is no good evidence for establishing BCAA tolerance levels in humans [16, 17], while their potential clinical use is currently proposed both in healthy individuals and in several metabolic and neurological human diseases (see [10] for a comprehensive review). Furthermore, there is increasing concern on the abuse of BCAA-containing dietary integrators and energetic drinks, particularly in adolescents and young adults [18, 19].

In this chapter we will briefly present four studies aimed at investigating the potential neurotoxicity of BCAAs and their possible implication in neurodegenerative risk, carried out by different research groups at the National Institute of Health (Istituto Superiore di Sanità) with the support of the Italian Ministry of Health. These studies were performed in either in vitro or in vivo models, ranging from the analysis of cell cultures derived from embryonic rat brain [20, 21] to the in vivo evaluation of chronic BCAA supplementation effects on gene expression [22], behavioral phenotype and electrophysiology in wild-type mice and in the transgenic model of ALS [23].

BCAA Neurotoxicity in Rat Cortical Cultures

Contrusciere et al. [20] attempted to verify if BCAAs could induce toxicity or exacerbate excitotoxic cell damage in cell cultures derived from embryonic rat brain. They treated pure neuronal, mixed astrocytic/neuronal or pure astrocytic cultures derived from embryonic rat cortex and hippocampus at gestational day 18, with valine, isoleucine, or BCAA mixture (2.5 mM Valine, 2.5 mM Leucine, 2.5 mM Isoleucine), in the presence or not of glutamate receptor agonists, such as glutamate and α-amino-3-hydroxyl-5-methyl-4-isoxazole-propionate (AMPA). In addition, Contrusciere and coworkers examined the effect of MK-801, a specific NMDA channel blocker, to ascertain if BCAA toxicity could involve NMDA receptor overstimulation.

The effects of BCAAs were then evaluated by: (1) examination of apoptotic cells identified by the condensation of chromatin and fragmentation of the nuclei, (2) immunocytochemistry, and (3) flow cytometry. Results showed that treatment with 25 mM valine for 48 h, in absence of any additional excitotoxic stimulus, induced a pro-apoptotic effect on cell cultures. When administered for 7 days, 2.5 mM valine was neurotoxic and increased glutamate toxicity. The BCAA isoleucine, at the same concentration of valine (25 mM), was not toxic when administered alone, nor did it increase glutamate-induced apoptosis. However, treatment with a BCAA mixture (2.5 mM valine, 2.5 mM leucine, 2.5 mM isoleucine) had a significant pro-apoptotic effect. Valine toxicity was completely counteracted by pretreatment with MK-801, a specific NMDA channel blocker, suggesting an involvement of glutamate receptor overstimulation in valine toxicity. Notably valine exerted pro-apoptotic effects only in mixed astrocytic/neuronal cortical culture and not in hippocampal-derived cultures, thus indicating that astrocytes were involved in BCAA effects. In addition, neurotoxicity was brain area specific, being detected in cortical, but not in hippocampal neurons (Fig. 19.1).

These results indicate that BCAAs are neurotoxic in vitro through a mechanism that involves overstimulation of NMDA receptors, requires the presence of astrocytes and occurs in distinct neuronal populations. Although based on an experimental model that is very far from the complexity of the human brain, these data provide a possible mechanistic link between an exogenous factor (high intake of BCAAs) and the increased risk for ALS in specific population subgroups.

BCAAs Alter the Immune Properties of Microglial Cells

Besides neurons and astrocytes, high BCAA levels could influence the functional activities of other types of brain cells, among which microglia, the main macrophage population of brain parenchyma. Microglial cells actively survey the brain parenchyma [24] to readily respond to signals released from

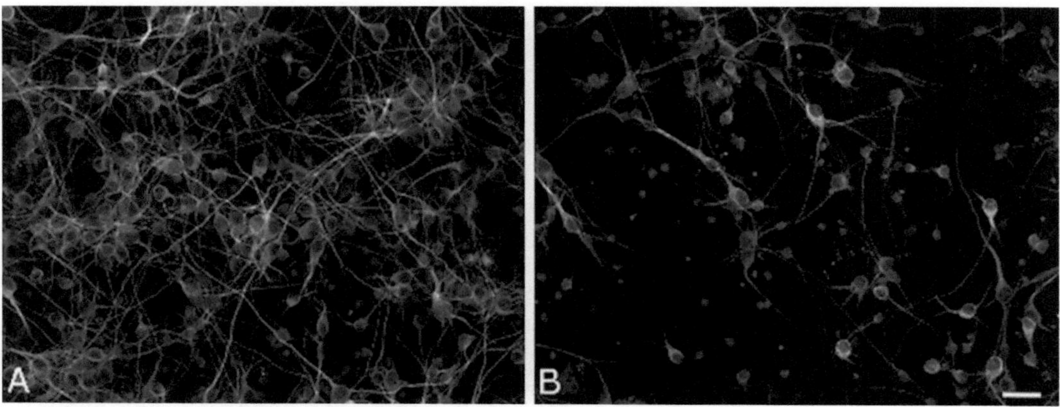

Fig. 19.1 In vitro valine neurotoxicity. Astrocytic/neuronal cortical cultures were treated at DIV 14 with 25 mM valine for 48 h and immunostained for MAP2. Compared to control cultures (**a**), valine-treated cultures (**b**) showed loss of neurons and severe damage of the dendritic network. Bars = 50 μm

damaged cells or pathogens. The rapid response of microglia to the altered microenvironment—known as microglial activation—can result in either protective or toxic activities. As described for peripheral macrophages, the functional phenotype of activated microglia is strictly stimuli-dependent and related to the specific pathological condition examined. The repertoire of activated microglial phenotypes can span from classical M1 inflammatory phenotype to the M2 anti-inflammatory and protective phenotype [25]. It is important to stress that M2 properties, generally associated with the resolution of inflammation, do not necessarily lead to positive outcomes as inappropriate downregulation of inflammatory response can be detrimental as well as an excessive inflammatory response.

In a recent study, De Simone and colleagues evaluated the effects of BCAAs on microglial reactivity through the analysis of several genes and markers associated with either M1 or M2 phenotypes. They used primary microglial cells harvested from mixed glial cultures that had been cultivated in normal or high BCAA medium (H-BCAA, 1–10 mM valine, leucine, and isoleucine 1:1:1) for 8 days. They also compared specific functional properties of normal and H-BCAA microglial cultures, including migration, phagocytosis and reaction to a typical pro-inflammatory challenge such as bacterial endotoxin (lipopolysaccharide, LPS). They were able to show that H-BCAAs significantly influence microglia gene expression profile and functions, leading to an intermediate phenotype characterized by a partial skewing toward the M2 state. They reported that microglial cells cultured in high BCAA-containing medium under unstimulated conditions exhibit a lower expression of the M1 genes IL-1β, TNF-α, and iNOS and a concomitant higher expression of the M2 gene MRC-1. These features would suggest a less reactive phenotype. However, these cells are also characterized by an increased production of reactive oxygen species, as indicated by the higher levels of the lipid peroxidation product 15-F_{2t}-IsoP, and by a lower production of the neuroprotective factor IGF-1, suggesting an unbalance in favor of neurotoxic activities. H-BCAA microglial cultures also show an increased capacity to phagocyte opsonized fluorescent latex beads, in agreement with the M2-like phenotype, which is typically associated to sustained phagocytic activity. Other microglial functions, such as basal motility and migration in response to a well-defined microglial chemoattractant such as ADP, were comparable in control and H-BCAA cultures (Table 19.1).

Finally, in H-BCAA cultures, the levels of NO, IL-1β, and TNF-α induced by LPS were significantly decreased, suggesting a milder inflammatory reaction. On the other hand, M2-associated genes such as IL-10 and MRC-1 were increased, suggesting a pattern of activation skewed towards the M2 phenotype. Nonetheless, both Arginase-1 mRNA expression and arginase activity were substantially lower than in control cultures, further supporting the failure of H-BCAA cultures in acquiring a full M2 phenotype. Arginase-1 competes with iNOS for the common substrate L-arginine and transforms

Table 19.1 High levels of BCAA on microglial gene profile and functions

M1 genes	IL-1β	↓
	iNOS	↓
	TNFα	↓
M2 genes	IL-10	=
	Arg-1	=
	MRC-1	↑
	IGF-1	↓
Functions	ROS production	↑
	Motility	=
	Phagocytosis	↑

Microglial cells were maintained in high BCAA medium and then analyzed for the expression of M1 and M2 markers, ROS production, microglial motility, and phagocytosis. H-BCAA microglial cultures exhibited an intermediate M1/M2 phenotype characterized by a partial skewing toward the M2 state, with enhanced phagocytic activity but also increased free radical generation and decreased neuroprotective functions

Table 19.2 Effects of high levels of BCAAs on microglial activation by LPS

M1 genes	IL-1β	↓
	iNOS	↓
	TNFα	=/↓[a]
M2 genes	IL-10	↑
	Arg-1	↓
	MRC-1	↑

Microglial cells that had been cultivated in normal or high BCAA medium were stimulated with LPS. M1 or M2 markers were analyzed and compared to activated cultures maintained in normal medium. H-BCAA microglial cultures had a milder inflammatory reaction with a pattern of activation skewed towards the M2 phenotype

[a]Gene/protein expression

it into urea and ornithine, the precursor of a family of small molecules termed the polyamines with neuroprotective functions, suggesting that H-BCAA exposed microglia might be deficient in their neurotrophic activities (Table 19.2).

On the basis of these findings, De Simone et al. [21] hypothesize that the unique functional phenotype induced in microglia by BCAA could result in a low-grade inflammatory state and a less efficient microglial response to local damage, two features that might increase the susceptibility to neurodegenerative processes. Notably, loss of microglial surveillance function has been recently described in a mouse model of amyotrophic lateral sclerosis, during the clinical phase of disease.

BCAA Dietary Supplementation Induces Expression of Oxidative Stress Genes in Wild-Type Mice

The discovery of disease-associated mutations in the superoxide dismutase gene (SOD1) in a subset of cases of familial ALS (FALS) [27, 28] prompted extensive research into the role of oxidative stress in the pathogenesis of FALS cases and in animal models expressing these mutations [29–31].

Table 19.3 Genes modulated by dietary BCAAs supplementation in murine cortex tissues

Upregulated	Downregulated
– *Als2*: Amyotrophic lateral sclerosis 2 (juvenile)	– *ApoE*: Apolipoprotein E
– *Cygb*: Cytoglobin	– *Ccs*: Copper chaperone for superoxide dismutase
– *Ercc2*: Excision repair cross-complementing rodent repair deficiency 2	– *Park7*: Parkinson disease 7
– *Fancc*: Fanconi anemia, complementation group C	– *Prdx6*: Peroxiredoxin 6
– *Gpx2*: Glutathione peroxidase 2	– *Prdx6-rs1*: Peroxiredoxin 6, related sequence 1
– *Hbq1*: Hemoglobin theta 1	– *Sod1*: Superoxide dismutase 1
– *Ngb*: Neuroglobin	– *Tmod1*: Tropomodulin 1
	– *Txnrd1*: Thioredoxin reductase 1
	– *Txnrd3*: Thioredoxin reductase 3

As a matter of fact, oxidative stress may play different roles, acting both directly, through damage of crucial molecules, and indirectly, modulating the level of expression and/or the bioavailability of other important players in the mechanisms of signal transduction that dictate cell survival [32]. In such a framework, Piscopo et al. [22] investigated the effect of a BCAA diet, at doses comparable to human usage, on the expression profile of a set of genes involved in oxidative stress. Male C57BL6/J mice received either a basic standard diet or the same standard diet enriched by 2.5 % in BCAAs, Val–Leu–Ileu, 1:2:1 from postnatal day 21 till the age of 3 months. The BCAA concentration in the diet was selected on the basis of sport medicine indications [33], clinical protocols [11], and reported concentration of BCAAs in commercial products. At 3 months of age, mice were sacrificed, brains removed, and cortex analyzed with a quantitative RT-PCR array for 84 mRNA transcripts that included genes encoding catalase, glutathione peroxidases, and peroxiredoxins neutralizing cellular H_2O_2, along with genes involved in superoxide metabolism such as superoxide.

Some key genes that are involved in oxidative stress pathways were found to be regulated in different way by BCAA treatment. In particular genes encoding antioxidant proteins were downregulated, whereas genes encoding oxygen transporters were upregulated (see Table 19.3). The first downregulated genes codify for superoxide dismutase 1 (SOD), which is associated with familial ALS and for the copper chaperone for SOD, which regulates SOD activity by acquiring copper and consequently activating its function.

Another important gene downregulated by BCAA diet is Park 7 (also named DJ1 gene), involved in the protection of cells from oxidative stress and chaperone activity; Park 7 has been involved in early onset of Parkinson's disease. The ApoE ε4 allele, a well-known genetic risk factor for the sporadic form of Alzheimer's disease, is also associated with the age of onset of sporadic ALS, suggesting a correlation between these diseases.

With regard to upregulated genes, findings revealed a disturbance of molecular processes underlying oxygen transport. Piscopo et al. reported altered regulation of four oxygen transporters: neuroglobin, cytoglobin, hemoglobin theta 1, and Fanconi anemia complementation group C (FANCC); all these carriers are expressed in the brain and, intriguingly, induced by conditions of neuronal hypoxia and cerebral ischemia [34–36]. Another upregulated gene following BCAA diet supplementation was Als2 encoding alsin. Notably, autosomal recessive mutations in this gene lead to a clinical spectrum of motor dysfunction, including juvenile-onset alS (ALS2), primary lateral sclerosis, and hereditary spastic paraplegia [37, 38]. This set of results show that BCCA dietary supplementation can significantly alter some oxidative stress pathways in the brain (Fig. 19.2 and Table 19.3). So far, the functional meaning of BCCA-induced gene expression alteration remains unclear, nevertheless, considering the role of oxidative stress in the pathogenesis of ALS and of neurodegenerative diseases in general, the authors suggest that caution should be applied in the widespread use of BCAAs as dietary integrators in sports.

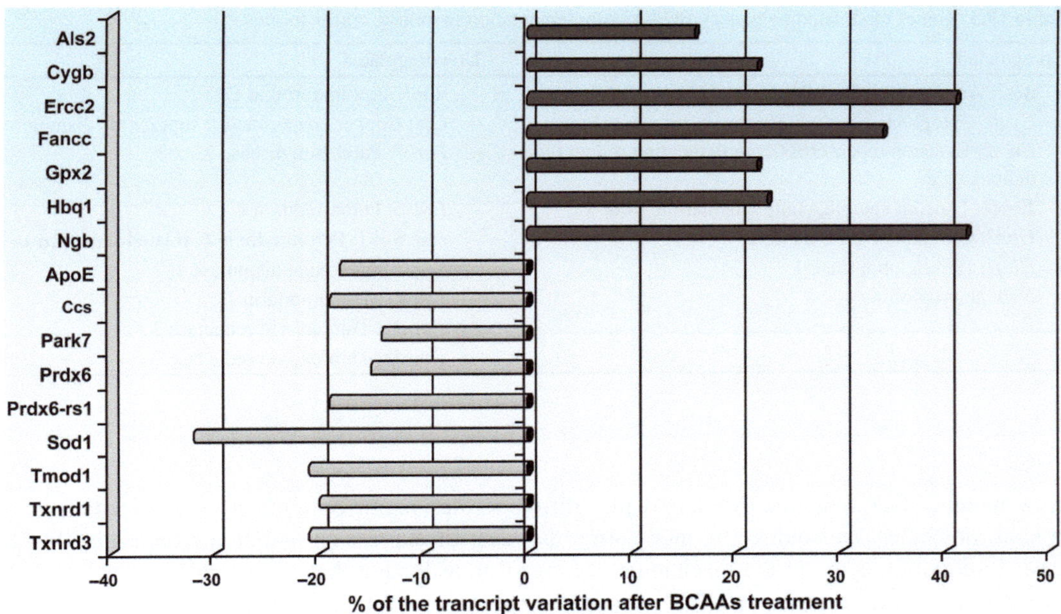

Fig. 19.2 Graphical representation of the effects of enriched BCAAs diet on gene expression. The figure shows the percentage of transcriptional variation obtained in a set of genes after BCAAs dietary supplementation compared with the control group. The treatment upregulates particularly some oxygen transporters (*black bars*), while downregulates antioxidant genes (*gray bars*) [22]. Abbreviations are listed in Table 19.3

Dietary BCAA Supplementation Influences Behavior and Synaptic Transmission in a Mouse Model of ALS

The transgenic mice carrying the human mutated Sod1 gene with a glycine/alanine substitution at codon 93 (G93A) are the most widely used model for investigating the mechanism of ALS. Acting as a powerful antioxidant and anti-inflammatory agent, SOD-1 plays a critical role in preventing toxic effects of free radicals on cells. The expression of the mutant human SOD1 in mice leads to an ALS-like neurodegenerative phenotype [2, 3]. This mutation is linked to selective impairment of the EAAT2/glutamate transporter-1 (GLT-1) and altered expression of metabotropic glutamate receptors 5 (mGluR5). The altered levels of extracellular glutamate could contribute to neurotoxicity [23].

Animal studies evaluating a specific vulnerability to dietary BCAAs in both normal and vulnerable gene backgrounds represent a critical tool to verify their potential toxicity. To this aim, Venerosi and coworkers [23] investigated the hypothesis that chronic dietary supplementation with BCAA could worsen ALS-like symptoms in the G93A transgenic model, bearing the mutated human superoxide dismutase 1 (SOD1). G93A or control mice (WT) received from the 4th to the 16th week of life dietary supplementation with BCAA (same concentration as in the paper by Piscopo et al. [22]). The final concentration of BCAA provided in the food was 2.5 % wet/weight (2:1:1 ratio of leucine, isoleucine, and valine). Mice consumed about 4 g/kg/day, equivalent to the maximum recommended daily dose taken by professional or non-professional athletes. After BCAA supplementation motor coordination, motor activity and pain threshold were evaluated at 8, 9, and 12 weeks of age and when motor ALS-like symptoms appeared, mice were sacrificed and synaptic transmission in motor cortex, electrophysiological response to glutamatergic stimulation and brain expression of the glial glutamate transporter-1 (GLT-1) were evaluated.

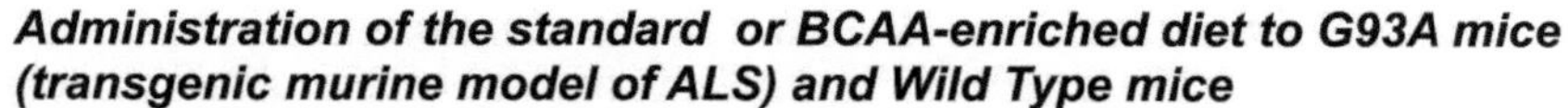

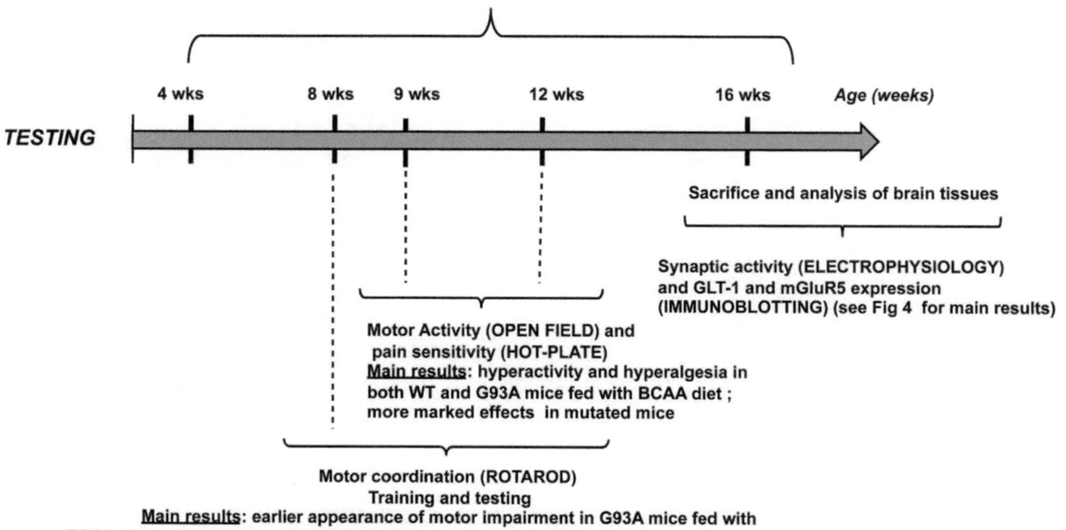

Fig. 19.3 Behavioral effects of chronic BCAA (from 4 weeks to 16 weeks of age) diet supplementation in WT and G93A-SOD1 mice. BCAAs caused motor hyperactivity in WT and more markedly in G93A mice, worsened the motor coordination performances of G93A mutated mice on both 9 and 12 weeks of age, and lowered pain threshold in both genotypes

The results indicated that at 2 months of age G93A mice weighed significantly less than both WT and SOD1 in the standard diet condition, while BCAAs contribute to a higher intake of food in G93A. G93A mice that consumed BCAA-enriched diet had higher body weight than WT mice, probably due to enhanced basal metabolism and higher energetic need, but this data were not related with any improvement of motor coordination. In the rotarod test, assessing muscular fatigue and motor coordination, G93A mice showed significant impaired performance at 8–9 weeks of age, irrespectively of the diet received. At 12 weeks of age there is a significant decline in motor coordination in G93A mice, with BCAA supplementation significantly worsening the motor performance of mutated mice. Spontaneous behavior was assessed by open-field test at 9 weeks of age. BCAA supplementation significantly enhanced motor activity more markedly in G93A mice. Finally, pain sensitivity was evaluated by hot-plate test at 9 and 12 weeks. Hot-plate data further confirmed the hyperactive profile of G93A mice, while BCAAs per se induced hyperalgesia in all genotypes. At 12 weeks of age, G93A-STD showed a very high pain threshold and BCAA supplementation restored control-like pain sensitivity in G93A mice too (Fig. 19.3).

At 16 weeks of age, at the time of detection of the first ALS-like symptoms (tremors, spontaneous epilepsy, and clasping of the hind limbs), mice were sacrificed to analyze BCAA effects on glutamatergic neurotransmission. Electrophysiology and immunoblotting studies performed in the full symptomatic stage highlighted a rather complicated picture, where BCAA supplementation either had no effect or apparently counteracted some of the abnormalities in glutamatergic function already reported for G93A mice (Fig. 19.4). In ALS pathogenesis, an abnormal increase in the glutamatergic tone is believed to occur due to the downregulation/inactivation of glutamate transporters, and of the GLT-1 subtype in particular [39, 40]. In agreement, Venerosi et al. found a significant reduction in GLT-1 expression in the brain and the spinal cord of G93A mice, and increased vulnerability to application of glutamate in cortical slices. The reduction of GLT-1 was completely prevented in the cortex and significantly attenuated in the striatum after BCAA supplementation. Furthermore, the increased

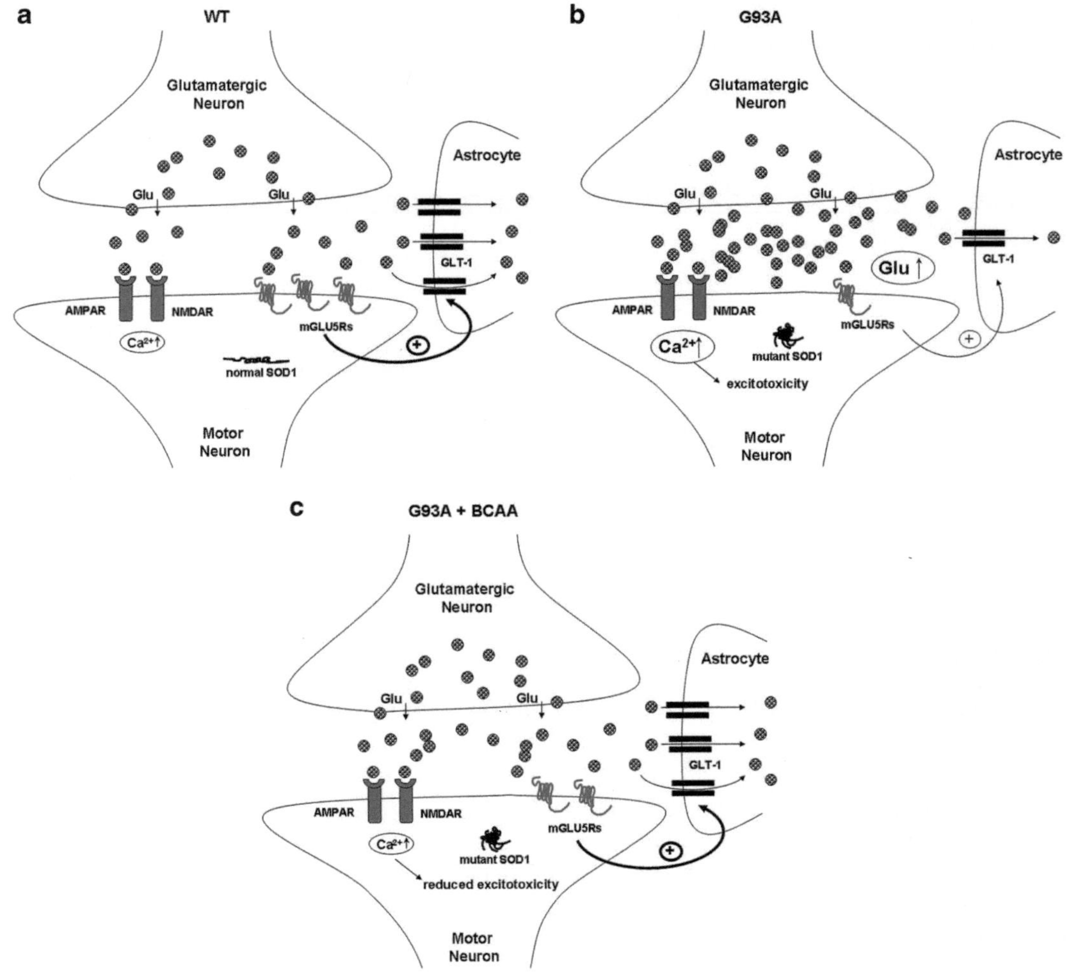

Fig. 19.4 BCAA supplementation counteracts some of the abnormalities in glutamatergic function of G93A mice. A quite complicated picture arises from electrophysiology and immunoblotting studies performed in G93A mice in a full symptomatic stage [23]; despite a general worsening of the behavioral profile, BCAA supplementation apparently reduces some of their molecular and synaptic abnormalities in glutamatergic function, shown as follows. (**a**) In WT mice, in the presence of normal SOD1, the level of extracellular glutamate (Glu), released from presynaptic neuron, is mainly regulated by GLT-1, a glutamate transporter present on astrocytes; also mGlu5Rs, on the membrane of the downstream motor neuron, contribute to modulate glutamate uptake by activating GLT-1; such a tight control on extracellular glutamate avoids the possible toxic effects of the excitatory neurotransmitter on motor neurons. (**b**) In G93A mice, in the presence of mutant SOD1, such a fragile balance is impaired; both GLT-1 on astrocytes and mGlu5R on neurons are strongly reduced in motor cortex; as a consequence the clearance of extracellular glutamate is severely compromised and an excessive amount of neurotransmitter remains in the extrasynaptic space, overstimulating AMPA and NMDA receptors on motor neurons and leading to an abnormal increase of intracellular Ca^{2+} and to cell toxicity. (**c**) In G93A mice under BCAA diet supplementation both GLT-1 and mGlu5R expression in motor cortex is almost completely restored; such an effect could explain the decreased glutamate toxicity on synaptic transmission observed in cortical slices from G93A-BCAA mice

glutamate-induced toxicity was no longer observed in cortical slices of G93A-BCAA mice. On their whole, electrophysiology results would suggest a beneficial influence of BCAA diet on the perturbation of glutamate uptake (Fig. 19.4).

Interestingly, all the effects elicited by the BCAA diet were specific for the G93A genotype, since no differences were observed in WT or SOD1 mice. To summarize, BCAA supplementation seems to exacerbate the behavioral profile typical of G93A mice, including significant worsening in motor performances and enhancement of hyperactivity, whereas it appears to counteract some of the effects of SOD1 mutation on glutamatergic neurotransmission. This seems to exclude a direct correlation between the behavioral changes observed in G93A mice and the altered glutamatergic neurotransmission. However, a complex brain functional pathway controls motor activity in rodents, involving the interplay among glutamate, GABA, and dopamine neurotransmission. As BCAAs compete with serotonin and catecholamine precursors to enter the brain through the same transporter [10], it is also possible that a rise in BCAAs plasma levels interferes with the maturation and function of other neurochemical systems, besides the glutamatergic one.

Altogether, the behavioral results did not confirm a causative role of BCAA supplementation in inducing or accelerating onset of ALS-like pathology in a transgenic mouse model, but G93A mice showed a specific susceptibility to BCAA supplementation.

Conclusions

Overall, increasing experimental evidence indicates that BCAAs at doses corresponding to human exposure may promote neurotoxic conditions either in vitro or in vivo models. BCAAs have pro-apoptotic effects in cell cultures likely by inducing an overstimulation of the NMDA glutamate receptors. The neurotoxic effects of BCAAs require the presence of astrocytes; in addition chronic exposure of microglia cells to BCAAs alters their immune properties by increasing the production of reactive oxygen species, and inducing a partial skewing toward the M2 state and a less efficient microglial response to inflammatory stimuli. These events would promote the establishment of a low-grade chronic inflammation and increase the likelihood of neurodegeneration.

In in vivo models chronic supplementation of BCAAs at doses comparable to reported human usage markedly modulate the expression of genes linked to oxidative stress (i.e., Sod1), as well as of genes possibly involved in neurodegenerative processes and ALS aetiology (i.e., Als2) suggesting that chronic dietary supplementation of BCAAs alter different oxidative stress or pathological pathways in the brain with potentially dangerous effects. Behavioral and electrophysiology studies carried out in a transgenic mouse model of ALS, the G93 SOD1 mutated mouse, indicate that BCAA exposure is able to worsen the neurological phenotype typical of this strain of mice, while reducing the susceptibility of mutated mice to glutamate toxicity in electrophysiology and immunoblotting assays.

These results are based on experimental models that do not fully reproduce the complexity of the human brain. It should also be considered that G93A SOD1 mice models familiar ALS, which accounts only for 10 % of total ALS cases. However, these data provide a possible link between an exogenous factor (high intake of BCAAs) and the increased risk of neurotoxic/neurodegenerative damage especially in individuals with a genetic predisposition. These findings are also supported by the observation that hyperexcitability of cortical neurons, which could facilitate excitotoxic events, has been demonstrated by electrophysiological recordings in both cultured neurons treated with BCAAs and in motor cortex slices from mice fed with BCAA-enriched diet [41]. Furthermore, recent evidence demonstrates that BCAAs modulate nerve growth factor expression [42] and this could suggest that an excess of these compounds may interfere with key processes such as cell growth, differentiation and repair.

Considering that BCAAs are largely used in general population as dietary integrators to improve mental and physical performances, the finding that they can induce significant CNS effects raises concerns over their spread, chronic use, and points to the need of further investigation to clarify the pros and cons of their uptake in health and disease.

References

1. Robberecht W, Philips T. The changing scene of amyotrophic lateral sclerosis. Nat Rev Neurosci. 2013; 14:248–64.
2. Bendotti C, Carri MT. Lessons from models of SOD1-linked familial ALS. Trends Mol Med. 2004;10:393–400.
3. Gurney ME, Pu H, Chiu AY, Dal Canto MC, et al. Motor neuron degeneration in mice that express a human Cu, Zn superoxide dismutase mutation. Science. 1994;264:1772–5.
4. Vucic S, Kierna MC. Pathophysiology of neurodegeneration in familial amyotrophic lateral sclerosis. Curr Mol Med. 2009;9:255–72.
5. Shaw CA, Höglinger GU. Neurodegenerative diseases: neurotoxins as sufficient etiologic agents? Neuromolecular Med. 2008;10:1–9.
6. Belli S, Vanacore N. Proportionate mortality of Italian soccer players: is amyotrophic lateral sclerosis an occupational disease? Eur J Epidemiol. 2005;20:237–42.
7. Vanacore N, Binazzi A, Bottazzi M, Belli S. Amyotrophic lateral sclerosis in an Italian professional soccer player. Parkinsonism Relat Disord. 2006;12:327–9.
8. Abel EL. Football increases the risk for Lou Gehrig's disease, amyotrophic lateral sclerosis. Percept Mot Skills. 2007;104:1251–4.
9. Davis JM, Alderson NL, Welsh RS. Serotonin and central nervous system fatigue: nutritional considerations. Am J Clin Nutr. 2000;72(2 Suppl):573S–8.
10. Fernstorm JD. Branched-chain amino acids and brain function. American Society of Nutritional Sciences, 4th Amino Acid Assessment workshop, 2005.
11. Tandan R, Bromberg MB, Forshew D, Fries TJ, et al. A controlled trial of amino acid therapy in amyotrophic lateral sclerosis: I. Clinical, functional, and maximum isometric torque data. Neurology. 1996;47:1220–6.
12. Zinnanti WJ, Lazovic J. Interrupting the mechanisms of brain injury in a model of maple syrup urine disease encephalopathy. J Inherit Metab Dis. 2012;35:71–9.
13. Wajner M, Coelho DM, Barschak AG, Araujo PR, Pires F, Lulhier FLG, Vargas CR. Reduction of large neutral amino acid concentrations in plasma and CSF of patients with maple syrup urine disease during crises. J Inherit Metab Dis. 2000;23:505–12.
14. Fernstrom JD, Fernstrom MH. Tyrosine, phenylalanine, and catecholamine synthesis and function in the brain. J Nutr. 2007;137(6 Suppl 1):1539S–47.
15. Lieth E, LaNoue KF, Berkich DA, Xu B, Ratz M, Taylor C, Hutson SM. Nitrogen shuttling between neurons and glial cells during glutamate synthesis. J Neurochem. 2001;76:1712–23.
16. Baker DH. Tolerance for branched-chain amino acids in experimental animals and humans. J Nutr. 2005;135:1585S–90.
17. Hutson SM, Sweatt AJ, Lanoue KF. Branched-chain [corrected] amino acid metabolism: implications for establishing safe intakes. J Nutr. 2005;135:1557S–64.
18. Pencharz PB, Elango R, Ball RO. Determination of the tolerable upper intake level of leucine in adult men. J Nutr. 2012;142:2220S–4.
19. Seifert SM, Schaechter JL, Hershorin ER, Lipshultz SE. Health effects of energy drinks on children, adolescents, and young adults. Pediatrics. 2011;127:511–28.
20. Contrusciere V, Paradisi S, Matteucci A, Malchiodi-Albedi F. Branched-chain amino acids induce neurotoxicity in rat cortical cultures. Neurotox Res. 2010;17:392–8.
21. De Simone R, Vissicchio F, Mingarelli C, De Nuccio C, Visentin S, Ajmone-Cat MA, Minghetti L. Branched-chain amino acids influence the immune properties of microglial cells and their responsiveness to pro-inflammatory signals. Biochim Biophys Acta. 2013;1832(5):650–9.
22. Piscopo P, Crestini A, Adduci A, Ferrante A, Massari M, Popoli P, Vanacore N, Confaloni A. Altered oxidative stress profile in the cortex of mice fed an enriched branched-chain amino acids diet: possible link with amyotrophic lateral sclerosis? J Neurosci Res. 2011;89(8):1276–83.
23. Venerosi A, Martire A, Rungi A, Pieri M, Ferrante A, Zona C, Popoli P, Calamandrei G. Complex behavioural and synaptic effects of dietary branched chain amino acids in a mouse model of amyotrophic lateral sclerosis. Mol Nutr Food Res. 2011;55(4):541–52.
24. Nimmerjahn A, Kirchhoff F, Helmchen F. Resting microglial cells are highly dynamic surveillants of brain parenchyma in vivo. Science. 2005;308:1314–8.
25. Colton CA. Heterogeneity of microglial activation in the innate immune response in the brain. J Neuroimmune Pharmacol. 2009;4:399–418.
26. Dibaj P, Steffens H, Zschüntzsch J, Nadrigny F, Schomburg ED, Kirchhoff F, Neusch C. In vivo imaging reveals distinct inflammatory activity of CNS microglia versus PNS macrophages in a mouse model for ALS. PLoS One. 2011;6:e17910.

27. Rosen DR, Siddique T, Patterson D, Figlewicz DA, Sapp P, Hentati A, Donaldson D, Goto J, Oregan JP, Deng HX, Rahmani Z, Krizus A, Mckennayasek D, Cayabyab A, Gaston SM, Berger R, Tanzi RE, Halperin JJ, Herzfeldt B, Vandenbergh R, Hung WY, Bird T, Deng G, Mulder DW, Smyth C, Laing NG, Soriano E, Pericakvance MA, Haines J, Rouleau GA, Gusella JS, Horvitz HR, Brown RH. Mutations in Cu/Zn superoxide-dismutase gene are associated with familial amyotrophic-lateral-sclerosis. Nature. 1993;362:59–62.
28. Deng HX, Hentati A, Tainer JA, Iqbal Z, Cayabyab A, Hung WY, Getzoff ED, Hu P, Herzfeldt B, Roos RP, et al. Amyotrophic lateral sclerosis and structural defects in Cu, Zn superoxide dismutase. Science. 1993;261:1047–51.
29. Andrus PK, Fleck TJ, Gurney ME, Hall ED. Protein oxidative damage in a transgenic mouse model of familial amyotrophic lateral sclerosis. J Neurochem. 1998;71:2041–8.
30. Pasinelli P, Brown RH. Molecular biology of amyotrophic lateral sclerosis: insights from genetics. Nat Rev Neurosci. 2006;7:710–23.
31. Barber SC, Mead RJ, Shaw PJ. Oxidative stress in ALS: a mechanism of neurodegeneration and a therapeutic target. Biochim Biophys Acta. 2006;1762:1051–67.
32. Cozzolino M, Ferri A, Carrì MT. Amyotrophic lateral sclerosis: from current developments in the laboratory to clinical implications. Antioxid Redox Signal. 2008;10:405–43.
33. De Lorenzo A, Petroni ML, Masala S, Melchiorri G, Pietrantuono M, Perriello G, Andreoli A. Effect of acute and chronic branched chain amino acids on energy metabolism and muscle performance. Diabetes Nutr Metab. 2003;2003(16):291–7.
34. Khan AA, Wang Y, Sun Y, Mao XO, Xie L, Miles E, Graboski J, Chen S, Ellerby LM, Jin K, Greenberg DA. Neuroglobin-overexpressing transgenic mice are resistant to cerebral and myocardial ischemia. Proc Natl Acad Sci U S A. 2006;103:17944–8.
35. Mammen PP, Shelton JM, Ye Q, Kanatous SB, Mc Grath AJ, Richardson JA, Garry DJ. Cytoglobin is a stress-responsive hemoprotein expressed in the developing and adult brain. J Histochem Cytochem. 2006;54:1349–61.
36. He Y, Hua Y, Liu W, Hu H, Keep RF, Xi G. Effects of cerebral ischemia on neuronal hemoglobin. J Cereb Blood Flow Metab. 2009;29:596–605.
37. Hadano S, Hand CK, Osuga H, Yanagisawa Y, Otomo A, Devon RS, Miyamoto N, Showguchi-Miyata J, Okada Y, Singaraja R, Figlewicz DA, Kwiatkowski T, Hosler BA, Sagie T, Skaug J, Nasir J, Brown Jr RH, Scherer SW, Rouleau GA, Hayden MR, Ikeda JE. A gene encoding a putative GTPase regulator is mutated in familial amyotrophic lateral sclerosis 2. Nat Genet. 2001;29:166–73.
38. Rowland LP. Primary lateral sclerosis, hereditary spastic paraplegia, and mutations in the alsin gene: historical background for the first International Conference. Amyotroph Lateral Scler Other Motor Neuron Disord. 2005;6:67–76.
39. Sala G, Beretta S, Ceresa C, Mattavelli L, et al. Impairment of glutamate transport and increased vulnerability to oxidative stress in neuroblastoma SH-SY5Y cells expressing a Cu, Zn superoxide dismutase typical of familial amyotrophic lateral sclerosis. Neurochem Int. 2005;46:227–34.
40. Trotti D, Rolfs A, Danbolt NC, Brown Jr RH, Hediger MA. SOD1 mutants linked to amyotrophic lateral sclerosis selectively inactivate a glial glutamate transporter. Nat Neurosci. 1999;2:427–33.
41. Carunchio I, Curcio L, Pieri M, Pica F, Caioli S, Viscomi MT, Molinari M, Canu N, Bernardi G, Zona C. Increased levels of p70S6 phosphorylation in the G93A mouse model of amyotrophic lateral sclerosis and in valine-exposed cortical neurons in culture. Exp Neurol. 2010;226:218–30.
42. Scaini G, Mello-Santos LM, Furlanetto CB, Jeremias IC, Mina F, Schuck PF, Ferreira GC, Kist LW, Pereira TC, Bogo MR, Streck EL. Acute and chronic administration of the branched-chain amino acids decrease nerve growth factor in rat hippocampus. Mol Neurobiol. 2013;48(3):581–9.

Chapter 20
Leucine and Ethanol Oxidation

Hitoshi Murakami and Michio Komai

Key Points

- Leucine accelerates ethanol clearance.
- Leucine enhances alcohol-metabolizing enzyme activities, such as ADH and low Km ALDH.
- In contrast, valine has no effect on ethanol oxidation.
- The effect of leucine on ethanol oxidation is not exerted directly in the liver.
- Leucine treatment before alcohol intake may be important for enhancing ethanol oxidation.

Keywords Leucine • BCAAs • Ethanol clearance • ADH • ALDH

Abbreviations

MEOS	Microsomal ethanol oxidation system
ADH	Alcohol dehydrogenase
ALDH	Aldehyde dehydrogenase
NAD	Nicotinamide adenine dinucleotide
PPARα	Peroxisome proliferator-activated receptor-α
AMPK	AMP-activated protein kinase
IGF-1	Insulin like growth factor-1
4E-BP1	Eukaryotic initiation factor-binding protein 1
S6K	Ribosomal protein S6 kinase
mTOR	Mammalian target of rapamycin
BCAAs	Branched chain amino acids
KIC	α-Keto-isocaproic acid
KMV	β-Methyl isovalerianic acid
AST	Aspartate aminotransferase

H. Murakami, Ph.D. (✉)
Institute for Innovation, Ajinomoto, Co. Inc., Kawasaki-ku, Kawasaki, Kanagawa, Japan
e-mail: hitoshi_murakami@ajinomoto.com

M. Komai, Ph.D.
Tohoku University, Aoba-ku, Sendai, Miyagi, Japan
e-mail: mkomai@biochem.tohoku.ac.jp

R. Rajendram et al. (eds.), *Branched Chain Amino Acids in Clinical Nutrition: Volume 1*, Nutrition and Health, DOI 10.1007/978-1-4939-1923-9_20,

TBARS	Thiobarbituric acid reactive substances
eIF	Eukaryotic initiation factor
IL-6	Interleukin-6
TNF-α	Tumor necrosis factor-α

Introduction

Ethanol Metabolism

Ethanol is metabolized to acetaldehyde by catalase in the peroxisome, by the microsomal ethanol oxidation system (MEOS) containing cytochrome P-450, which is located in the endoplasmic reticulum, and by the alcohol dehydrogenase (ADH) pathway of cytosol or the soluble fraction of the cell [1, 2]. Among these metabolic processes, ADH is a key enzyme for metabolizing ethanol, and the role of non-ADH enzymes in systemic alcohol metabolism is unclear [3]. Ethanol absorption is primarily controlled by gastric emptying because the primary region of ethanol absorption is the small intestine, and ethanol is absorbed via simple diffusion [4]; for example, slow gastric emptying leads to increases in the first-pass metabolism of ingested ethanol. Ingested ethanol is primarily metabolized in the stomach and liver. In the stomach, ADH, including the isoenzymes ADH I, III, and IV, oxidizes the ingested ethanol, but most of the ethanol (approximately 90 %) is absorbed in the small intestine. Absorptive ethanol is metabolized by ADH, including the isoenzymes ADH I, II, and III, and primarily oxidizes in the liver. In comparison, the MEOS shares many properties with other microsomal drug-metabolizing enzymes, such as P-450; therefore, MEOS pathway activity increases under high ethanol concentrations and/or after chronic ethanol consumption. Ethanol and acetaldehyde are toxic and damage cells. Catalase can oxidize ethanol in vitro with H_2O_2, but it may not be a major system for ethanol oxidation [2]. ADH and/or catalase, MEOS enzyme pathways metabolize ethanol to produce acetaldehyde. Acetaldehyde is a harmful compound that causes hangovers and flush. It is detoxified by aldehyde dehydrogenase (ALDH), which is primarily composed of mitochondrial low Km ALDH, including the isoenzymes ALDH I, II, III, and IV, in the liver. The acetaldehyde metabolized by ALDH is converted to acetic acid and water, resulting in detoxification. The enzymatic activities of both of ADH and ALDH are regulated by the availability of the coenzyme nicotinamide adenine dinucleotide (NAD). NAD is reduced to NADH when ethanol and acetaldehyde are oxidized by ADH and ALDH, and NADH oxidation is the limiting factor of ethanol metabolism [2]. In addition to metabolizing ethanol, NAD and NADH regulate several metabolisms as coenzymes. However, ethanol metabolism is preferable to other metabolic pathways in terms of the resulting toxicity of ethanol and its metabolite, acetaldehyde; therefore, ethanol and acetaldehyde intake affect some metabolic pathways, such as lipid, hormone, protein, and amino acid metabolism, and induce disorders such as liver failure and muscle wasting.

Metabolic Changes Resulting from Ethanol Consumption

Ethanol metabolism causes changes in redox homeostasis and induces metabolic disorders, such as hepatitis and liver cirrhosis. Lieber reviewed the adverse effects of ethanol intake [2, 5]. The change in redox state is associated with hyperlactacidemia, which contributes to acidosis. Hyperlactacidemia reduces the kidney's capacity to excrete uric acid, thus inducing hyperuricemia. Furthermore, ethanol induces ketosis and purine breakdown, which are associated with hyperuricemia. In addition, enhanced purine breakdown increases the oxygen species resulting from xanthine oxidase. The CYP2E1 form of cytochrome P450 enzymes generates several reactive oxygen species [5]. Chronic ethanol intake induces cytokine formation in the liver and promotes oxidative stress via both the increased formation of ROS and the depletion of oxidative defenses in the cell. Liver cells from ethanol-treated animals are more susceptible to the cytotoxic effects of TNF-α and other cytokines than cells from control animals [6, 7]. These ethanol-induced changes in redox state, oxidative stress, and cytokines are associated with the metabolic disturbances described below.

Ethanol and Lipid Metabolism

The change in the redox state (i.e., the increase in the NADH/NAD ratio) generates the mobilization of peripheral triglyceride from the adipose tissue. The NADH accumulation inhibits β-oxidation by inhibiting the mitochondrial fatty acid-oxidizing dehydrogenase. Additionally, the ethanol intake decreases the fatty acid oxidation by decreasing the expression of peroxisome proliferator-activated receptor-α (PPARα)-regulated genes. PPARα knockout mice develop worse hepatomegaly and hepatocyte damage with ethanol consumption as a result of increased exposure to oxidative stress via the decrease in glutathione peroxidase, superoxide dismutase, and catalase [8]. In addition, ethanol consumption decreases the activity of AMP-activated protein kinase (AMPK), which is a key regulator of lipid metabolism, resulting in the decrease of fatty acid oxidation and increase of fatty acid synthesis through the activation of sterol regulatory element-binding protein 1 and acetyl-CoA carboxylase. These lipid metabolism impairments caused by ethanol consumption induce alcoholic fatty liver [8].

Ethanol and Protein Metabolism

Ingested ethanol is a potent inhibitor of hepatic regeneration [9]. Ethanol suppresses growth hormone-induced signal transduction, resulting in a decrease in insulin like growth factor-1 (IGF-1) gene expression [10]. In humans, acute ethanol ingestion impairs the postprandial protein synthesis of hepatic proteins, such as albumin and fibrinogen [11]. Albumin synthesis is impaired, even when alcohol intake is moderate [12]; this impairment is associated with limited ATP availability, which results from the change in redox state [13]. In addition, ethanol blunts protein synthesis in skeletal muscle by decreasing the phosphorylation of eukaryotic initiation factor-binding protein 1 (4E-BP1) and ribosomal protein S6 kinase (S6K) [14]. Generally, leucine stimulates protein synthesis by activating 4E-BP1 and S6K via the mTOR pathway; however, ethanol impairs leucine's anabolic effect on protein metabolism, an impairment that is not attributable to differences in the plasma concentration of insulin, IGF-1, and leucine [15].

Ethanol and Amino acid Metabolism

Ethanol affects amino acid metabolism. Amino acids are well known as protein component and regulators of the signaling pathway of protein metabolism; therefore, ethanol-related changes in amino acid metabolism are also associated with impaired liver regeneration. Plasma amino acid concentration is affected by ethanol consumption. In rats, acute ethanol intake decreases alanine, arginine, aspartic acid, β-alanine, glycine, phenylalanine, and serine, but the plasma levels of other amino acids are not significantly changed [16]. Ethanol ingestion alters the metabolism of sulfur amino acids, such as cysteine and taurine, in the liver, resulting in a decrease in glutathione level via a decrease in glutathione synthesis [17]. Chronic ethanol consumption increases the concentration of branched chain amino acids (BCAAs) in plasma, the liver, skeletal muscle, and the jejunal mucosa in rats through a decrease in protein synthesis and increase in protein breakdown [18]. In addition, chronic ethanol intake reduces the flux of a substrate of leucine (α-ketoisocaproate; KIC) via a decrease in the basal and total activity of branched chain keto acid (BCKA) dehydrogenase and non-increase in the conversion of the enzyme to its inactive form [19]. Furthermore, it is well known that plasma BCAA concentrations are reduced in cases of liver cirrhosis [20]. It has been indicated that the predominant mechanism for the decreases in plasma leucine levels in cirrhosis is an increase in the oxidized leucine fraction associated with decreased leucine turnover [21].

Ethanol also has a direct and/or selective inhibitory effect on amino acid transport into the cells. Ethanol inhibits sodium-dependent alanine and cysteine uptake, but no effect has been observed on sodium-independent alanine and cysteine transport in basolateral rat liver plasma membranes [22]. In comparison, ethanol exposure stimulates leucine uptake in rat fetal hepatocytes via sodium independent-system L [23].

Therefore, ethanol exposure changes several metabolic pathways, resulting in liver failure via fatty liver and liver cirrhosis. Thus, the acceleration of ethanol oxidation is very important to prevent liver disease and the metabolic anabolism impairment induced by alcohol intake.

Amino Acid Supplementation Improves the Metabolic Changes Caused by Ethanol Consumption

The development of ethanol-induced fatty liver, alcoholic hepatitis, and cirrhosis has been partially attributed to nutritional deficiencies in compounds such as amino acids, lipids, and vitamins [24]. Some studies have indicated that amino acids reduce the liver damage and oxidative stress caused by chronic ethanol consumption. Beauge et al. showed that only the amino acids that are precursors of pyruvate, alanine, aspartate or glutamate are able to activate ethanol oxidation in rat hepatocytes [25]. These compounds supply malate-aspartate shuttle components and stimulate the oxidation of NADH generation via ethanol consumption, resulting in enhanced ethanol clearance. Yang et al. demonstrated that a diet supplemented with amino acids (alanine, glutamine, glutamate, protein, and ten essential amino acids) reduced plasma aspartate aminotransferase (AST) levels in rats that chronically consume ethanol [26]. Glutamine pretreatment suppresses plasma inflammation in chronically ethanol-fed rats [27]. Treatment with histidine or carnosine after liver injury from chronic ethanol exposure increases the mRNA expression levels of catalase and glutathione peroxidase and downregulates the mRNA expression levels of IL-6 and TNF-α in mice livers [28]. Aspartate attenuates ethanol-induced oxidative stress, such as thiobarbituric acid reactive substances (TBARS), and reduces glutathione transferase activity in rats [29]. Torii et al. indicated that a preference for both alanine and glutamine was observed in preference tests when alcoholic rats developed hepatic disorder [30], and supplementation with these amino acids prevented the ethanol-induced inhibition of liver regeneration [31]. In addition, treatment with d/l-cysteine and alanine before ethanol loading accelerates ethanol and acetaldehyde clearance [32].

It is well known that BCAA supplementation improves liver failures, such as liver cirrhosis. A BCAA-enriched diet partly prevents the morphological ultrastructural changes in the liver caused by chronic ethanol consumption and improves positive body weight gain in rats [33]. Leucine administration improves myocardial protein synthesis, which is impaired by ethanol intoxication, by enhancing anabolic signaling pathways, such as eukaryotic initiation factor 4G (eIF4G) [34]. In addition, we recently reported the effect of each BCAA on ethanol clearance after acute ethanol loading in rats [35].

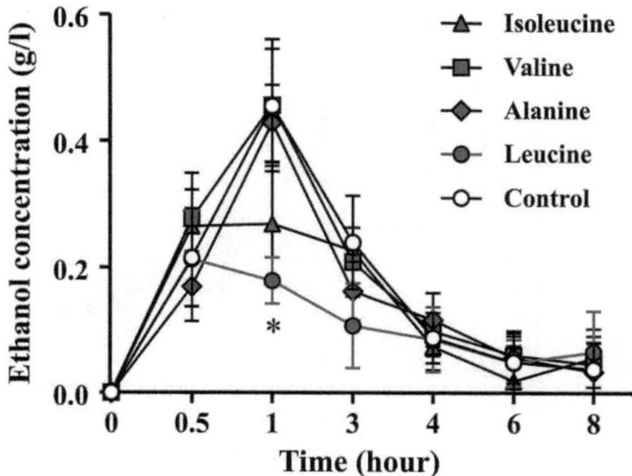

Fig. 20.1 The effects of orally administered amino acids on blood ethanol clearance in rats. Leucine significantly decreased the blood ethanol concentration 1 h after ethanol administration, but the other amino acids did not significantly decrease the blood ethanol values. The values are presented as the mean ± SEM of four rats in each group. Comparisons with the control group at each time point were performed with Bonferroni's post-test following a two-way ANOVA for multiple comparisons (*$P < 0.05$). With permission from Amino Acids, 2012, 43, 2545–51 [35]

Leucine Accelerates Ethanol Oxidation

The effect of pretreatment with each BCAA via oral administration or tail-vein infusion on ethanol clearance was investigated after acute ethanol loading in rats. First-pass ethanol metabolism in the stomach and liver is important for alcohol metabolism [4]; therefore, to understand the effects of amino acids in the digestive tract on ethanol absorption and metabolism, each amino acid was administered either orally or via the tail vein. Leucine significantly decreased the blood ethanol levels after oral ethanol administration, and the individuals in the isoleucine group tended to have reduced ethanol levels (Fig. 20.1). In contrast, alanine treatment, which is known as amino acid-enhancing ethanol oxidation [31], slightly decreased the ethanol concentrations, but the reduction was not significant. Valine did not affect blood ethanol oxidation. The amount of orally administered alanine in this study was low compared with the levels reported in a previous study (0.084 g/kg body weigh in the present study [35] and 1 g/kg BW in the previous study [31]). Therefore, the effect of orally administered alanine on ethanol clearance in rats may be weak in this study [35]. Furthermore, leucine also significantly decreased the blood ethanol concentration after acute ethanol intake in rats who received infusions of each amino acid into the tail vein (Fig. 20.2). The infused alanine significantly decreased the blood ethanol levels after ethanol intake. However, isoleucine and valine did not decrease the blood ethanol concentrations. Leucine is the only amino acid that has been found to enhance ethanol clearance via both administration routes.

Leucine induces an increase in ethanol clearance by enhancing ethanol enzyme activity. ADH activity tended to be increased by pretreatment with leucine and alanine before ethanol intake, and these amino acids also significantly increased ADH activity after ethanol intake. In contrast, ADH activity was lower in the valine group before and after ethanol administration (Table 20.1). The low Km ALDH activity was significantly increased in the leucine group (Table 20.2). Liver ADH is the primary enzyme responsible for ethanol oxidation, and low Km ALDH encodes the mitochondrial enzyme that is primarily responsible for oxidizing the ethanol-derived acetaldehyde [36–38]. Low Km ALDH impairment negatively affects the capacity for acetaldehyde metabolism, which is one of the primary causes of the alcohol-flush reactions that have been observed in Asian individuals [38]. Therefore, these results indicate that pretreatment with leucine accelerates ethanol clearance and

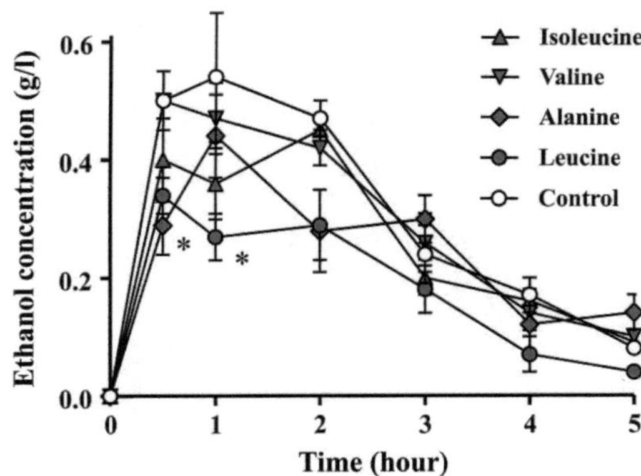

Fig. 20.2 The effects of amino acids infused into the tail vein on ethanol oxidation in rats. Leucine significantly decreased the blood ethanol concentration 1 h after ethanol administration, and the ethanol concentration in the leucine group was lower for 5 h after the ethanol administration. In addition, alanine significantly decreased the blood ethanol level 0.5 h after acute ethanol administration. However, isoleucine and valine did not decrease the blood ethanol concentrations. The values are presented as the mean ± SEM of six rats in each group. Comparisons with the control group at each time point were performed using Bonferroni's post-test following a two-way ANOVA for multiple comparisons (*$P<0.05$). With permission from Amino Acids, 2012, 43, 2545–51 [35]

Table 20.1 Changes in alcohol dehydrogenase activity after each amino acid and ethanol administration

	(U/mg protein)	Control	Leucine	Isoleucine	Valine	Alanine
	0 min	0.058±0.015	0.088±0.018	0.030±0.012	0.034±0.018	0.108±0.026
ADH	15 min	0.041±0.015	0.095±0.006	0.086±0.013	0.038±0.026	0.120±0.027*
	30 min	0.042±0.014	0.090±0.016*	0.042±0.008	0.022±0.010	0.070±0.016
	60 min	0.090±0.016	0.125±0.014	0.045±0.017	0.035±0.010	0.075±0.030

Alanine significantly increased alcohol dehydrogenase (ADH) activity 15 min after ethanol administration, and leucine increased ADH activity 30 min after ethanol intake. The values are presented as the mean ± SEM of five rats in each group. The results were compared with those of the control group, which was given distilled water, using Dunnet's test following an ANOVA for multiple comparisons (*$P<0.05$)
With permission from Amino Acids, 2012, 43, 2545–51 [35]

Table 20.2 Changes in acetaldehyde dehydrogenase activity after each amino acid and ethanol administration

	(U/mg protein)	Control	Leucine	Isoleucine	Valine	Alanine
	0 min	0.446±0.499	1.884±1.227	2.649±1.091	0.112±0.125	4.067±2.312
low Km	15 min	0.511±0.626	3.719±0.657*	1.255±0.887	N.D.	1.859±1.573*
ALDH	30 min	0.945±0.672	5.249±3.903	0.223±0.249	2.154±1.544	2.283±2.184
	60 min	0.772±0.485	4.057±2.584	1.435±1.284	N.D.	0.801±0.576
low Km	0 min	0.544±0.216	0.634±0.251	0.606±0.227	0.589±0.184	0.690±0.263
ALDH	15 min	1.899±0.414	2.820±0.651	2.215±0.410	2.186±0.472	2.215±0.338
	30 min	0.732±0.372	0.766±0.360	0.828±0.413	0.603±0.204	0.831±0.391
	60 min	1.000±0.380	0.926±0.264	0.978±0.357	0.923±0.404	0.942±0.351

Low Km aldehyde dehydrogenase (ALDH) activity was significantly increased in the leucine group 15 min after ethanol administration and continued to increase during 60 min. High Km ALDH activity did not differ between the groups. The values are presented as the mean ± SEM of five rats in each group. Comparisons with the control group, which was given distilled water, were determined using Dunnet's test following an ANOVA for multiple comparisons (*$P<0.05$)
With permission from Amino Acids, 2012, 43, 2545–51 [35]

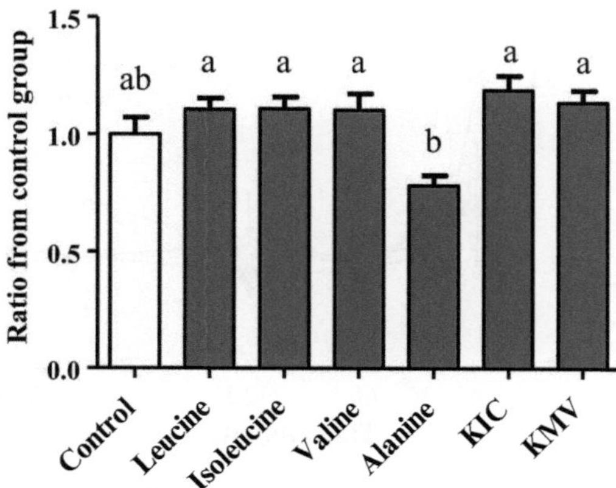

Fig. 20.3 The effects of amino acids on ethanol oxidation in isolated rat hepatocytes. Alanine significantly decreased the ethanol concentration in the medium 90 min after ethanol injection. However, the BCAAs and their metabolites did not decrease the ethanol concentrations in the isolated rat hepatocytes. The values are presented as the mean ± SEM of six set in each group. Comparisons of each group were performed with a Bonferroni's post-test following a two-way ANOVA for multiple comparisons (*$P<0.05$). With permission from Amino Acids, 2012, 43, 2545–51 [35]

enhances acetaldehyde oxidation by activating the alcoholic enzymes in the liver. In contrast, low Km ALDH was lower in individuals pretreated with valine (Table 20.2).

BCAAs are primarily metabolized in skeletal muscles and the brain by branched chain amino acid aminotransferase. This process produces branched chain α-keto acids, such as α-keto isocaproic acid from leucine and β-methyl isovalerianic acid (KMV) from isoleucine. Branched chain α-keto acids are metabolized in the liver. Some studies have demonstrated that leucine directly affects hepatocytes. Leucine has been shown to inhibit proteolysis in isolated-rat hepatocytes [39] and stimulate hepatic growth factor production in hepatic stellate cells [40]. Therefore, leucine or its metabolite, α-keto isocaproic acid, may directly affect ethanol clearance in the liver. To understand the effect of leucine or its metabolite on ethanol oxidation in the hepatocytes, this study investigated changes in the media ethanol levels of isolated rat hepatocytes incubated with ethanol after preincubation with each amino acid or BCAA metabolites. Interestingly, leucine, other branched amino acids and α-keto acids did not accelerate ethanol oxidation (Fig. 20.3). In contrast, alanine significantly increased ethanol oxidation in the hepatocytes (Fig. 20.3). Each BCAA also accelerated ethanol clearance in the liver perfusion performed using a modification of the method described by Mortimore [41] (data not shown). NADH reoxidation is believed to be the primary rate-determining step in hepatic ethanol oxidation, and the malate-aspartate shuttle is an important rate determinant in hepatic ethanol oxidation [42]. Precursors of malate-aspartate shuttle components, such as alanine, increased ethanol metabolism in hepatocytes [25]. In addition, it has been suggested that alanine stimulates the glucose-alanine cycle, which improves the NADH redox state and results in increased ethanol elimination [43]. Therefore, the effect of leucine on ethanol clearance may not be attributable to changes in the redox state via leucine metabolism in the liver, and leucine and its metabolite will not directly affect ethanol oxidation in the liver.

Alcohol-metabolizing enzymes are activated by hormonal and nutritional factors. ADH has been shown to be increased by insulin in vivo and by IGF-1 and growth hormone via an increase in cyclic AMP in hepatocytes [44–46]. Insulin also enhanced the effect of IGF-1 on ADH activation [45]. In addition, insulin secretion has been shown to be stimulated by amino acids, such as leucine [47], but ethanol has been shown to inhibit insulin secretion [48]. As shown in Fig. 20.4, pretreatment with leucine significantly increased insulin secretion just prior to ethanol administration, but alanine did not confer this effect. This suggests that the stimulation of hormone secretion, such as insulin secretion, by leucine may contribute to enhanced alcohol-metabolizing enzyme activities. The mechanism of leucine's effect on

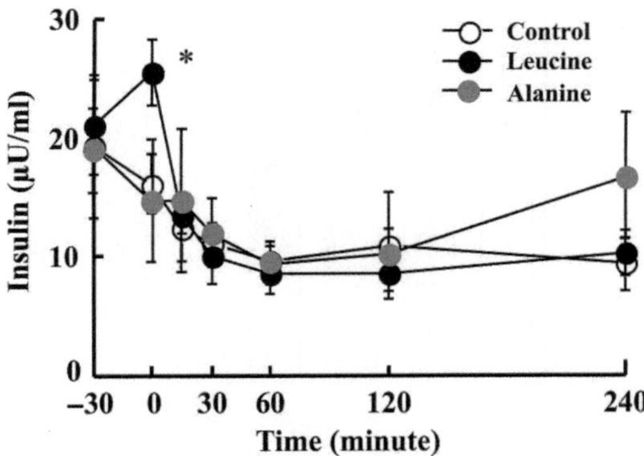

Fig. 20.4 The effects of leucine or alanine on insulin secretion after the administration of each amino acid and ethanol. Leucine significantly increased the insulin concentrations 30 min after administration, but alanine did not confer this effect. The values are presented as the mean ± SEM of five rats in each group. Comparisons with the control group at each time point were performed with Bonferroni's post-test following a two-way ANOVA for multiple comparisons (*$P<0.05$). With permission from Amino Acids, 2012, 43, 2545–51 [35]

ethanol clearance may be associated with the change in liquid factors (e.g., insulin). However, leucine's mechanism of action on the acceleration of ethanol clearance is unclear. Further studies are required to investigate this mechanism.

Estimation of Effective Amount of Leucine for Humans

The amount of leucine used in the study performed by Murakami et al. was 0.125 g/kg body weight in rats [35]. In addition, two-thirds of this amount (0.083 g/kg body weight) of leucine had the same effect on ethanol clearance in rats (data not shown). The increase of BCAAs in plasma depends on the BCAA intake [49]. In an animal study, the plasma leucine concentration after BCAA administration (the amounts of leucine, isoleucine, and valine were 0.5, 0.25, and 0.25 g/kg body weight, respectively) was 12.6-fold higher than that of the control group, which was administered water [50]. Thus, plasma leucine levels will increase approximately 2.1-fold when leucine is administered at 0.083 g/kg body weight in rats. Plasma leucine concentration increases 3.80-fold over the basal level with an intake of 5 g of BCAAs (the leucine content was 2.56 g) in young Japanese men (22–25 years old) [49]. Therefore, we estimate that 1.41 g of leucine is needed to accelerate ethanol oxidation in humans.

Conclusions

Leucine accelerates ethanol clearance after acute ethanol administration by enhancing alcohol-metabolizing enzyme activities, such as alcohol and aldehyde dehydrogenase. In contrast, valine has no effect on ethanol oxidation. Effect of leucine on ethanol oxidation is not exerted directly in the liver. The increase in alcohol dehydrogenase activity resulting from leucine treatment may be associated with the induction of insulin secretion. In addition, leucine treatment before alcohol intake may be important to enhance alcoholic enzyme activities and accelerate ethanol oxidation.

Chronic alcohol intake leads to liver failures, such as hepatic inflammation and fatty liver, and induces liver cirrhosis. Therefore, it is important to accelerate ethanol oxidation after alcohol intake to prevent these liver failures. Leucine may be a powerful solution for preventing liver failure resulting from acute and chronic alcohol intake.

References

1. Lieber CS. Metabolism of alcohol. Clin Liver Dis. 2005;9(1):1–35.
2. Lieber CS. Ethanol metabolism, cirrhosis and alcoholism. Clin Chim Acta. 1997;257(1):59–84.
3. Haseba T, Ohno Y. A new view of alcohol metabolism and alcoholism–role of the high-Km Class III alcohol dehydrogenase (ADH3). Int J Environ Res Public Health. 2010;7(3):1076–92.
4. Oneta CM, Simanowski UA, Martinez M, et al. First pass metabolism of ethanol is strikingly influenced by the speed of gastric emptying. Gut. 1998;43(5):612–9.
5. Lieber CS. ALCOHOL: its metabolism and interaction with nutrients. Annu Rev Nutr. 2000;20:395–430.
6. Hoek JB, Pastorino JG. Ethanol, oxidative stress, and cytokine-induced liver cell injury. Alcohol. 2002;27(1):63–8.
7. Koch OR, Pani G, Borrello S, et al. Oxidative stress and antioxidant defenses in ethanol-induced cell injury. Mol Asp Med. 2004;25(1–2):191–8.
8. Sozio M, Crabb DW. Alcohol and lipid metabolism. Am J Physiol. 2008;295(1):E10–6.
9. Wands JR, Carter EA, Bucher NL, Isselbacher KJ. Inhibition of hepatic regeneration in rats by acute and chronic ethanol intoxication. Gastroenterology. 1979;77(3):528–31.
10. Xu X, Ingram RL, Sonntag WE. Ethanol suppresses growth hormone-mediated cellular responses in liver slices. Alcohol Clin Exp Res. 1995;19(5):1246–51.
11. De Feo P, Volpi E, Lucidi P, et al. Ethanol impairs post-prandial hepatic protein metabolism. J Clin Invest. 1995;95(4):1472–9.
12. Volpi E, Lucidi P, Cruciani G, et al. Moderate and large doses of ethanol differentially affect hepatic protein metabolism in humans. J Nutr. 1998;128(2):198–203.
13. Deaciuc IV, D'Souza NB, Lang CH, Spitzer JJ. Effects of acute alcohol intoxication on gluconeogenesis and its hormonal responsiveness in isolated, perfused rat liver. Biochem Pharmacol. 1992;44(8):1617–24.
14. Vary TC, Deiter G, Goodman SA. Acute alcohol intoxication enhances myocardial eIF4G phosphorylation despite reducing mTOR signaling. Am J Physiol. 2005;288(1):H121–8.
15. Lang CH, Frost RA, Deshpande N, et al. Alcohol impairs leucine-mediated phosphorylation of 4E-BP1, S6K1, eIF4G, and mTOR in skeletal muscle. Am J Physiol. 2003;285(6):E1205–15.
16. Abdel-Nabi R, Milakofsky L, Hofford JM, Hare TA, Vogel WH. Effect of ethanol on amino acids and related compounds in rat plasma, heart, aorta, bronchus, and pancreas. Alcohol. 1996;13(2):171–4.
17. Jung YS, Kwak HE, Choi KH, Kim YC. Effect of acute ethanol administration on S-amino acid metabolism: increased utilization of cysteine for synthesis of taurine rather than glutathione. Adv Exp Med Biol. 2003;526:245–52.
18. Bernal CA, Vazquez JA, Adibi SA. Leucine metabolism during chronic ethanol consumption. Metab Clin Exp. 1993;42(9):1084–6.
19. Bernal CA, Vazquez JA, Adibi SA. Chronic ethanol intake reduces the flux through liver branched-chain keto-acid dehydrogenase. Metab Clin Exp. 1995;44(10):1243–6.
20. Morgan MY, Marshall AW, Milsom JP, Sherlock S. Plasma amino-acid patterns in liver disease. Gut. 1982;23(5):362–70.
21. Holecek M, Tilser I, Skopec F, Sprongl L. Leucine metabolism in rats with cirrhosis. J Hepatol. 1996;24(2):209–16.
22. Moseley RH, Murphy SM. Effects of ethanol on amino acid transport in basolateral liver plasma membrane vesicles. Am J Physiol. 1989;256(3 Pt 1):G458–65.
23. Henderson GI, Frosto TA, Heitman DW, Schenker S. Ethanol stimulates leucine uptake by rat fetal hepatocytes via trans-stimulation. Am J Physiol. 1989;256(2 Pt 1):G384–9.
24. Leevy CM, Moroianu SA. Nutritional aspects of alcoholic liver disease. Clin Liver Dis. 2005;9(1):67–81.
25. Beauge F, Mangeney M, Nordmann J, Nordmann R. Comparative study of the effect of amino acids on ethanol oxidation in isolated hepatocytes from starved and fed rats. Adv Exp Med Biol. 1980;132:393–402.
26. Yang SC, Ito M, Morimatsu F, Furukawa Y, Kimura S. Effects of amino acids on alcohol intake in stroke-prone spontaneously hypertensive rats. J Nutr Sci Vitaminol. 1993;39(1):55–61.
27. Peng HC, Chen YL, Chen JR, et al. Effects of glutamine administration on inflammatory responses in chronic ethanol-fed rats. J Nutr Biochem. 2011;22(3):282–8.
28. Liu WH, Liu TC, Yin MC. Beneficial effects of histidine and carnosine on ethanol-induced chronic liver injury. Food Chem Toxicol. 2008;46(5):1503–9.
29. Oh SI, Lee MS, Kim CI, Song KY, Park SC. Aspartate modulates the ethanol-induced oxidative stress and glutathione utilizing enzymes in rat testes. Exp Mol Med. 2002;34(1):47–52.
30. Torii K. A new pharmacological and physiological aspects of L-amino acids. Nihon Yakurigaku Zasshi. 1997;110 Suppl 1:28P–32.
31. Tanaka T, Imano M, Yamashita T, et al. Effect of combined alanine and glutamine administration on the inhibition of liver regeneration caused by long-term administration of alcohol. Alcohol Alcohol. 1994;29(1):125–32.

32. Tsukamoto S, Kanegae T, Nagoya T, et al. Effects of amino acids on acute alcohol intoxication in mice–concentrations of ethanol, acetaldehyde, acetate and acetone in blood and tissues. Arukoru Kenkyuto Yakubutsu Izon. 1990;25(5):429–40.
33. Farbiszewski R, Chyczewski L, Holownia A, Pawlowska D, Chwiecko M. Branched-chain amino acid enriched diet given simultaneously with ethanol partially prevents morphological and biochemical changes in the liver. Patologia Polska. 1990;41(4):183–6.
34. Vary T. Oral leucine enhances myocardial protein synthesis in rats acutely administered ethanol. J Nutr. 2009;139(8):1439–44.
35. Murakami H, Ito M, Furukawa Y, Komai M. Leucine accelerates blood ethanol oxidation by enhancing the activity of ethanol metabolic enzymes in the livers of SHRSP rats. Amino Acids. 2012;43(6):2545–51.
36. Deitrich RA, Petersen D, Vasiliou V. Removal of acetaldehyde from the body. Novartis Found Symp. 2007;285: 23–40. discussion 40-51, 198–199.
37. Yin SJ. Alcohol dehydrogenase: enzymology and metabolism. Alcohol Alcohol. 1994;2:113–9.
38. Ehrig T, Bosron WF, Li TK. Alcohol and aldehyde dehydrogenase. Alcohol Alcohol. 1990;25(2–3):105–16.
39. Venerando R, Miotto G, Kadowaki M, Siliprandi N, Mortimore GE. Multiphasic control of proteolysis by leucine and alanine in the isolated rat hepatocyte. Am J Physiol. 1994;266(2 Pt 1):C455–61.
40. Tomiya T, Nishikawa T, Inoue Y, et al. Leucine stimulates HGF production by hepatic stellate cells through mTOR pathway. Biochem Biophys Res Commun. 2007;358(1):176–80.
41. Mortimore GE, Khurana KK, Miotto G. Amino acid control of proteolysis in perfused livers of synchronously fed rats. Mechanism and specificity of alanine co-regulation. J Biol Chem. 1991;266(2):1021–8.
42. Sugano T, Handler JA, Yoshihara H, Kizaki Z, Thurman RG. Acute and chronic ethanol treatment in vivo increases malate-aspartate shuttle capacity in perfused rat liver. J Biol Chem. 1990;265(35):21549–53.
43. Cunningham CC, Preedy VR, Paice AG, et al. Ethanol and protein metabolism. Alcohol Clin Exp Res. 2001;25 (5 (Suppl ISBRA)):262S–8.
44. Lakshman MR, Chambers LL, Chirtel SJ, Ekarohita N. Roles of hormonal and nutritional factors in the regulation of rat liver alcohol dehydrogenase activity and ethanol elimination rate in vivo. Alcohol Clin Exp Res. 1988;12(3): 407–11.
45. Mezey E, Potter JJ, Mishra L, Sharma S, Janicot M. Effect of insulin-like growth factor I on rat alcohol dehydrogenase in primary hepatocyte culture. Arch Biochem Biophys. 1990;280(2):390–6.
46. Mezey E, Potter JJ, Rhodes DL. Effect of growth hormone on alcohol dehydrogenase activity in hepatocyte culture. Hepatology. 1986;6(6):1386–90.
47. Milner RD. Stimulation of insulin secretion in vitro by essential aminoacids. Lancet. 1969;1(7605):1075–6.
48. Singh SP, Patel DG, Snyder AK. Ethanol inhibition of insulin secretion by perifused rat islets. Acta Endocrinol (Copenh). 1980;93(1):61–6.
49. Zhang Y, Kobayashi H, Mawatari K, et al. Effects of branched-chain amino acid supplementation on plasma concentrations of free amino acids, insulin, and energy substrates in young men. J Nutr Sci Vitaminol. 2011;57(1): 114–7.
50. Murakami H, Shimbo K, Inoue Y, Takino Y, Kobayashi H. Importance of amino acid composition to improve skin collagen protein synthesis rates in UV-irradiated mice. Amino Acids. 2012;42(6):2481–9.

Chapter 21
Isoleucine, Leucine and Their Role in Experimental Models of Bladder Carcinogenesis

Min Wei, Xiao-Li Xie, Shotaro Yamano, Anna Kakehashi, and Hideki Wanibuchi

Key Points

1. There is no epidemiological data relevant to an association of dietary branched chain amino acids (BCAA) with the risk of bladder cancer.
2. A few experimental studies evaluated the effects of L-isoleucine and L-leucine on bladder carcinogenesis using a two-stage (initiation–promotion) carcinogenesis protocol with male F344 rats.
3. The effects of dietary BCAA on rat bladder carcinogenesis is dependent on the type of basal diet used: supplementation of the CE2 and AIN-93G diets, but not the MF diet, with L-leucine or L-isoleucine has tumor-promoting activity on bladder carcinogenesis in rats.
4. L-leucine or L-isoleucine themselves do not induce toxicity and therefore regenerative proliferation is not likely to be involved in their tumor promoting activity.
5. L-leucine and L-isoleucine increased expression of amino acid transporters in bladder carcinogen-initiated urothelial cells; but in hyperplasias and bladder tumors, increased expression of these transporters became independent of L-leucine and L-isoleucine.
6. Long-term use of BCAAs as a dietary supplement should be avoided until more is known about their effects on carcinogenesis in humans. This is particularly applicable to patients with bladder cancer.

Keywords L-leucine • L-Isoleucine • Bladder carcinogenesis • Tumor-promoting activity • Male F344 rats

Abbreviations

BBN	*N*-butyl-*N*-(4-hydroxybutyl)nitrosamine
BCAA	Branched chain amino acids
PN hyperplasia	Papillary and nodular hyperplasia
SEM	Scanning electron microscopy.

M. Wei, M.D., Ph.D. (✉) • X.-L. Xie, M.D., Ph.D. • S. Yamano, M.S.
A. Kakehashi, Ph.D. • H. Wanibuchi, M.D., Ph.D.
Department of Pathology, Osaka City University Graduate School of Medicine, Osaka, Japan
e-mail: mwei@med.osaka-cu.ac.jp; xiexiaoli1999@126.com; s.yamano@med.osaka-cu.ac.jp; anna@med.osaka-cu.ac.jp; wani@med.osaka-cu.ac.jp

R. Rajendram et al. (eds.), *Branched Chain Amino Acids in Clinical Nutrition: Volume 1*, Nutrition and Health, DOI 10.1007/978-1-4939-1923-9_21,

Introduction

Branched chain amino acids (BCAA) may facilitate tumor growth; consequently, it is reasonable to assume that the dietary intake of BCAA may play a role in bladder carcinogenesis. To date, however, there is no epidemiological data relevant to a relationship of dietary BCAA with the risk of bladder cancer. On the other hand, there are a few experimental studies reported in the literature. This chapter focuses on the studies evaluating the effects of L-isoleucine and L-leucine on bladder carcinogenesis in experimental animals. These three studies employed a two-stage (initiation–promotion) carcinogenesis protocol with male F344 rats. The results show that the effects of dietary BCAA on rat bladder carcinogenesis is dependent on the type of basal diet used.

Studies in Experiment Animals

Effects of Supplementing the CE-2 Diet with L-leucine or L-isoleucine on Rat Bladder Carcinogenesis

In the study using CE-2 (CLEA, Tokyo, Japan) as the basal diet, the effects of diet supplementation with L-leucine and L-isoleucine on *N*-butyl-*N*-(4-hydroxybutyl)nitrosamine (BBN)-initiated rat bladder carcinogenesis were evaluated in one 40-week and one 60-week experiment [1].

In the 40-week experiment, male F344 rats were divided into BBN-initiation and non-initiation groups. The BBN-initiated groups were administered 0.05 % BBN in the drinking water for 4 weeks to initiate bladder carcinogenesis, and then fed basal diet alone, basal diet supplemented with 2 % L-leucine, or basal diet supplemented with 2 % L-isoleucine for 36 weeks (Table 21.1). The non-initiation groups were fed basal diet alone, basal diet supplemented with 2 % L-isoleucine, or basal diet supplemented with 2 % L-leucine for 36 weeks, from the beginning of week 5 through week 40. In the BBN-initiation groups, bladder carcinoma was observed in the groups administered L-leucine and L-isoleucine, but not in the BBN alone group (Table 21.1). The incidence and number of carcinomas were significantly increased in the group administered L-isoleucine compared to the BBN alone group. In the non-initiation groups, no bladder lesions were observed in the basal diet group or the 2 % L-isoleucine or 2 % L-leucine groups.

In the 60-week experiment, rats were divided into BBN-initiation and non-initiation groups, as in the 40-week experiment. The BBN-initiation groups were fed basal diet alone, basal diet supplemented with L-isoleucine at doses of 2 % or 4 % or basal diet supplemented with L-leucine at doses of 2 % or 4 % for 56 weeks after the 4-week BBN-initiation treatment (Table 21.1). The non-initiation groups were fed basal diet alone or basal diet supplemented with 4 % L-leucine from the beginning of week 5 through week 60. At the end of the week 60, the incidence and number of bladder urothelial carcinomas was significantly increased in a dose-dependent manner in groups administered BBN followed by L-isoleucine or L-leucine compared to the BBN-alone group. The incidence and number of papilloma were also increased by L-isoleucine and L-leucine treatment, although the increase was not reported to be significant. No abnormality of the bladder epithelium was noted in the non-initiation groups.

These results demonstrate that L-leucine and L-isoleucine have tumor promotion effects on bladder carcinogenesis in rats when supplemented to the CE-2 basal diet.

Table 21.1 Effects of L-leucine and L-isoleucine on BBN-initiated bladder carcinogenesis in male F344 rats

	Treatment duration with BCAA (weeks)	No. of rats	Incidence (%)			Multiplicity[a]		
			Papilloma	Carcinoma	Total (papilloma+carcinoma)	Papilloma	Carcinoma	Total (papilloma+carcinoma)
CE-2 basal diet (Nishino et al. 1986. [1])								
BBN initiation (4 weeks) →								
CE-2 alone	36	30	5 (17)	0	NA	0.16±0.06	0	NA
CE-2 + 2% L-leucine	36	30	7 (23)	4 (13%)	NA	0.29±0.1	0.11±0.05	NA
CE-2 + 2% L-isoleucine	36	27	3 (11)	5 (19)[b]	NA	0.14±0.07	0.17±0.07[b]	NA
Non-initiation (4 weeks) →								
CE-2 + 2% L-leucine	36	17	0	0		0	0	
CE-2 + 2% L-isoleucine	36	18	0	0		0	0	
BBN initiation (4 weeks) →								
CE-2 alone	56	31	0	7 (23%)	NA	0	0.21±0.07	NA
CE-2 + 2% L-leucine	56	31	15 (48)	16 (52)[b]	NA	0.68±0.15	0.86±1.18[b]	NA
CE-2 + 4% L-leucine	56	31	16 (52)	23 (74)[b]	NA	0.58±0.11	1.18±0.16[b]	NA
CE-2 + 2% L-isoleucine	56	31	18 (58)	14 (45)	NA	0.83±0.14	0.65±0.14[b]	NA
CE-2 + 4% L-isoleucine	56	31	19 (61)	24 (77)[b]	NA	0.77±0.13	1.06±0.15[b]	NA
Non-initiation (4 weeks) →								
CE-2 + 4% L-leucine	56	21	0	0		0	0	
AIN-93G basal diet (Xie et al. 2012. [4])								
BBN initiation (4 weeks) →								
AIN-93G alone	25	30	10 (33)	7 (23)	16[c] (53)	0.5±0.9	0.33±0.66	0.83±1.09
AIN-93G + 2% L-leucine	25	30	16 (53)	15 (50)[b]	22[d] (73)	0.73±0.83	0.5±0.51	1.23±1.07
AIN-93G + 2% L-isoleucine	25	30	17 (53)[b]	11 (37)	24[e] (80)	0.7±0.7	0.57±0.77	1.27±0.98
MF basal diet (Xie et al. 2012. [4])								
BBN initiation (4 weeks) →								
MF alone	25	10	4 (40)	0	4 (40)	0.5±0.71	0	0.5±0.71
MF + 2% L-leucine	25	10	3 (30)	2 (20)	5 (50)	0.5±0.97	0.2±0.42	0.7±0.95
MF + 2% L-isoleucine	25	10	3 (30)	0	3 (30)	0.3±0.48	0	0.3±0.48

(continued)

Table 21.1 (continued)

	Treatment duration with BCAA (weeks)	No. of rats	Incidence (%)			Multiplicity[a]		
			Papilloma	Carcinoma	Total (papilloma+carcinoma)	Papilloma	Carcinoma	Total (papilloma+carcinoma)
BBN initiation (4 weeks) →								
MF alone	32	20	8 (40)	4 (20)	11[f] (55)	0.4±0.5	0.2±0.41	0.6±0.6
MF + 2% L-leucine	32	20	8 (40)	8 (40)	13[g] (65)	0.55±0.76	0.5±0.69	1.05±1.1
MF + 2% L-isoleucine	32	20	10 (50)	4 (20)	12[h] (60)	0.7±0.8	0.2±0.41	0.9±0.91

NA not available

[a]Numbers of bladder lesions are presented as mean ± standard error of the mean per 10 cm of basement membrane in the CE-2 diet experiment, and mean ± standard deviation of the mean per rat in the AIN-93G and MF diet experiments

[b]Significantly different from the respective basal diet alone control group

[c]One rat had both papillomas and carcinomas

[d]Nine rats had both papillomas and carcinomas

[e]Four rats had both papillomas and carcinomas

[f]One rat had both papillomas and carcinomas

[g]Three rats had both papillomas and carcinomas

[h]Two rats had both papillomas and carcinomas

Effects of Supplementing AIN-93G and MF Diets with L-leucine or L-isoleucine on Rat Bladder Carcinogenesis

In the study using AIN-93G and MF (Oriental Yeast Co., Tokyo, Japan) as the basal diets, the effects of dietary supplementation with L-leucine and L-isoleucine on BBN-initiated bladder carcinogenesis and the effects of these amino acids on the urine were evaluated [2].

In the experiment evaluating the promotion effect of BCAA supplementation on BBN-initiated carcinogenesis, male F344 rats were divided into AIN-93G diet and MF diet groups. All rats were administered 0.05 % BBN in the drinking water for the first 4 weeks and then were fed basal diet alone, or basal diet supplemented with 2 % L-leucine or 2 % L-isoleucine (Table 21.1). All rats in the AIN-93G diet groups and 10 rats of each MF diet group were killed at week 29 and the remaining rats in the MF diet groups were killed at week 36. The incidence of tumors with volumes greater than 15 mm^3 at week 29 in the AIN-93G diet group supplemented with L-leucine and L-isoleucine was increased compared to the control, and the increase in the L-leucine supplemented group was significant (see Fig. 21.1). The incidence of all carcinomas at week 29 in the AIN-93G diet group supplemented with L-leucine was also significantly increased compared to the control (Table 21.1). In addition, L-isoleucine supplementation significantly increased the incidence of papillomas and total tumors (papilloma and carcinoma) in the AIN-93G diet group. In contrast, neither L-leucine nor L-isoleucine supplementation had an affect on the incidence of papillomas, carcinomas, or total tumors in the MF diet groups at either week 29 or week 36. These results show that the basal diet had a marked effect on bladder cancer promotion by L-leucine and L-isoleucine.

Urine is of fundamental importance in bladder cancer research. High urinary pH caused by certain chemicals can contribute to their tumor promotion activity [3, 4], and urinary microcrystals and other solid precipitates dispersed in the urine are also relevant to rodent bladder carcinogenesis [4]. In the experiment determining the effects of L-isoleucine and L-leucine on the urine, rats were fed basal diet (MF diet or AIN-93G) or basal diet supplemented with 2 % L-leucine or 2 % L-isoleucine for 2 weeks. Rats fed the MF diet had urinary pH values as high as 7.5 while rats fed AIN-93G diet had urinary pH values less than or equal to 6.5. No significant differences in pH values were detected between the amino acid-supplemented groups and the basal diet control groups. These results indicate that in the AIN-93G diet groups, urinary pH does not mediate tumor promotion by L-leucine or L-isoleucine. Analysis of urine sediment by scanning electron microscopy (SEM) showed that in the MF diet groups several kinds of urine microcrystals and aggregates were present, however, no significant differences were observed between the amino acid supplemented groups and the basal diet control groups. In the AIN-93G diet groups, there were no obvious urine sediments. These results demonstrate that in the AIN-93G diet groups, urinary sediments do not mediate tumor promotion by L-leucine or L-isoleucine. Importantly, no pathological lesions were observed in the bladder urothelium by light microscope or SEM in any of the groups, indicating these two amino acids themselves do not induce toxicity and therefore regenerative proliferation is not likely to be involved in the their tumor promoting activity.

Effect of L-leucine and L-isoleucine Supplementation to the AIN-93G Diet on the Expression of Amino Acid Transporters During Rat Bladder Carcinogenesis

To evaluate the expression of L-leucine and L-isoleucine related transporters and tumorigenesis-associated genes during the early responses of BBN-initiated bladder urothelium to L-leucine and L-isoleucine dietary supplementation, rats were administered BBN for 4 weeks and then fed AIN-93G basal diet or diet supplemented with 2 % L-leucine or 2 % L-isoleucine for 8 weeks followed by basal diet for another 8 weeks [5].

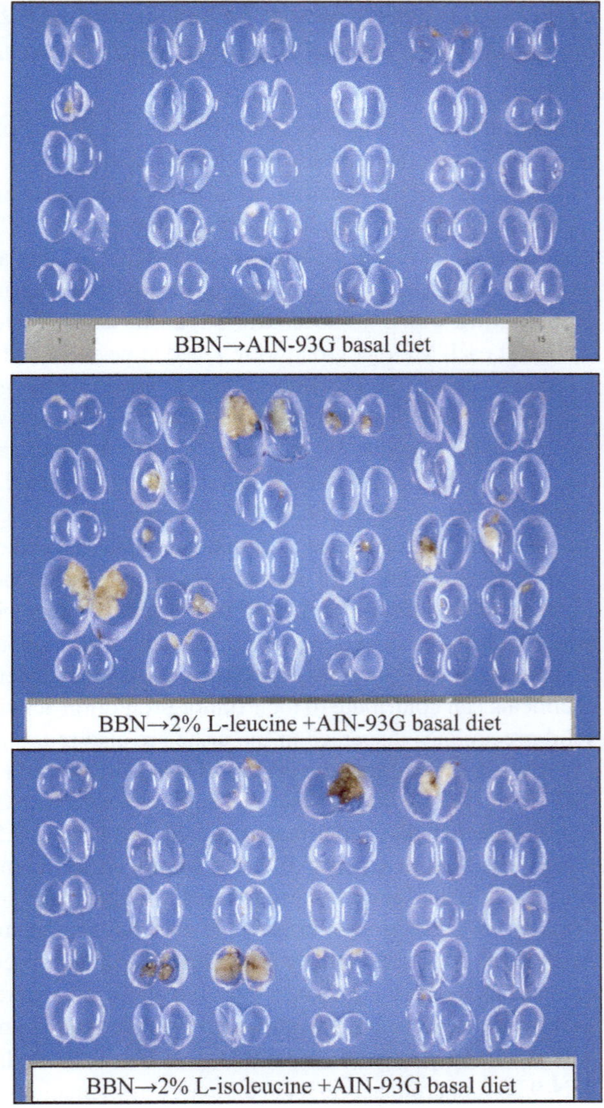

Fig. 21.1 Macroscopic images of rat bladders of rats fed AIN-93G basal diet without and with BCAA supplementation at week 29

At the end of the experiment, week 20, the multiplicity of papillary and nodular (PN) hyperplasia, which is a preneoplastic lesion rat bladder carcinogenesis, was significantly increased in the L-leucine and L-isoleucine-treated groups compared to the BBN control group. In addition, the incidences and multiplicities of papillomas, urothelial carcinomas, and total tumors in the bladder were elevated in the L-isoleucine supplemented group; although, due to the short treatment duration, the increases were not significance.

Eight weeks supplementation with L-leucine and L-isoleucine elevated expression of the amino acid transporters 4F2hc, y+LAT1, LAT2, LAT3, and LAT4, and the tumorigenesis-associated genes TNF-α, c-fos, and β-catenin in normal-appearing bladder urothelium. After removal of supplemental levels of L-leucine and L-isoleucine from the diet, expression of 4F2hc, y+LAT1, LAT2, and LAT4 remained elevated whereas expression of LAT3 returned to baseline levels. Expression of LAT1 became elevated in normal-appearing bladder urothelium during the 8 weeks after removal of supplemental levels of L-leucine and L-isoleucine from the diet. In contrast, while expression of 4F2hc, LAT2, and LAT4 was also elevated in PN hyperplasias and/or bladder tumors, expression of amino acid transporters in these lesions was not affected by amino acid supplementation. These data suggest that in rats administered

BBN to initiate bladder carcinogenesis, L-leucine and L-isoleucine increased expression of amino acid transporters in bladder epithelial cells, but that in PN hyperplasias and bladder tumors, increased expression of these transporters became independent of L-leucine and L-isoleucine.

The main transporters responsible for the uptake of BCAAs are LAT1, y+LAT1, LAT2, LAT3, and LAT4 [6–8]. Differences in the expression of amino acid transporters between tumor cells and normal cells has been proposed as a basis of differential amino acid uptake by tumor tissue and nonmalignant adjacent tissue [9], and upregulated expression of amino acid transporters is important for tumor cell growth and, consequently, is related to the progress of carcinogenesis [10–12]. In addition, LAT1, LAT2, and y+LAT1 are catalytic 4F2hc light chains [6, 7], and 4F2hc has been shown to interact with β1 integrin [13]: This interaction may play an important role in cell adhesion during metastasis. Notably, 4F2hc expression in resected non-small-cell lung cancer with lymph node metastases is associated with poor prognosis [14]. Taken altogether, the data suggest that L-leucine and L-isoleucine mediated enhancement of tumorigenic transformation of BNN-initiated urothelium was likely instituted by upregulating the expression of 4F2hc, LAT1, LAT2, and LAT4, and possibly y+LAT1 and LAT3, in BBN-initiated bladder urothelium, and advancement through the process of carcinogenesis resulted in elevated expression of tumorigenesis-associated genes such as TNF-α, c-fos, and β-catenin.

Conclusions

The results of the animal studies discussed here indicate that long-term supplementation of the CE2 and AIN-93G diets, but not the MF diet, with L-leucine or L-isoleucine has tumor-promoting activity on bladder carcinogenesis in rats. An important point is that the amino acid contents of L-leucine and L-isoleucine in the AIN-93G basal diet (1.73 % leucine, 0.96 % isoleucine), MF basal diet (1.78 % leucine, 0.89 % isoleucine), and CE-2 diet (1.8 % leucine, 1.03 % isoleucine) are comparable. The factors in these basal diets which affect bladder cancer promotion by L-leucine and L-isoleucine remain to be investigated.

The World Health Organization recommended daily amount of leucine and isoleucine is 39 mg and 20 mg, respectively, per kilogram body weight [15]. Rats fed AIN-93G basal diet (1.73 % leucine, 0.96 % isoleucine) supplemented with 2 % L-leucine or 2 % L-isoleucine ingested approximately 1400 mg leucine or 1100 mg isoleucine/kg body weight/day. This is 36 and 55 times the dietary requirement for L-leucine and L-isoleucine, respectively. While much higher than the recommended daily amount of L-leucine and L-isoleucine, the total BCAA level in the supplemented diet was similar to that in BCAA-enriched (1.2 g protein/kg body weight/day with 50 % of the total amino acid content being BCAA) total parenteral nutrition. This high level of BCAA was shown to be beneficial for cancer cachexia in a prospective randomized crossover trial [16]. However, given the results of the animal experiments discussed here, long-term use of BCAAs as a dietary supplement should be avoided until more is known about their effects on carcinogenesis in humans. This is particularly applicable to patients with bladder cancer. Further investigations are needed to determine the precise mechanism of the tumor promoting activity of BCAAs and to ascertain safe levels of BCAA supplementation for bladder cancer patients.

References

1. Nishio Y, Kakizoe T, Ohtani M, Sato S, Sugimura T, Fukushima S. L-isoleucine and L-leucine: tumor promoters of bladder cancer in rats. Science. 1986;231:843–5.
2. Xie XL, Wei M, Yunoki T, Kakehashi A, Yamano S, Kato M, Wanibuchi H. Long-term treatment with L-isoleucine or L-leucine in AIN-93G diet has promoting effects on rat bladder carcinogenesis. Food Chem Toxicol. 2012;50:3934–40.

3. Fukushima S, Shibata MA, Kurata Y, Hasegawa R. The roles of L-ascorbic acid, urinary pH, and Na or K ion concentration in rat bladder epithelial cell proliferation with special reference to tumor promotion. Prog Clin Biol Res. 1991;369:195–208.
4. Clayson DB, Fishbein L, Cohen SM. Effects of Stones and Other Physical Factors on the Induction of Rodent Bladder-Cancer. Food Chem Toxicol. 1995;33:771–84.
5. Xie XL, Kakehashi A, Wei M, Yamano S, Takeshita M, Yunoki T, Wanibuchi H. L-Leucine and L-isoleucine enhance tumor growth of BBN-induced urothelial tumors in the rat bladder by modulating expression of amino acid transporters and tumorigenesis-associated genes. Food Chem Toxicol. 2013;59:137–44.
6. Babu E, Kanai Y, Chairoungdua A, Kim DK, Iribe Y, Tangtrongsup S, Jutabha P, Li Y, Ahmed N, Sakamoto S, Anzai N, Nagamori S, Endou H. Identification of a novel system L amino acid transporter structurally distinct from heterodimeric amino acid transporters. J Biol Chem. 2003;278:43838–45.
7. Bodoy S, Martin L, Zorzano A, Palacin M, Estevez R, Bertran J. Identification of LAT4, a novel amino acid transporter with system L activity. J Biol Chem. 2005;280:12002–11.
8. Del Amo EM, Urtti A, Yliperttula M. Pharmacokinetic role of L-type amino acid transporters LAT1 and LAT2. Eur J Pharm Sci. 2008;35:161–74.
9. Baracos VE, Mackenzie ML. Investigations of branched-chain amino acids and their metabolites in animal models of cancer. J Nutr. 2006;136:237S–42.
10. Luo X, Yin P, Reierstad S, Ishikawa H, Lin Z, Pavone ME, Zhao H, Marsh EE, Bulun SE. Progesterone and mifepristone regulate L-type amino acid transporter 2 and 4 F2 heavy chain expression in uterine leiomyoma cells. J Clin Endocrinol Metab. 2009;94:4533–9.
11. Wang Q, Bailey CG, Ng C, Tiffen J, Thoeng A, Minhas V, Lehman ML, Hendy SC, Buchanan G, Nelson CC, Rasko JE, Holst J. Androgen receptor and nutrient signaling pathways coordinate the demand for increased amino acid transport during prostate cancer progression. Cancer Res. 2011;71:7525–36.
12. Kobayashi H, Ishii Y, Takayama T. Expression of L-type amino acid transporter 1 (LAT1) in esophageal carcinoma. J Surg Oncol. 2005;90:233–8.
13. Miyamoto YJ, Mitchell JS, McIntyre BW. Physical association and functional interaction between beta1 integrin and CD98 on human T lymphocytes. Mol Immunol. 2003;39:739–51.
14. Kaira K, Oriuchi N, Imai H, Shimizu K, Yanagitani N, Sunaga N, Hisada T, Kawashima O, Kamide Y, Ishizuka T, Kanai Y, Nakajima T, Mori M. CD98 expression is associated with poor prognosis in resected non-small-cell lung cancer with lymph node metastases. Ann Surg Oncol. 2009;16:3473–81.
15. Protein and amino acid requirements in human nutrition. World Health Organ Tech Rep Ser. 2007: 1–265, back cover.
16. Tayek JA, Bistrian BR, Hehir DJ, Martin R, Moldawer LL, Blackburn GL. Improved protein kinetics and albumin synthesis by branched chain amino acid-enriched total parenteral nutrition in cancer cachexia. A prospective randomized crossover trial. Cancer. 1986;58:147–57.

Index

R. Rajendram et al. (eds.), *Branched Chain Amino Acids in Clinical Nutrition: Volume 1*, Nutrition and Health, DOI 10.1007/978-1-4939-1923-9,

MIX
Papier aus verantwortungsvollen Quellen
Paper from responsible sources
FSC® C105338

If you have any concerns about our products,
you can contact us on
ProductSafety@springernature.com

In case Publisher is established outside the EU,
the EU authorized representative is:
Springer Nature Customer Service Center GmbH
Europaplatz 3, 69115 Heidelberg, Germany

Printed by Libri Plureos GmbH
in Hamburg, Germany